Ancient Nets and Fishing Gear

Proceedings of the International Workshop on "Nets and Fishing Gear in Classical Antiquity: A First Approach"

Cádiz, November 15-17, 2007

Ancient Nets and Fishing Gear

Proceedings of the International Workshop on "Nets and Fishing Gear in Classical Antiquity: A First Approach"

Cádiz, November 15-17, 2007

Edited by

Tønnes Bekker-Nielsen
and
Darío Bernal Casasola

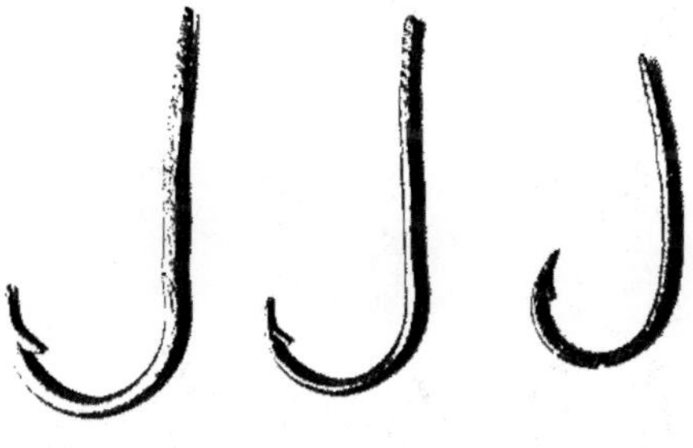

Universidad de Cádiz, Servicio de Publicaciones
and Aarhus University Press

Cover image:	Fishing with casting-net – Rota, Cádiz, November, 2007 – (D. Bernal)
	Fishing scene from Gallic pottery (Hermet, 1934, pl. 28)
Rear cover:	Mosaic from Thugga (Bardo Museum, Tunis)
	Detail of the dragnet at Conil (Hoefnagel sixteenth century, facsimile)

Published by:
Servicio de Publicaciones de la Universidad de Cádiz
C/ Doctor Gregorio Marañón, 3, 11002 Cádiz (Spain)
www.uca.es/publicaciones
publicaciones@uca.es

Aarhus University Press
Langelandsgade 177
DK-8200 Aarhus N (Denmark)
www.unipress.dk

Layout: Trébede Ediciones, S.L.
Cover: Trébede Ediciones, S.L.
Impreso en España/Printed in Spain by: Pedro Cid, S.A.

ISBN: 978-84-9828-302-0
Depósito Legal: M-31.159-2010

For Lourdes and Marit,
and for Africa, Emilie, Kirstine and Rebeca

This volume is a product of the Research Project SAGENA (HUM-03015), financed by the "Consejería de Innovación, Ciencia y Empresa" of the Regional Government of Andalusia (www.sagena.es), and forms part of the activities of the Research Team HUM-440 of the IV PAIDI of the "Junta de Andalucía".

CONTENTS

PART II. POSTERS

PART III. PERSPECTIVES

Contributors

Carmen Alfaro Giner
Universidad de Valencia
Departamento de Historia de la Antigüedad y de la Cultura Escrita. Avda. Blasco Ibáñez 28, E-46010 Valencia
carmen.alfaro@uv.es

Tønnes Bekker-Nielsen
University of Southern Denmark
Department of History and Civilization. Engstien, 1, DK-6000 Kolding
tonnes@hist.sdu.dk

Carlo Beltrame
Università Ca' Foscari di Venezia
Dipartimento di Scienze dell'Antichità e del Vicino Oriente. Dorsoduro 3484/D, I-30123 Venezia
beltrame@unive.it

Darío Bernal Casasola
Universidad de Cádiz
Departamento de Historia, Geografía y Filosofía. Avda. Dr. Gómez Ulla s.n., E-11003 Cádiz
dario.bernal@uca.es

Giulia Boetto
Centre National de la Recherche Scientifique
Centre Camille Jullian (CNRS - Université de Provence)
MMSH. 5, rue Château de l'Horloge, BP 647, F-13094 Aix-en-Provence
boetto@mmsh.univ-aix.fr

Macarena Bustamante Álvarez
Universidad de Cádiz
Departamento de Historia, Geografía y Filosofía. Avda. Dr. Gómez Ulla s.n., E-11003 Cádiz
macarena.bustamante@uca.es

Juan Jesús Cantillo Duarte
Universidad de Cádiz
Departamento de Historia, Geografía y Filosofía. Avda. Dr. Gómez Ulla s.n., E-11003 Cádiz
juanjesuscantillo@yahoo.es

Ignacio Clemente Conte
Institución Milá i Fontanals (CSIC)
Departamento de Arqueología y Antropología. c/ Egipcíaques 15, E-08001 Barcelona
ignacio@imf.csic.es

Daniela Cottica
Università Ca' Foscari di Venezia
Dipartimento di Scienze dell'Antichità e del Vicino Oriente. Dorsoduro 3484/D, I-30123 Venezia
cottica@unive.it

Milagrosa Soriguer Escofet
Universidad de Cádiz
Departamento de Biología. Polígono Río San Pedro, E-11510 Puerto Real (Cádiz)
mila.soriguer@uca.es

José Juan Díaz Rodríguez
Universidad de Cádiz
Departamento de Historia, Geografía y Filosofía. Avda. Dr. Gómez Ulla s.n., E-11003 Cádiz
josejuan.diaz@uca.es

Salvador Domínguez Bella
Universidad de Cádiz
Departamento de Ciencias de la Tierra. Campus Universitario de Puerto Real, E-11510 Puerto Real (Cádiz)
salvador.dominguez@uca.es

David Florido del Corral
Universidad de Sevilla
Departamento de Antropología Social. c/ Doña María de Padilla s/n, E-41004 Sevilla
dflorido@us.es

Virginia García Díaz
Universiteit Leiden
Archaeology. Reuvensplaats 3, NL-2311 BE Leiden
v.garcia.diaz@arch.leidenuniv.nl

Enrique García Vargas
Universidad de Sevilla
Departamento de Prehistoria y Arqueología. c/ Doña María de Padilla s/n, E-41004 Sevilla
egarcia@us.es

José Antonio Hernando Casal
Universidad de Cádiz
Departamento de Biología. Polígono Río San Pedro, E-11510 Puerto Real (Cádiz)
joseantonio.hernando@uca.es

José Lagóstena Gutiérrez
Universidad de Cádiz
Departamento de Historia, Geografía y Filosofía. Avda. Dr. Gómez Ulla s.n., E-11003 Cádiz
jose.lagostena@uca.es

Juan José López Amador
Museo Municipal
c/ Pagador 1, E-11500 El Puerto de Santa María (Cádiz)
lopezama@ono.com

Guadalupe López Monteagudo
Consejo Superior de Investigaciones Científicas
Departamento de Historia Antigua y Arqueología. c/ Albasanz 26-28, E-28037 Madrid
guadalupelopez@ceh.csic.es

Thijs Maarleveld
University of Southern Denmark
Maritime Archaeology Programme. Niels Bohrs vej 9, DK-6700 Esbjerg
t.maarleveld@hist.sdu.dk

Atle Ove Martinussen
Museum Centre in Hordaland
Salhusvegen 201, N-5107 Salhus
atleovemartinussen@muho.no

Arturo Morales Muñiz
Universidad Autónoma de Madrid
Departamento de Biología. c/ Darwin, 2, Campus de Cantoblanco, E-28049 Madrid
arturo.morales@uam.es

Manuela Pérez Rodríguez
Universidad de Cádiz
Departamento de Historia, Geografía y Filosofía. Avda. Dr. Gómez Ulla s.n., E-11003 Cádiz
manuela.perez@uca.es

José Ramos Muñoz
Universidad de Cádiz
Departamento de Historia, Geografía y Filosofía. Avda. Dr. Gómez Ulla s.n., E-11003 Cádiz
jose.ramos@uca.es

Jose Antonio Ruiz Gil
Universidad de Cádiz
Departamento de Historia, Geografía y Filosofía. Avda. Dr. Gómez Ulla s.n., E-11003 Cádiz
jantonio.ruiz@uca.es

Antonio Manuel Sáez Romero
Universidad de Cádiz
Departamento de Historia, Geografía y Filosofía. Avda. Dr. Gómez Ulla s.n., E-11003 Cádiz
antonio.saez@uca.es

Ross Iain Thomas
University of Southampton
School of Humanities. Avenue Campus, Highfield, Southampton, UK-SO17 1BF
ross@soton.ac.uk

Athena Trakadas
University of Southampton
School of Humanities. Avenue Campus, Highfield, Southampton, UK-SO17 1BF
trakadas@morocco-group.org

Eduardo Vijande Vila
Universidad de Cádiz
Departamento de Historia, Geografía y Filosofía. Avda. Dr. Gómez Ulla s.n., E-11003 Cádiz
eduardo.vijande@uca.es

Cristina Zabala Giménez
Universidad de Cádiz
Departamento de Biología. Polígono Río San Pedro, E-11510 Puerto Real (Cádiz)
cristina.zabala@uca.es

INTRODUCTION

An interdisciplinary workshop in Cádiz

Darío Bernal Casasola & Tønnes Bekker-Nielsen

The archaeological visibility of ancient fishing, typically carried out by the less privileged social groups in the Graeco-Roman world, is very limited. Research on ancient fishing relies on a very limited number of pictorial representations – mostly in mosaics, particularly abundant in *Africa Proconsularis,* and coins – a few literary texts and a countless number of scattered references and pieces of evidence, generally difficult to interpret. Among the latter are archaeological remains of fishing gear, very common on coastal sites of the Mediterranean, the Atlantic, the Black Sea and the Red Sea, and hitherto little studied by scholars.

The neglect of this source material among historians of the ancient economy prompted the universities of Southern Denmark and Cádiz, involved in maritime history and archaeology projects in the Black Sea and the Strait of Gibraltar respectively, to join forces and provide a general overview of the topic with the organisation of an international workshop entitled "Artes de Pesca en la Antigüedad Clásica. Un primer balance" ("Fishing Gear in Classical Antiquity. A First Approach"), held in the Faculty of Philosophy and Letters of Cádiz University, from November 15 to 17, 2007 (figure 1).

A better knowledge of the evolution of fishing technologies in Antiquity is crucial for an understanding of the nutritional history of Europe, the Maghreb and the Levant, as well as of the human impact on the maritime resources of the Mediterranean and Black Sea. Although we possess a significant amount of evidence on Graeco-Roman fishing technologies and on the organisation of fisheries, it had never been systematically organised and analysed by an international, interdisciplinary circle of researchers. On the contrary, there has been a tendency to approach the topic on a local or, at best, a regional level and with a diachronic perspective (that is, comparing ancient techniques with those of later periods in the same geographical context) rather than developing a synchronic view (relating to evidence from the same period found in distant geographical locations). The only exception is the study of fish processing, for which a rich literature including systematic supra-regional studies exists.

Figure 1. Poster for the international workshop held in Cádiz.

The main aim of the workshop was to lay the foundations for a systematic study of fishing technology and organisation in the Mediterranean from the Phoenician period to the Islamic expansion (800 BC-700 AD). It is only by combining the evidence of textual, pictorial and archaeological source material that we can hope to draw a comprehensive picture of ancient fishing. By clarifying the current state of research, a new focus for the debate and a unified methodology for future work can be defined. A group of scholars from several countries (Denmark, Spain, France, Italy, Portugal, Morocco and the UK) were invited to take a fresh look at the physical remains of ancient fishing tackle and relate it to pictorial representations, literary and legal texts, and fishing practices of later periods, inferring relationships between fishing techniques and animal remains found on archaeological sites and drawing implications for the use of vessels for sea fishing.

According to a famous passage from the *Corpus Iuris Civilis* that was quoted more than once in the course of the workshop, the "sea is common to all". The sea is the common perspective of all the papers in this volume; on the other hand, the authors' approaches to the sea, its fauna and its exploitation are many and varied, reflecting the interdisciplinary nature of the meeting and of the research field itself.

Fishing is essentially a contest of wits between land-living humans and the creatures of the sea. In the first paper of this volume, Arturo Morales considers nets and other fishing gear from the perspective of the fish they are intended to catch, while the following paper by Carmen Alfaro takes the reader on a guided tour through the history of net fishing technology from prehistoric Bohemia via Lithuania, Santorini, Oropos and *Augusta Raurica* to the western Mediterranean.

The contribution by Darío Bernal represents an ambitious step forward towards the creation of a coherent taxonomy and typology of fishing instruments, primarily hooks and sinkers. It highlights the potential but also the problems of distinguishing between different net types on the basis of their non-organic remains. As Ross Thomas demonstrates in the following paper, in exceptional environments such as Egypt, remains of the net fabric itself may permit a classification of nets by function.

Pictures were the point of departure for the paper by Guadalupe López, in which she demonstrates the richness and variety of the pictorial record preserved in the mosaics, but also the challenge of their interpretation within a context that is not only technological, but literary and mythological. Likewise, the contributions by Tønnes Bekker-Nielsen, Enrique García and David Florido attempt to place ancient fishing technology in a wider context: Bekker-Nielsen in relation to social and economical factors, García and Florido as part of a *longue durée* of beach fishing in the western Mediterranean, a theme also touched upon in several of the poster presentations.

With the contributions by Carlo Beltrame and Giulia Boetto, we move off the beach and on to the water. Both authors focus on the use of tanks for preserving live fish on board, a key aspect of ancient fishing technology which one hopes will be documented by further finds of well-boats, vessels with well-boxes, or fish tanks in the years to come. Being constructed mainly of wood, ancient boats have not often been preserved, but as pointed out by Thijs Maarleveld, the physical and chemical deterioration of the wooden structure is only one among many factors working against the preservation of ancient boats; equally important are regimes of preservation and levels of awareness. It is to be hoped that, given the increasing interest in marine archaeology, the future will reveal more ancient wrecks, especially from soft coastal sediments such as harbours and estuaries.

The papers presented at the conference were supplemented by a series of shorter poster presentations and a round table discussion. The two posters on lithic tools from La Esparragosa and on the SAGENA project demonstrate the richness of the southern Spanish archaeological record for ancient fishing, and point to the potential of typological and comparative analysis for its interpretation. The poster paper on *terra sigillata* by Macarena Bustamante Álvarez takes up the thread of Guadalupe López' paper on the pictorial evidence, relating changes in consumption patterns and iconographical preferences to the social and political history of the early imperial period, while in the following poster presentation, Athena Trakadas discusses a series of archaeological finds on and off the Moroccan coast as evidence for fixed-net fishing in Antiquity. The last three poster papers by Daniela Cottica, Atle Ove Martinussen, J.J. López and J.A. Ruiz share an ethno-archaeological approach, demonstrating how the accumulated knowledge of traditional fishermen may be used to improve our understanding of ancient fishermen developed their fishing implements.

The workshop was concluded by a summing-up of the main themes and questions in current research into ancient fishing technology.

A large number of interesting questions were taken up in the discussions following the presentations and in the lively round table debate (from which one contribution, by Athena Trakadas, has been included in this volume as an epilogue) and it may seem unfair to mention only a few:

First, the value of diachronic comparison, which is at the core of the contributuions by Martinussen, López and Ruiz, García and Florido, and Boetto, and also plays an important role in the analysis of Beltrame. As Beltrame points out, however, terminology can be ambiguous: a boat used for fishing is not necessarily a "fishing boat"; and as García and Florido demonstrate, *almadraba* may be used both to describe a specific fishing implement and a wider category of coast-based fishing activities. To the fisherman working in a coastal fishing team or to the observer on the shore, the social, economical and entrepreneurial aspects of the operation were more important than the finer technical distinctions, and distinct fishing technologies may have been subsumed under one and the same term (just as the modern word "car" may denote either an automobile or a railway carriage).

Second, the problems of the iconographical evidence. The north African mosaics provide a valuable and attractive pictorial record, but Darío Bernal points to the danger of taking visual representations from one area as evidence for fishing practices in another part of the Mediterranean, as techniques and forms of organization may have varied from place to place. A related problem is that mosaic patterns and even *terra sigillata* stamps travelled with their users from one place of production to another.

Third, and perhaps the most crucial problem, the relation between form and function and the associated problems of taxonomy and terminology. Viewed in isolation, many objects have more than one potential function: a net weight is difficult to distinguish from a loom weight, while the thousands of aquatic species far outnumber the different types of hooks and fishing implements: in other words, hook types tend to be multifunctional, targeting more than one species. Nets, too, are multifunctional; Guadalupe López reminded us of the place of nets in the context of the games, being used by the gladiators themselves and for catching the vast numbers of wild animals required for the amphitheatre. In Oppian's *Halieutika* we have a detailed list of ancient net types, but we do not know if it is exhaustive, and on the basis of the preserved remains alone – typically net weights – we cannot hope to achieve a similarly detailed subdivision into functional categories. As Daniela Cottica pointed out in the round table discussion, viewing the objects in their context and in the light of modern parallels may bring us closer to an understanding of their function; as shown in Atle Ove Martinussen's presentation, this requires a serious commitment to preserving the traditional craft knowledge of today's fishermen.

With the participation of more than fifty scholars, mostly Spanish but also Gibraltarian, Italian, Danish, Portuguese, British, Dutch, Norwegian and Moroccan, the working environment was active and interdisciplinary. Apart from the academic sessions and the round table ("think tank session"), the programme also included visits to the Cádiz Archaeological Museum and to the Roman fish processing plant on the site of the "Teatro Andalucía". To gain an impression of traditional fishing methods, the group visited the fishing settlement at Rota with its stone traps known as "corrales", followed by a social session at the Salina San Vicente, in San Fernando (figure 2), in which all participants took part. All this would not have been possible without the enormous effort of the two secretaries of the work-

Figure 2. Participants observing fishing techniques at the Salina San Vicente.

shop, José Juan Díaz Rodríguez and José Lagóstena Gutiérrez, from the Archaeology Area of the University of Cádiz, whom we wish to thank warmly for their effort and dedication. The financial support and sponsorship of the Ministry of Education and Science, the Regional Government of Andalusia (through the "Consejería de Innovación, Ciencia y Empresa"), the HMAP (History of Marine Animal Populations) Project and the Pouroulis Foundation, along with the two organising universities, were instrumental for the successful completion of the workshop.

Publication of the workshop's results

The high scientific level attained during the sessions, the number of contributions presented and the general agreement reached after the Round Table entitled "Fishing Tackle in the Ancient World: The Issue of Nomenclature and Research Perspectives", made it clear that the scientific results reached during the workshop should be published in book form. This was made possible thanks to the financial support of the SAGENA Project (HUM-03015), of the Department of Innovation, Science and Private Enterprise, Regional Government of Andalusia (www.sagena.es), the Cádiz University Press (Mr Gonzalo Butrón Prida) and Aarhus University Press (Ms Sanne Lind Hansen and Mr Claes Hvidbak), for whose wholehearted support we are sincerely grateful.

Future perspectives: toward regional *corpora* and studies

Despite the novelty of the topics discussed and the disparate nature of of the evidence, the workshop produced significant results and raised a series of exciting new questions for future research. Did the Phoenicians introduce metal hooks in the west? Are the tuna fishing nets known as *almadrabas* in the Cádiz area of Phoenician origin? Can the mosaic iconography or the descriptions in Oppian's *Halieutika* be used to draw general inferences about the fishing equipment used in the Atlantic, the Mediterranean and the Black Sea? Were fisheries truly over-exploited during the Republic and the early Roman Empire? These questions, with their wide-ranging historical implications, can only be answered through an intensive and detailed analysis of fishing equipment found on archaeological sites, hence the necessity of pursuing this line of research in future years.

Until 2008 no *corpus* of fishing equipment anywhere in the ancient world had been published. Since then, a few preliminary proposals have been suggested for *Mauretania Tingitana* (the doctoral thesis of A. Trakadas), ancient *Baetica* (ongoing within the framework of the SAGENA Project) and the Vesuvius-Pompeii-*Herculaneum* area (under the aegis of a Spanish-Italian project sponsored by UCA and UNIVE). A future extension of this approach to the Atlantic, the Mediterranean, the Black Sea and the Red Sea will make it possible to identify distinctive regional variations in fishing practices. The development of general models of artefactual behaviour from a diachronic and/or a geographical point of view, allowing the drawing of conclusions about influences and technological transfers between ancient societies and the survival of local traditions in a Mediterranean, will not, however, be possible in the short term. To be able to approach these fascinating topics in the future we will have to continue with the historical-archaeological analysis of fishing equipment, of which the contributions to this volume mark the first steps.

Darío Bernal Casasola and Tønnes Bekker-Nielsen
Cádiz and Kolding, February 2010

Detail of a Tunisian mosaic at Bardo Museum, Tunisia.

Papers

Fishing for plaice at low tide with leisters taken from Sáñez-Reguart, 1988, Vol. II, 110 (detail).

1. Inferences about Prehistoric Fishing Gear based on Archaeological Fish Assemblages

Arturo Morales Muñiz

Introduction

Though the archaeological record is far from straightforward on the subject, the evidence suggests that among the hominines, only modern man has learned to fish (Morales & Roselló, 2005-2006). Whether this reveals a fundamental aspect of the primeval human niche or not, most scholars concede that the confirmation of such a hypothesis requires evidence beyond the mere presence of fish remains in archaeological contexts. Among this evidence, only the retrieval of fishing gear could take us from the realm of educated speculation into that of objective data-testing. As it turns out with so many evolutionary processes, matters were probably rather fuzzy in the transition from hunting to fishing as both proximal (i.e., subsistence practices) and distal causes, such as cognitive developments allowing people access to resources in an alien environment, played roles that will probably remain unknown forever.

What seems reasonable to assume is that the introduction of fishing was linked with the increasing ability of *Homo sapiens* to capture an ever wider array of terrestrial species through time. Indeed, both the Upper Palaeolithic period of western Europe and the Later Stone Age of Africa testify to modern man's incorporation of a large number of small, swift and otherwise difficult to catch animals, such as birds and lagomorphs, in an apparent broadening of Paleolithic diets first noted by Lewis Binford (1968) that Kent Flannery later labelled the Broad Spectrum Revolution (Flannery, 1969). Although fishes were specifically mentioned as part of this broadening of diets (Flannery, 1969, 77), at the time of his writing these consisted mainly of freshwater species and then only in quite low numbers.

Up until that moment, marine fishes had rarely been documented on archaeological sites due to retrieval biases and to the submergence of the continental platforms that took place after the onset of the Holocene (Wheeler & Jones, 1989; Morales & Roselló, 2008). As fishing and related activities were mostly carried out at the shore itself, the flooding of formerly terrestrial environments erased most of the evidence, a contingency that archaeologists are now slowly acknowledging.

If fishing is indeed a rather recent innovation within the niche of *Homo sapiens*, one must presume that the origins of fishing gear lie in the realm of hunting instruments. It is thus in the hunter's toolkit that one should look for the ancestors of later fishing tackle.

A second question to be discussed at some length in the following lines has to do with the biological features of the fishes and their ability to escape capture. Despite a rather conservative tendency for fishing strategies and fishing implements to remain unaltered, the past fifty millennia witnessed changes at the level of fish-fisherman co-evolution that one can loosely use to interpret the evidence of archaeological fish assemblages. It seems therefore sensible to open this discussion by specifying what makes a fish vulnerable to a non-aquatic predator.

A final matter for concern that will also be dealt with below is the fact that interpretation of fish assemblages in terms of fishing technology leans heavily upon ethnographic inferences and analogies, extrapolating the present into the past. This is obviously unavoidable, but fraught with difficulties given that present-day analogies, whether these concern human groups, fishing tackle or fish behaviour, are embedded in totally different contexts – due partly to technological innovation, but mainly to problems such as pollution, overfishing and land degradation that were only minor or anecdotal constraints during prehistoric and early historic times.

Restricting the inferential universe

What kinds of fishing tackle were deployed by people up until classical antiquity? This is a difficult, yet pivotal question, since the answer would assist us in our interpretation of archaeological fish collections. To answer it, one cannot restrict the investigation to fishing gear proper but must extend it to include auxiliary elements, such as vessels, the primeval forms of fishing gear that may have looked very different from the final product, and the places where fishing actually took place.

In the same way that we know electrical fishing to be a twentieth-century development, offshore fishing (i.e., fishing in the open sea beyond the sight of the coast) appears to have been an essentially European Middle Age event (e.g., Barrett *et alii*, 2004). Restricting fishing to inshore waters would have limited the range of available fishes, and the ichthyoarchaeological data thus far gathered for Europe appears to confirm this. Thus, the major catch in Northern Europe consisted of the gadiforms (codlike fishes), whereas in the Mediterranean the Sea breams *(Sparidae)* dominated and taxa such as groupers (*Epinephelus* sp.), grey *(Mugilidae)* and red mullets *(Mullidae)* represented items of secondary importance (Bødker Enghoff, 1999, 2000; Morales & Roselló, 2008).

Clupeids of various kinds (i.e., herring (*Clupea harengus*) in the NW Atlantic, sardine (*Sardina pilchardus*) in the more temperate latitudes), appear to have been on the increase from classical antiquity onwards but were rarely main objectives before the Middle Age (Sahrhage & Lundbeck, 1992, 57ff). Other secondary catches in the NW Atlantic included euryhaline flatfishes such as plaice (*Platichthys flessus*) and several families of perciforms (e.g., *Sciaenidae* (meagre), *Carangidae* (jacks), etc.) in the Mediterranean. In the Black Sea, the catch from pre-classical to classical times was dominated by sturgeons *(Acipenseridae)*, along with carp *(Cyprinidae)*, sheat-fish *(Siluridae)* and a few percids (i.e., the perch (*Perca fluviatilis*) and the pike perch (*Stizostedion lucioperca*); Morales *et alii*, 2007). The impression of littoral fishing is compelling in all cases.

In fresh waters, on the other hand, all of the European taxa would have been within the reach of any littoral fishery from the beginning, thus one may surmise that the specific abundance of certain groups at specific sites – unlike the situation in the case of sea fishing – might have been dictated as much by choice as by availability, with crucial implications for the deployment of specific gear.

A couple of other points to be borne in mind have to do with boats and crews. From what one can gather from complementary information on this subject (see, e.g., Guadalupe López Monteagudo, this volume), it appears that up until classical antiquity fishing boats were small and slow for the most part, functioning more as convenient plattforms from which a couple of fishermen operated than as elements for the active pursuit and capture of fishes. If this was indeed the case, then trawling must have been a rare phenomenon (many scholars place the earliest use of trawl fishing in the sixteenth century AD or even later; T. Bekker-Nielsen, pers.comm.) and the capture of swift fishes by active pursuit essentially precluded. The corollary of such a hypothesis is that fast pelagics such as jacks and scombrids (i.e., tunas and mackerels), must have been taken by interception methods rather than by active netting. This, in turn, could only have occurred when they were in very shallow waters, a phenomenon documented through different lines of evidence and that, again must have restricted fishing operations to the shoreline zone (Cleyet-Merle, 1990; Sahrhage & Lundbeck, 1992; Merino, 1997; Sáñez-Reguart, 1988, 1993).

The number of people involved in fishing operations is well documented from classical antiquity onwards though some Old Kingdom Egyptian reliefs already show teams of fishermen working a fish net (Brewer & Friedman, 1990, 39). Ethnographic evidence hints that, even when practised for subsistence, whenever fishes were abundant communal fishing ensued (Stewart, 1977; Gallant, 1985; Bekker-Nielsen, 2005). Sheer availability could have determined what kinds of fishes the fishing at a particular place would have concentrated on, thus what specific subset of fishing tackle one should be more likely to encounter – or try to infer – in an archaeological assemblage.

For most pelagic (i.e., open water) fishes, the early stages of development are more restricted by environmental parameters than is the case for the inshore species (Gil de Sola, 1999; Helfmann *et alii*, 1998; Margalef, 1989). The planktonic stages are particulary vulnerable to unsuitable water temperatures. For such reasons, pelagic fishes often produce very different recruitments on an annual basis so that years of glut alternate with years of dearth. This phenomenon has been documented since ancient times (Gallant, 1985; Horden & Purcell, 2000); the eighteenth-century fishing records from the Gulf of Cádiz also reveal highs and lows of devastating consequences for people relying exclusively on pelagic resources (Burgos-Madroñero, 2003).

For this reason, even when set at far lower levels of average annual yields, the exploitation of the littoral, non-migratory species was probably a safer bet at all times for local fishermen stationed at the coast, in particular whenever fishing involved small groups of people and buffering losses through alternative strategies was either restricted or non-existent (Arbex, 1990; Brandt, 1984). As a corollary of such limitations, "specialized" fishing of pelagic taxa must have been a risky choice before the existence of exogenous buffers at the onset of Classical Antiquity in Europe.

With these provisos in mind, we now turn to the fishes themselves, reviewing data on their biology that could help us explain why the choice of specific fishing tackle should vary from group to group.

The fishermen's companion to the biology of fishes

Obviously fishes, like other aquatic organisms, are poorly designed to avoid terrestrial and aerial predators that operate in ways different from the aquatic counterparts with which they have been coping for aeons. Still, most terrestrial and aerial predators are only effective either in very shallow waters or in closed bodies of water, where fishes have their escape abilities restricted. To birds, fishes are most vulnerable when they swim close to the surface (figure 1). From this standpoint, marine, stricly benthonic (i.e., bottom-dwelling) and pelagic species would be in principle less vulnerable to human predation than freshwater and littoral demersal (i.e., swimming close to the bottom) taxa. This sort of evolutionary trend in fishing, leading from the more to the less accessible species, appears to be in agreement with the data thus far gathered from the archaeoichthyological record and, again, confirms that throughout Prehistory and well into Classical Antiquity, fishing was an essentially "coastal", meaning shallow water, endeavour (Cleyet-Merle 1990; Merino 1997; Morales & Roselló, 2008).

Even under optimal conditions, fishing is a far from straightforward task. Throwing a spear, for example, with just enough power to impale a fish, but not enough to break the gear, requires the fisherman to estimate the size of the animal

Figure 1. Mosaic of Oceanus and Tethys (Yakto Complex Turkey, mid-fourth century AD) where hands are used to catch a marine invertebrate (lobster?). Fishing with bare hands implies either grabbing or using the hands to deliver a fast blow from below the fish that will throw it out of the water and onshore.

and the depth of the water, correcting for distortions due to light refraction. If the surface of the water is rough, further corrections need to be made in order to neutralize a blurred image, and if the fish is moving, its speed also needs to be taken into account. No easy task by any means.

The archaeoichthyological record does not always confirm the trend of less accessible fishes becoming more common with time. This indicates that other factors, including availability of alternative resources, differences in technological development, cultural choices and even taboos, need to be taken into account as well (Leach, 2006). Elements of particular interest in our case would be the sensory abilities of fishes, as many types of gear have been specifically designed to neutralize or circumvent these.

One of the keys to effective fishing, as with any kind of predatory activity, is to capture the animals before they are given a chance to react. As fishermen know from experience, this element of surprise varies from species to species due to their different sensory capabilities.

Generally speaking, among the two major kinds of fishes, bony fishes are far more visual than cartilaginous ones (i.e., sharks and skates) and these, in turn far

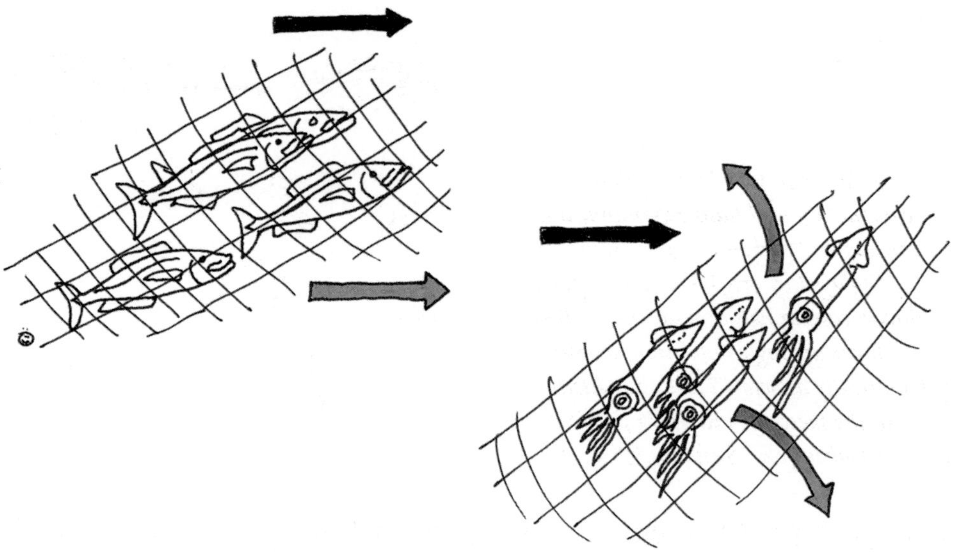

Figure 2. Why cannot one catch squids with nets? Surrounded by a trawling net that proceeds faster than they can, squids, provided with good sight but no lateral line system, believe their eyes and escape through the mesh when its size allows. Fishes under stress, on the other hand do not trust their eyes as much as they trust their lateral line which informs them of a moving wall, rather than a multiperforated device, next to them. As a result, they keep on swimming until they are engulfed by the trawl.

more olfactive. Obviously, there are exceptions (e.g., bony fishes living in muddy waters often have poor sight), but the "rule" has had many practical implications for the development of fishing gear (Brandt, 1984). In addition, all fishes respond to certain stimuli that are meaningless to terrestrial animals but which fishermen must have known intuitively since time immemorial. This would be the case of electrical impulses, which travel well in water and allow the fishes to detect changes in electrical fields, as well as pressure waves, that also travel well in an essentially incompressible medium. These stimuli allow fishes to detect obstacles and prey at a distance. Although the electrical "nature" of some fishes has been known for more than a thousand years, the discovery of the "attraction" of fishes to the anode pole of an electrical field and the development of electric gear are very recent (i.e., twentieth-century) events that need not concern us here (Brandt, 1984). The application of knowledge of the intuitive functioning of the lateral line system (i.e., the sonar-like sense of fishes), on the other hand, has guided the development of certain gear, in particular trawling nets (figure 2).

If the animals cannot be taken by surprise, they can still be captured without major problems when their reactions can be anticipated. Most fishes react instantaneously to sudden stimuli, be these visual, auditive or other. These alarm reactions are for the most part innate, thus highly stereotyped, meaning that fishermen know what

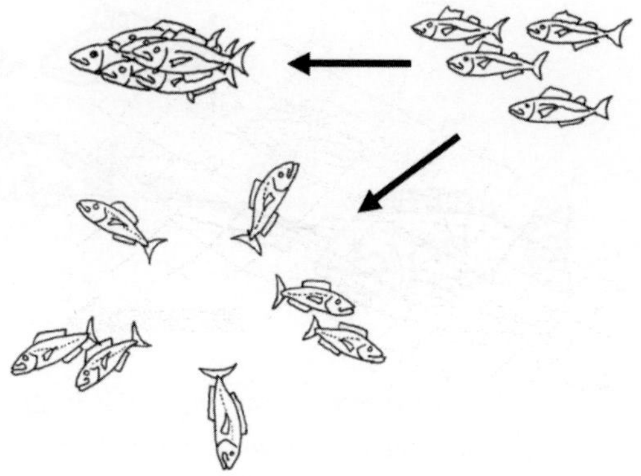

Figure 3. Of the two behaviours, compaction and flash expansion, that initiate and end the chain of responses displayed by gregarious fishes under threat, casting nets aim at intercepting fishes during compaction, as this is systematically associated with a reduction in the speed of travel that facilitates capture (taken from Stewart, 1977, 96).

is likely to happen next (figure 3). Anticipation is therefore critical: when tuna fishermen beat the waters with their oars, such loud noises will make the schools swim away, allowing people to connect previously independent sections of the seine nets to prevent escape and to funnel the fishes into the death chamber of the madrague. This is but a single instance where human knowledge about fish behaviour is used to trick the animals and facilitate their capture.

Another realm where anticipation works wonders is to know precisely what attracts fishes. Many fishes have a favourite prey, with which fishermen are keen to bait their hooks and traps, but often the attractants are of a more general nature and again attuned to specific sensorial abilities. For attracting large, swift pelagics like scombrids, for example, hooks do not even need to be baited provided they send the right visual signals to the predators. Most of these fishes are visually attracted to their prey, small countershaded pelagics that shine as mirrors in the water. Almost any small object that shines in the water is therefore taken as prey and attacked. Since the Late Stone Age of Southern Arabia, people have been making mother-of-pearl hooks with pearl mussels *(Pinctada sp.)* whose shine even to this day lures scombrids into taking unbaited hooks (Roselló *et alii*, 2005). For olfactive animals, such as sharks, blood and rotten meat works best whereas for fishes that respond more to movement or sound, various devices from coconut shells to dead fishes jerking through the water can do the same job.

Many other innate aspects of fish behaviour are exploited to the fisherman's advantage (figure 4). One such case is the tendency for many gregarious species to become more densely packed and to reduce their speed when frightened, something that allows for casting nets to be thrown over them and seine nets to encircle the schools. The relentless urge of migratory species to swim upstream or downstream during their spawning runs, together with the flow and ebb movements of the tides, ensured the success of weirs and traps of all kinds (see below). Other exam-

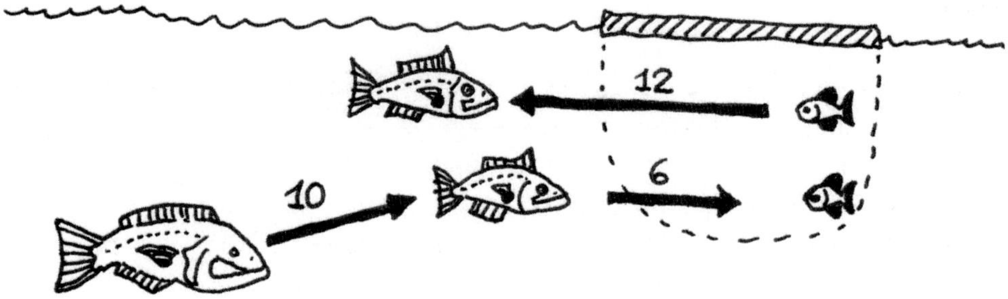

Figure 4. The advantages of hovering in the shade. On sunny days in clear water, a shaded observer can detect sunlit objects 12m away, which represents a 20% advantage over the 10m a sunlit observer can manage for sunlit objects. More critical, this sunlit observer would not be able to detect the fish in the shade until it was 6m away from it. Although such advantage decreases under different conditions, such as clouds and turbidity of the water, it helps explain why so many fishes, whether open water species or not, love to hover in the shade, a trait that fishing gear such as stone traps exploit (taken, with modifications, from Helfman et alii, *1997, 339).*

ples are the attraction that many fishes with nocturnal feeding peaks (e.g., sardines) exhibit towards light, and the curiosity and hunger that fishermen know work so well in the case of fish traps.

As for hooks and baits, multiple factors affect the way fishes interact with them. Species with diurnal or nocturnal feeding patterns, for example, exhibit different reactions through their daily cycles (Løkkeborg & Bjordal, 1989) and for many species these behaviours vary through the tidal range, seasonally, or simply when the weather is dry or rainy (Leach, 2006, 101). The presence of other species likewise affects the degree of interaction with this gear making some species feel more wary, others more aggressive. Eventually, a deep acquaintance with such details turned fishermen into the most specialized and skilled among the human predators.

A brief survey of fishing gear

As stated earlier, one key assumption throughout this paper is that it only took time and ingenuity before some of the instruments in the toolkit of the prehistoric hunter were transformed into efficient fishing implements. Based on this obviously debatable premise, what follows is a cursory survey of the features and limitations of the main categories of fishing gear in terms of the fishes they can potentially catch, thus the fish assemblages each one of them is likely to produce in the archaeological record. The main ethnographic references here have been the native American cultures of the northwest Pacific coast, superbly documented by Stewart (1977).

Hands

As the basic "tool" of the hominines, one can postulate hands to have been the primeval fishing gear.[1] Fishing with one's bare hands is no easy task except under very special circumstances, such as when animals are very slow or when their movements are severely limited due to the spatial features of the waters where they are found.[2] The two main biotopes where fishes are so restricted in their movements are tidal pools in rocky shores and equivalent areas of very shallow depth in freshwaters. In the former instance, fishing would better qualify as gathering and the main targets would be small sized fishes such as sculpins (Cottidae), blennies (Blennidae) and gobids (Gobiidae) with little meat on them and quite territorial, thus found at low densities and unlikely to have been of any significance as a source of food (though perhaps as bait!). In fresh water, depending on the circumstances (e.g., spawning runs, rapids and falls, etc.), the potential fishermen are likely to find animals of far larger size and at far higher densities. Throughout the northern hemisphere, most of these would belong to one of two families, the salmonids (trouts and salmon) and the ciprinids (carpfishes).

Fishing with hands would leave no external traces on the animals and this obviously applies also to their skeletons. The inference of this method of fishing in archaeological sites would solely rest on the taxonomic composition of the samples but whereas a collection of rock-pool fishes could theoretically be taken as evidence of hand deployment, the presence of the aforementioned freshwater groups could always be more probably explained by a wealth of alternative methods.

Ichtiotoxins

Ethnographic data indicate that knowledge about poisons must have been a basic element in the toolkit of the hunter and gatherer, and the large number of plants whose toxins have been traditionally used to catch fishes in places such as Spain

---

1 What about feet? When I was a child in Asturias (northern Spain) I used to go catch small brill *(Scophthalmus rhombus)* at low tide by treading on them. Other European fishermen have been doing so for millennia. Species such as dab *(Limanda limanda)* are still caught in this way on the North Sea coast (T. Bekker-Nielsen, pers.comm.).

2 There exists one further instance where hands were used to collect fishes, namely the mass mortalities that ocassionally washed vast numbers of animals ashore. Documented ethnographically on the beaches of South Africa, Patagonia, New Zealand, and Peru, but perhaps a widespread phenomenon before the introduction of fishing, such mass mortalities might have been one of the evolutionary driving forces that led hominines to develop a taste for fish meat.

demonstrates that it probably did not take long before people realized their value as "fishing gear" (Álvarez, 1999).

The main limitations of these venoms, whether used to kill fishes or simply stun them, are spatial, since dilution in water rapidly diminishes their effects. From that perspective, one may take it that the limitations mentioned in the case of hand fishing, also apply to ichtiotoxins; thus the range of taxa may in many instances be quite similar. The main difference with hand fishing lies in the fact that venoms stun fishes large and small alike, thus the samples found in an archaeological collection would not be so restricted in terms of size when compared to those in a collection produced by hand fishing. Also, poisons would not be worth employing in tidal pools; the marine settings where they have been traditionally used concentrate in tropical waters and include tidal flats along the shore and mangrove swamps where the variety of fishes likely to be encountered is high and includes both benthonic, demersal and pelagic taxa (eg., Soleidae (soles), Mugilidae, Centracanthidae (mojarras), Ariidae (catfishes), etc.) that live over sandy or muddy bottoms, not rocks.

A second difference setting these assemblages apart from those produced by hand fishing is that ichtiotoxins, like traps and nets, collect "populations", not individuals. In retrospect, in the absence of complementary evidence such as ethnographic data or the retrieval of toxic plants, the analyst may be at pains to detect the use of this technique on the basis of a particular taxonomic composition or the size distribution of fishes in a sample.

Spears

The spear is probably the oldest offensive weapon in the toolkit of the hominine and one may surmise that it was only a matter of time before it was deployed on aquatic targets, long before people introduced the changes in design that would turn it into a more effective fishing device (Merino, 1997, 97). One initial modification were probably the barbed points ("hooks") that helped the spear point hold the body of the animal (figure 5). In cases where the spear was used to stab the fish, the development of multiple prongs was a second modification that eventually led into the kind of gear we now call a leister (figure 6).

When spears began to be thrown at moving targets in the water, chances of hitting were always higher than chances of killing, if only because the target must always have been above a minimum critical size threshold, meaning medium to large fishes. In such situations, and in order not to lose the fish (and often the gear as well!) the harpoon was developed. Unlike the spear, where the point is firmly fastened to the shaft, the head of the harpoon becomes detached when it strikes the fish but remains attached to a retrieving line that allows the fisherman to follow the injured animal and haul it back. There were many modifications introduced to this basic

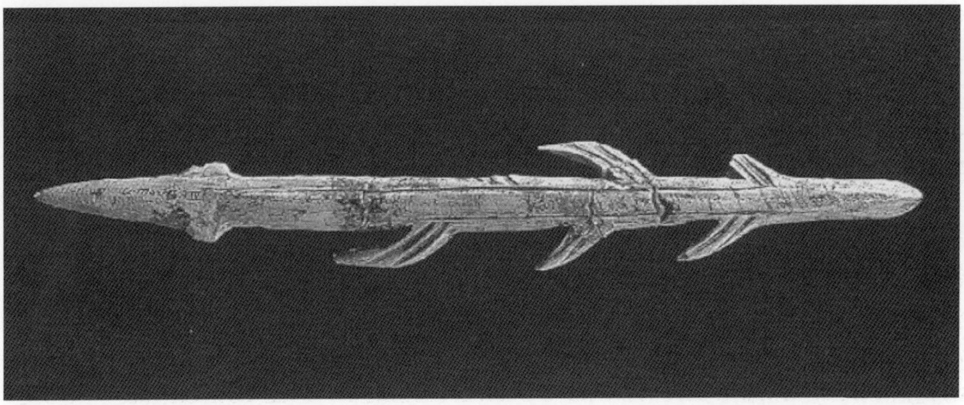

Figure 5. A Magdalenian harpoon with its characteristic barbed points meant to embed the projectile in the flesh of the fish.

design. Harpoons used by native people on the northwestern coast of America, for example, combined two or more heads on separate prongs (Stewart, 1977, 70), and a throwing harpoon often had a finger grip at the butt end from which the shaft was propelled towards the fish, which was afterwards hauled in using the attached retrieving line (Stewart, 1977, 65).

Whereas leisters can be used for fishes of sizes ranging from small (< 30cm SL[3]) to large (> 1m), they work best on medium sizes whereas harpoons, as mentioned previously, work better the larger the target. This is so because a large, struggling fish impaled on a fixed spear/leister can break either the shaft or point, or free itself by thrashing about. A detachable harpoon, on the other hand, not only allows the fish to move about in the water without placing excessive strain on the gear while it struggles, but while doing this the barbs are imbedded even deeper into its flesh.

Fishing with a leister requires less skill than using a harpoon. Indeed, spearing is often done on foot while walking over the very shallow waters of estuaries, tidal lagoons and river margins. In Europe the main targets, all marine, would have been flatfishes like the flounder (Pleuronectes platessa) and plaice as well as certain skates (Rajidae, Myliobatidae), with eels (Anguilla anguilla) and other sluggish moving, bottom-dwelling taxa that penetrate into shallow waters such as stargazers (Uranoscopus scaber) and weevers (Trachinidae) constituting marginal items. Leisters are far less effective for spearing moving targets, be these of medium or large size, although documentary, pictorial and ethnographic sources confirm that occassionally fishermen stationed on boats have been able to impale large prey, such as tunas and other scombrids like the wahoo *(Acantocybium solandri).*

---

3 SL = standard length, measured from the tip of the snout to the caudal peduncle.

When harpooning on foot or from boats, fishermen have preferred moving targets of the medium to large pelagics, with particular success at locations where these congregate. Such places include the mouths of rivers where spawning runs of fishes like salmon and shad (*Alosa* sp.) gather but, above all, bays and inlets where migration fluxes of pelagics like tuna come very close to the shore. So specialized was the art of harpooning, now an essentially lost tradition, that a vast array of types was developed for use in small streams, rivers with deep pools, rapids, etc. (Stewart, 1977, 65). A spectacular bifurcated harpoon that was developed by the natives of the northwestern Pacific coast for capturing sturgeon *(Acipenser tramontanus)* may have had its prehistoric counterpart in Europe at a time when sturgeon were common (Stewart, 1977, 69-71).

Archaeological assemblages produced by fishing with spears, leisters and harpoons, would incorporate the aforementioned taxa and probably exhibit a rather narrow range of sizes, with a preference for medium to large individuals, and a total absence of fishes below the critical 30cm threshold that we have arbitrarily defined as "small". In terms of traces on the bones, one would expect to find an abundance of punctures, notably on the vertebrae, as the signature of spear fishing.

Bow and arrow

What appears to have been a Late Upper Paleolithic development in the hunter's toolkit probably did not take long to be deployed as fishing gear, yet never became a truly successful fishing instrument (figure 7). The reason may be that bow and arrow combine several of the drawbacks with very few of the advantages of spears, leisters and harpoons. For one thing, their use requires as much skill as that required for throwing a harpoon, yet the arrow is easily lost if the fish is not killed instantaneously. In addition, due to the small mass of the arrow its force is rapidly neutralized by the viscosity of water, making a rapid kill unlikely. This forces the bowman to incorporate an additional calculation into his mental process and restricts arrow use to only the shallowest waters, for which more effective gear already existed by Paleolithic times. Finally, the small size of the arrow point means that a blow to a prey on the move is more likely to be lethal on a smaller target – that, however, is also far more difficult to hit. The alternative would be to aim a very long arrow at slow or static animals, permitting a lethal blow to be directed at the head of the fish even when lying in slightly deeper water (i.e., more than one metre).

The scarcity of the ethnographical evidence for bow and arrow fishing indicates that this alternative proved particulary unfruitful (Brandt, 1984). The few instances where bow and arrow have been systematically documented, occur in fresh water, preferably still and with soft rather than rocky bottoms. In all cases, the prey are medium to large, sluggish and bottom-dwelling fishes like the Pirarucu (*Arapaima*

Figure 6. Fishing for plaice at low tide with leisters (taken from Sáñez-Reguart, 1988, Vol. II, 110).

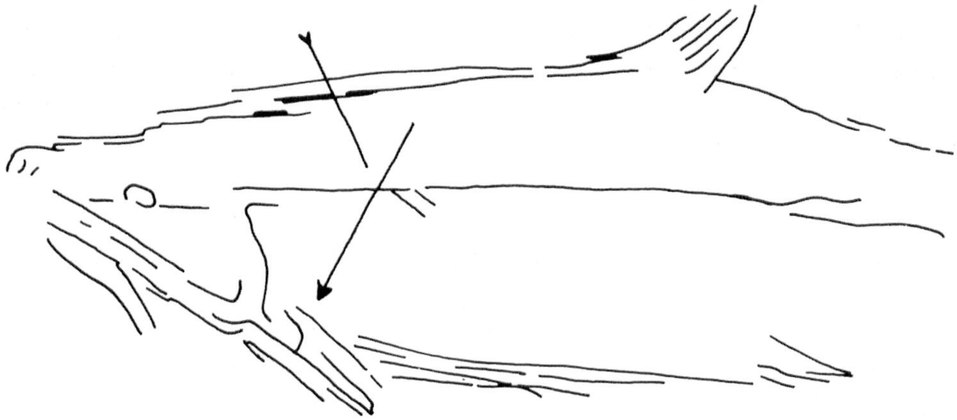

Figure 7. Magdalenian salmon from Gourdan cave (France). The projectiles depicted over its body have been interpreted as the oldest representation of arrows in parietal art. When confirmed, such a finding would not only place the origin of this weapon fully within the Upper Paleolithic, it would also evidence the practising of a kind of fishing that has not been documented ethnographically in Europe (taken from Cleyet-Merle, 1990, 63).

gigas) of the Amazon basin and other bonytongues (*Osteoglossiformes*) of Africa and tropical Asia as well as several kinds of catfishes (*Siluriformes*).

In the past, one may surmise that in Europe many of the fishes that could be taken with spear and leister could likewise have been killed with arrows. In addition, stalking predators that remain close to the surface, such as Pike (*Esox lucius*) in freshwater or groupers (*Epinephelus* sp.) in coastal waters could, when remaining static enough to assure a successful shot, constitute additional taxa. As was the case for "spears at large", fishing in this way would also tend to produce a catch excluding small individuals.

As for marks, traces of arrows on a fish skeleton would amount to punctures which, depending on the condition of the bone, could be evident or impossible to detect. Arrowheads imbedded in bones, as have been found in mammals and birds, would be almost impossible to retrieve due to the brittleness of fish skeletons. In terms of taxonomic variety, the most likely targets would be rather voluminous and quite static freshwater fishes. Arrows have, apparently, never been used in marine environments.

Traps and weirs

Traps and weirs, of which an infinite variety are documented ethnographically (Brandt, 1984; Stewart, 1977) remain poorly documented in the archaeological record due to the organic materials from which most of them were constructed (figure 8). Considered as the most productive of the fishing devices due to the combination of a passive fishing technique with the potential for catching large quantities of fish within a short time, most traps and weirs required a lot of investment and thus were likely to be deployed only in those instances where large catches were assured and groups of people could be gathered.

The two major principles responsible for the success of the largest traps and weirs were the relentless urge of certain migratory fishes to make their way to their spawning grounds, by swimming either upstream (e.g., salmon, sturgeon, shad) or downstream (e.g., eels), and the ebb and flow of the tide that affect many littoral groups, prominent among them the sea breams and grey mullets. Based on the ethnographic records of the northwest Pacific, the concentrations of some species must have been tremendous when the runs were at their peak (Stewart, 1977, 19).

Whereas weirs were mostly interception devices and worked best when built right across a river or angled to guide the fish into traps, the latter could be placed at any location that the fishermen knew from experience was going to produce a successful catch. In these latter instances, traps were normally baited. As for stone traps, wall-like rock alignments, these allowed fishes to swin over or around them at high tide, only to end up in the "corral" (the Spanish term is most appropriate here) when the waters receded. Many wooden traps were built on this same principle and, accordingly, would be placed in the same way as *corrales*: in rivers, at their estuaries or in bays drained at low tide.

Traps must have been part of the hunter's toolkit since time immemorial, and only a matter of time before they were turned into efficient fishing devices. In all cases they would be most effective in shallow waters and it is among the fishes living in these waters that one must look for evidence for their deployment in the past.

The most likely candidates would be the aforementioned amphidromous species, but any medium to large-sized species congregating in large numbers in the littoral waters, such as meagre (*Argyrosomus regius*), needlefish (*Belone belone*) or gilthead (*Sparus aurata*), would qualify as potential indicators of trap or weir fishing when found in large numbers. The smallest of the basket traps would often capture the juveniles of these same species and certain codfishes plus the small bottom-dwelling taxa such as wrasses, used mostly for bait. None of the cartilaginous fishes, on the other hand, would constitute major catches since these are mostly non-gregarious animals unlikely to reach the surf line or penetrate into brackish waters.

Taxa which could be captured by means of stone traps would include a variety of small benthonic taxa (i.e., blennies, gobies, soles, plaice, etc) as well as some of the loosely gregarious, littoral free-swiming taxa such as grunts (Haemulidae) and silversides (Atherinidae). Among the very large fishes, only sturgeons (Acipenseridae) constituted major catches of stone traps until recent times (Sáñez-Reguart, 1988).

Figure 8. Roman mosaic from Kélibia (Tunisia) depicting a net similar to a fish basket trap, meant for the capture of partridges and other small-sized terrestrial prey.

Hooks

Hooks are the archetypal fishing gear: in the view of many authors, we have here an instrument that, for once, did not originate in the toolkit of the palaeolithic hunter but was always designed specifically for fishing. The fact that in some places baited hooks were occassionally used to catch aquatic birds such as cormorants (Phalacrocoridae) and terns (Sternidae) may reveal a sort of "return trip" of fishing gear to the toolkit of the hunter that would constitute an exception to the rule we have been defending so far.

Whether such evolutionary speculations are true or not, the fact remains that a hook without accessories, lines in particular, is a useless device. And although certain of these accessories have certainly improved with time, ample documentary evidence testifies to the efficiency of the hooks used by aboriginal fishermen, such as the Maori or the native Americans, when the Europeans first reached their fishing grounds (Stewart, 1977, 25; Leach, 2006, 95-98).

One of the advantages that hooks possess over the other types of fishing gear is their versatility. Not only can they be made from a wide variety of materials, they likewise exhibit an almost infinite variety of accessories such as lines, floats and sinkers which extend the range of locations where the gear can be deployed. This allows hooks to be used at any depth, and, depending on the specific combination of accessories, to target superficial, mid-water and bottom-living fishes alike. How large a fish a given hook size will take also depends on the details of its jaw anatomy (Owen & Merrick, 1994).

For the archaeoichthyologist such versatility is a mixed blessing: since so many different kinds and sizes of fishes can be caught on a hook, the question to be answered would be what kinds of fishes are *unlikely* to be taken by hook. To start, microphagous species such as clupeids (sardine, anchovy, shad), which are for the most part filter feeders, and soleids (sole) whose tiny jaws are designed to feed on minute benthonic invertebrates. Of the large fishes, even the small-mouthed invertebrate feeders such as sturgeons are taken by hook as are all of the herbivorous species, including the carpfishes. The second group unlikely to be taken by hook would be small fishes (i.e., fishes below 30cm SL).

Although nowadays metal hooks can be made to target small animals, the materials from which hooks were traditionally made (i.e., bone, antler, wood, shell) did not permit so versatile designs; in any case, it was far more productive to target small fish with mass-collecting gear, such as nets.

A third group unlikely to be taken by hook, at least until recent times, were the very large fishes, meaning animals over 20-30kg; even though when hooks were able to catch and hold them, and people capable of hauling them in, the lines would have not withstood the strain imposed by the large weight of such animals.

Finally, sharks that sport large teeth could not have been taken because they could easily cut the line to which the hook was attached. For all practical purposes, then, except for these four groups, almost all fishes to be found in littoral waters, whether fresh or marine, could be potential targets for hook fishing.

Nets

In the layman's imagination, nets are the paradigm of fishing instruments, yet ample evidence testifies to their use for capturing of birds and mammals. Whether this is indicative of their origin as part of the hunter's toolkit or not remains open at this point, although it seems far more probable to envision their use in aquatic environments being secondary to terrestrial settings than viceversa.

The simplest kind of fishing net were dipnets, manual devices that would qualify as a trap, consisting of a small "bag" attached at the end of a shaft. Nowadays, this type of gear is mainly used to bag fishes after hooking them. Used in conjunction with other fishing devices or in isolation, dipnets are mostly operative in shallow waters, with the fishermen on foot or, more rarely, for hauling prey onto a boat. Dipnets allow for the capture of medium sized fishes in an essentially individual manner but both historical and ethnographic data reveal that, under circumstances of local abundance, people could walk along the surfline through swarms of small fishes dipping nets to capture them (Stewart, 1977, 88). In these cases the species would mostly be spawning shoals of smelts and clupeids (see Sahrhage & Lundbeck, 1992, figure 50 on page 72). The diversity of taxa that can be captured with dipnets would be far larger, however, and would include the salmonids, carpfishes and perch as the main groups in fresh water and estuaries, as well as many of the littoral pelagic and demersal species in the marine environment, for instance juvenile codfish in the Northern Atlantic and sea breams and grey mullets in the Mediterranean.

A second category of net operated by a single fisherman is the casting net (Brandt, 1984). The fishermen, again either on foot or operating from a vessel, can cast it over a group of fishes in shallow water and bag them before they are given a chance to escape (figure 3). The most common prey in these operations is of small size, either truly small fishes such as clupeids, smelts or silversides or the juveniles of larger-sized taxa, as is the case of the grey mullets. Since mainly gregarious fishes are taken, net casting, despite its small scale, can be an extremely productive activity (Bekker-Nielsen, 2005, 86). Documented since the Neolithic of Twann (Switzerland) these nets must have been used in earlier times (Cleyet-Merle, 1990, 147).

The two remaining categories of nets that one may assume have been used throughout Prehistory and well into historic times were the seine and the gill net. Both types require the coordination of groups of people and can be deployed over a wider range of settings, in particular deeper waters, than is the case for the individual nets. The range of taxa that they can catch is, accordingly, far wider.

Fixed seine nets operate in still and extremely shallow waters. By wading into the water and splashing it, people drive fishes into the net, much in the same way as they would drive ungulates into a corral. This method of fishing is particulary fruitful with clumsy swimmers such as flounders and other benthonic fishes. Due to the "mass mortality" it provokes, this gear tends to be non-selective in terms of size (table 1-3).

For beach seine nets, the gear requires a boat to deploy the gear in the water, and two crews of fishermen on the beach, each holding one of its two ends. Once the net has been correctly set, these crews push the lines in, closing the net while hauling it onto the beach. Again, clumsy swimmers of many sizes are taken but many other species of demersal or even pelagic taxa (table 3) can also be encircled before they have a chance to escape.

Beach seines of a wide variety and increasing complexity have been developed at different times and places (Brandt, 1984). Although they originally functioned as traps, some of the more complex beach seines, as is the case of the madrague ("almadraba"), eventually also became a weir of sorts, with nets substituting for stones and wooden implements. This means that what started as a trap for the capture of a wealth of littoral, medium to small-sized taxa, ended up as an interception device able to capture the largest of the most pelagic fishes that roam the ocean (tunas). Likewise, by virtue of fishes shoaling in schools of similar size, the originally wide range of sizes that beach seines trapped was restricted when interception was added to the original trapping function of the beach seine.

Scientific name	**Common name**
Lamna nasus	porbeagle shark
Isurus sp.	mako sharks
Sphyrna sp.	hammerheads
Clupeidae	sardines & shads
Salmonidae	salmons & trouts
Zeus faber	John Dory
Seriola sp.	Amber jack
Pomatomus saltator	Blue fish
Scombridae	mackerels & tunas
Sphyraenidae	barracudas
Stromateus fiatola	Pomfret

Table 1. Main pelagic fish taxa in iberian archaeological assemblages.

Gill nets are screens hung vertically in the water that rely on a fish penetrating partly into the mesh, so that the net becomes entangled behind its gill covers, preventing the animal from withdrawing. Gill nets are the most selective of fishing gear since, for any given mesh size, neither small fishes, able to pass right through the mesh, nor large ones, unable to penetrate far enough into it, will be caught. Narrow size ranges of a particular fish species may thus constitute one good indicator that gill nets have been deployed in an archaeological setting, provided these fishes possess the characteristics of the "ideal gill net target", namely: 1) pelagic/demersal (table 1-3), 2) schooling/gregarious, and 3) not too large (i.e., rarely in excess of 50/60cm, representing 1-3kg). Given that setting gill nets in offshore waters seems to be a rather recent phenomenon (Sahrhage & Lundbeck, 1992, 90), a fourth trait one may add for pre-industrial archaeological fish collections would be a dominance of inshore-dwelling species that may include offshore migrants, such as mackerels and certain jacks which at times get quite close to the shore.

Gill nets have been more rarely set in fresh water although the natives of the northwestern Pacific coast deployed them for the capture of salmon (Stewart, 1977, 86). A mere distribution of sizes consistent with the use of gill netting, however, does not warrant far-fetched conclusions such as the development of cordage and netting technology 25,000 years ago in the lower Darling region of New South Wales (Australia), based on the narrow size distribution of the golden perches *(Macquaria ambigua)* found there (Balme, 1983, 30).

An old saying holds that gill nets are so productive because they in fact, attract fishes. As with so many other cases of traditional wisdom, there may be an element of truth here. Gill nets work best when fishes meet them head on, a circumstance

Scientific name	Common name
Eugomphodus sp.	sand sharks
Acipenseridae	sturgeons
Epinephelus sp.	groupers
Sciaenidae	drums & meagre
Mullidae	red mullets

Table 2. Main demersal fish taxa in iberian archaeological assemblages.

Scientific name	Common name
Mustelus	smooth-hound
Squatina	Angel shark
Rhinobatidae	Guitarfishes
Scyliorhinidae	dogfishes
Rajidae	rays & skates
Muraenidae	moray eel
Congridae	conger eel
Halobatrachus sp.	Frogfish
Molva sp.	lings
Labridae	wrasses
Scorpaenidae	Scorpionfishes
Triglidae	Gurnards
Pleuronectiformes	flatfishes
Pleuronectidae	plaice &flounder
Scophtalmidae	Turbot & Brill
Bothidae	wide-eyed flounders
Soleidae	soles

Table 3. Main benthonic fish taxa in iberian archaeological assemblages.

that precludes the use of the lateral line system and where only sight can be of help (figure 2). The eye, however does not detect the net until it is very close to it, and the only way out is an abrupt change of direction. Such a rapid movement is extremely difficult, often impossible to perform if the fish is leading a shoal. In this case, the inertia to proceed in the same direction is so great that the leaders of the group are pushed into the net in much the same way that the Titanic's inertia forced the ship into the iceberg. This means that, among the creatures of the sea, pelagic, gregarious fishes are the most vulnerable to gill nets, a danger that increases with the speed and size of the shoal. No wonder that productivity here runs high.

Scientific name	Common name
Dasyatidae	Stingrays (benthonic-epipelagic)
Myliobatidae	sea eagles (demersal-pelagic)
Galeorhinus galeus	Tope shark (demersal-pelagic)
Squalus acanthias	Spiny fish (benthonic-pelagic)
Merluccidae	hakes (demersal-pelagic)
Gadus morhua	Cod (demersal-pelagic)
Melanogrammus aeglefinus	haddock (demersal-pelagic)
Pollachius pollachius	Pollack (demersal-pelagic)
Pollachius virens	Saithe (demersal-pelagic)
Trisopterus sp.	Pout & Poor cod (demersal-pelagic)
Cyprinidae	carpfishes (benthonic-pelagic)
Belonidae	needlefishes (demersal-pelagic)
Mugilidae	grey mullets (demersal-epipelagic)
Trachurus sp.	horse mackerels (demersal-pelagic)
Dicentrarchus sp.	sea basses (demersal-pelagic)
Serranus sp.	combers (demersal-benthonic)
Haemulidae	grunts (demersal-pelagic)
Sparidae	sea breams (demersal-pelagic)

Table 4. Common fish taxa from iberian archaeological assemblages that cannot be unequivocally assigned to a particular zone in the water colum.

The multiple problems of interpretation

The archaeologist faces several major problems when it comes to identifying the type of gear that may have produced a particular fish assemblage. The first, and foremost task would be to ensure that the assemblage reflects what was present in the sediments when these were excavated. Retrieval biases, in particular selective losses due to inadequate retrieval techniques, have long been known to alter drastically the original composition of the samples, debilitating the validity of the inferences made by the analyst (Wheeler & Jones, 1989). Up until now, it was believed that screen mesh sizes of 1mm were sufficient to insure representative fish samples at archaeological sites. More recently, authors such as Bødker Enghoff (2005) have demonstrated that for the retrieval of the smallest fishes such as the smelt *(Osmerus eperlanus)*, meshes of 0.8mm and 0.6mm are necessary. Given that such screen sizes are rarely used on archaeological excavations, and that the smallest fishes are often taken with gear that differs from that employed for the larger specimens, Bødker Enghoff's results cast a shadow of suspicion over many ichthyoarchaeological as-

Figure 9. Identical gear, such as similar-sized hooks, allow to fish a variety of size classes from many different species.

semblages studied to date and raise concern when it comes to making inferences about fishing tackle deployed in the past.

A second matter of interest has to do with the determination of the agent that produced a particular fish assemblage. This issue may be of lesser importance on historic sites where documentary evidence is available, but for the Paleolithic, the "fishing" behaviour of various predators may mislead researchers into incorrect thinking when dealing with mixed deposits (i.e., those that are not only the result of human activities), although criteria for discriminating anthropic activity from that of the animals would obviously be of help here (Erlandson & Moss, 2001; Roselló & Brinkhuizen, 1994; Van Neer & Morales, 1992).

These matters notwithstanding, the two most important issues in the context of the present discussion are those dealing with the problems of equifinality and the validity of analogies in general.

Equifinality

Even when considering a given size category, most fishes can be taken in a variety of ways. In fact, we can think of almost no instance where a particular species is fished exclusively with a single kind of gear. "V" hooks came closest to that strict corres-

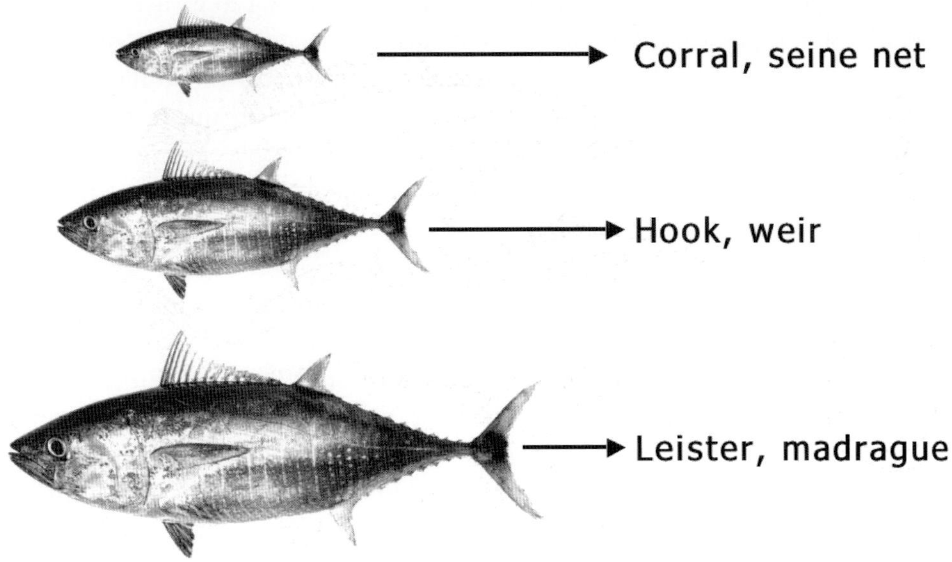

Figure 10. Size classes of species that reach to a large size are often taken through different methods employing different kinds of fishing gear. Estimating the size of the archaeological specimens may thus be as important as determining the species when it comes to inferring fishing tackle, yet such a procedure is not devoid of problems because tackle also changed through time, in different cultures, etc.

pondence in the case of the Pacific halibut (*Hipoglossus stenolepis*), yet were not the only tackle used to target this species since conventional hooks, and in shallower waters seine nets or even leisters could also be used; likewhise "V" hooks worked equally well for many other large-mouthed flatfish as is the case of the sand sole, *Psettichthys melanostictus*, in that same region (Stewart, 1977, 23). In a similar way, one may surmise that in pre-industrial times, the largest tunas could only have been appropriately fished with madragues, yet various Pacific cultures as well as people in the Canary islands, on the Black Sea and in the Strait of Gibraltar have traditionally captured these huge animals with harpoons and leisters (Merino, 1997, 623). Though most of the later catches would only affect individuals not "populations" (i.e., schools), the difference would be invisible to our eyes, since most archaeological deposits constitute accumulation horizons incorporating the products of a number of fishing episodes. These problems are compounded when confronting more versatile gear such as hooks (figure 9), more "versatile" species, i.e., taxa whose size range is very large (figure 10), or a high taxonomic diversity in an archaeological collection. Strict correspondence between a species and the fishing tackle that could bag it would then be at best an exception, at worst an illusion.

One way out of this ambiguity would be via context and complementary information. The best complementary information would of course be provided by the fishing gear itself or by documentary evidence testifying to its use but in that case, why should one care about the fish remains? The fact is that these types of evidence are not without interpretational problems, either. For one thing, most fishing gear until very recent times was made from perishable organic materials, with minimal chances of being retrieved. Even under the best possible circumstances, such evidence would neither be intact nor reflect the original variety of fishing tackle at a site. Furthermore, in the absence of ethnograhical analogies, one needs to consider the possibility that certain kinds of fishing gear passed unnoticed to the eye of the archaeologist. The heated debates over putative tackle, as for instance the Paleolithic "straight hooks" from Cueva de Nerja (Málaga, Spain) or the "arrow points" from the Paijian culture in the northern coast of Peru, constitute but two instances that reflect how ambiguous the complementary evidence might at times be (Aura & Pérez, 1998; Chauchat *et alii*, 2004; Gálvez & Queiroz, 2008). For all these reasons, the feedback from the faunal analyst to the archaeologist or historian cannot be dismissed as irrelevant. Even if fish remains cannot indicate what types of gear were actually deployed at a particular place or time, they may at least help decide what types were definitely not employed, focusing research on a narrower, more fruitful range of alternatives (see below).

As for context, a large variety of data, from coastal topography to sedimentary analyses, can hint at the types of gear likely to have been used at a particular site. Again, this evidence must be correctly contextualized in order to be of use. Coastal dynamics are intense in most areas, and coastlines in particular have changed dramatically since the onset of the Flandrian transgression during the Late Upper Paleolithic. Before trying to infer what kinds of fishing gear were deployed at a particular location, then, it is necessary to establish whether conditions at that location were different at the time when the fishing took place from those found nowadays. For answering this question, as in the case of complementary information, fish may prove instrumental, provided that their behaviour and habits have remained unchanged. This proviso leads us into the final issue of this section on the validity of present day fishes as analogues of their ancestors.

Analogies

Archaeozoology, like any discipline dealing with historical processes, draws heavily upon analogical inferences in order to explain data. The governing principle here is that of actualism (i.e., "the past can be interpreted in terms of how phenomena operate today"), whose roots go back to James Hutton's uniformitarianism (Gould, 1987, 66-67). Actualism has yielded uncounted benefits, yet it has sometimes been used due to the sheer lack of alternatives, as it is not devoid of problems (Morales, 1996). Fishes are a case in point.

It now seems undeniable that marine populations have changed dramatically over the past fifty years as a result of human activities, and also that their present status is but a pale reflection of their past. We have only recently realized, through archaeozoological studies among others, that this phenomenon is far from recent, its roots dating back to at least the time of the European colonial expansion (Jackson *et alii*, 2001). These same data reveal that many species, often key taxa within their ecological communities, became depleted or were hunted down to extinction at this time, bringing about radical re-arrangements in the ecosystems they were once a part of (Jackson *et alii*, 2001, 629).

One may surmise that radical diminutions in the abundance of certain commercial fishes may have rendered certain types of fishing tackle obsolete. Consider the case of the herring rake that Stewart so vividly describes in her book (1977, 76-77). This simple kind of gear, essentially a long stick with a series of points hafted to its end, was only effective when herring densities were so high that "paddling" the rake through the water impaled a large quantity of them. Once overfishing diminished the number of herring, the method was simply abandoned. Had it not been for the ethnographic accounts, all information concerning this kind of tackle would probably have vanished from sight as its perishable materials would not last for long, and the occasional findings of points at excavations would have been interpreted as remnants from other types of more conventional gear (e.g., hooks; *cf.* the discussion of equifinality, above). Similarly, without data on the herring rake, the faunal analyst would have been misled to interpret any large quantity of herring bones as evidence for the deployment of nets. In the ancient world, an equivalent case would be that of the multiple trident that Black Sea fishermen used for spearing young tunny (*cf.* Oppian 4.535-538). This device would only work in a dense shoal of fish and disappeared once the densities of tuna diminished (Bekker-Nielsen, 2005, 89). Perhaps one should consider the – presumably exaggerated – accounts of some classical authors, like Strabo and Gryllus, that referred to the fishing of tunas with bare hands and by stoning the densely packed schools that entered the Black Sea (Merino, 1997, 625) under the same light.

The point to stress is that present day analogies may not provide a reliable referential universe for interpreting the past, no matter how much one is forced to make use of them. The crucial question would be to know how many "herring rakes" have we lost thus far.

Another instance of the weaknesses posed by the use of recent analogies as a basis for inferences about the past concerns the biology of fishes. Not the fundamental features of their physiology or behaviour, which are genetically determined, and thus unlikely to change over a short time, but those more labile traits that influence habits or learning during a lifetime and could have had dramatic effects on the catches. One important case in point is how fishes interact with fishing gear. Different experiments over the past twenty years have revealed how fish react to hooks

(Fernö & Huse, 1983; Forster, 1973; Leach, 2006; Løkkeborg & Bjordal, 1989; Løkkeborg *et alii*, 1993; Orsi *et alii*, 1993). Among the many factors that influence the reaction of the fish to a hook, previous experience appears to be of critical importance as it makes the animals wary of further interaction. Beukema (1970) established that a carp (*Cyprinus carpio*) that escapes a hook does not bait for another year. Other studies reveal that, despite a tremendous amount of individual variation within species, painful experience with hooks makes fishes less susceptible of ever being hooked again (Fernö & Huse, 1983; Leach, 2006).[4]

From these data one may surmise that, although idiosyncratic responses and memory seem critical for the short-term survival of a particular fish, the large differences recorded in individual responses to hooks, undoubtedly a reflection of more fundamental differences, insure that a response with a genetic substrate, driven by natural selection, must have been an outcome of long-term interaction between fishes and hooks. In other words, through selective survival, fishes wary of hooks must have become more common with time.

Although our conclusions are based on scant data, thus need to be taken with caution, one may likewise surmise that hooks do not constitute an exception, and that a similar evolutionary rationale applies to other kinds of tackle (there are, however, no chances for fishes to "adapt" to techniques, such as electrical fishing or venoms, that kill them).

One corollary of the aforementioned coevolutionary process is that pristine populations of fishes, meaning those under little or no fishing pressure, must have been far less wary of fishermen and of fishing tackle, thus easier to capture. Ample documentary evidence from ethnographic records and reports by early European colonists testifies to how easy it was for the aboriginal people to fish (Jackson *et alii*, 2001). John Cabot's accounts of cods being fished by the thousands with baskets off the beaches of Newfoundland, for example, may seem far-fetched at present, and might indeed be taken as propaganda highlighting the bounties of the New World (Kurlansky, 1998, 48-49), yet may likewise contain an element of truth indicating how tame and easy to catch some of these commercial fishes were a mere 500 years ago (see above).

A second corollary of this coevolutionary trend is the way it determined fish distributions, thus the accessibility of these resources and the kinds of gear with which they could be harvested. As mentioned, cod in Newfoundland were reported being fished at the surf line where they are no longer found. Such reports are by no means exceptional, and shifts in the areas where fishes occur may in fact help explain some

---

4 This probably explains the awkward-looking, rotary hook of the Maories, designed to minimize painful sensations from the tip and to keep the fish unaware of the danger, postponing its aggressive reaction until it is too late (Leach, 2006, 96).

"aberrant" archaeological finds. At the cave of Santimamiñe (Basque country, Spain), for example, remains of hake (*Merluccius merluccius*) have been recently retrieved from Neolithic levels (Roselló & Morales, 2008). Dismissed as intrusive, given that hake is nowadays an offshore, mid-water species, further investigation revealed that these finds were indeed Neolithic. Santimamiñe lies on the margins of a river but its waters are close enough to the coast to be tidal, thus it is not unlikely that euryhaline species (i.e., those able to tolerate brackish waters) could penetrate way upriver. Whether hakes in the past could reach to very shallow waters or possessed a far greater tolerance to shifting salt levels we do not know, but reports of a flourishing hake fishery in the "rías" (estuaries) of Galicia (NW Spain) throughout the Middle Ages are numerous and have been compiled by Ferreira (1988, 58; see also Morales *et alii*, 2009). Biogeography tells us that fishes occupying a large latitudinal gradient tend to occur in shallower waters the higher the latitude and vice versa (Eckman, 1953), yet the key question here is: what would happen if populations from the shallower waters were fished down to extinction? Though, in general, consistent spearing would exert far lower pressure on local populations than netting and trapping, it could nevertheless substantially reduce the local numbers of large, infrequent species (e.g., top predators), provoking small-scale extinction events. For this reason, from an archaeological standpoint, sedentary, territorial species like pike, wrasses or groupers would always far better indicators of technological change or harvesting pressure than gregarious, migratory ones. At any rate, whether the present-day distribution reported by the various authors holds for the past we do not know, although one suspects that, being affected by so many factors, they must be undergoing perpetual change (Whitehead *et alii,* 1984-1989). What one should always keep in mind is that one major factor affecting the distributions, thus the availability, of certain species in the recent past must have been humans.

Thus, when one is attempting to explain why a species appears at a particular site, one needs to consider anthropogenic causes, that would need to be added to the natural ones, and would render inferences based on distributions of modern fishes meaningless (Rodrigo, 1994). From such a standpoint, ecological distributions over the water column would be far more reliable indicators than "horizontal" distributions despite the fact that many species may still be difficult to allocate to a specific category (*cf.* table 4).

A final instance where present-day analogies may be invalid concerns the dramatic size diminution trends documented for countless species, which often render comparisons between modern and archaeological specimens questionable.

One of the routines employed by the faunal analyst to distinguish natural from man-produced mortality is to compare the (observed) size distribution of the archaeological population with that of an (expected) present-day population of that same species, checking for discrepancies (Wheeler & Jones, 1989). The fact is that, whether showing a normal (i.e., bell-shaped) distribution or not, modern populations are most often truncated on the right (i.e., lacking the largest specimens) compared with their archaeological counterparts. In some cases the differences are dramatic, some archaeological samples featuring sizes that are impossible to estimate by sheer lack of equivalent specimens in the modern collections (Morales & Roselló, 2007, 2008). This recurrent "pattern" has been attributed either to selective fishing or to an overfishing that removes the largest and scarcest individuals from a population first (Leach, 2006, 205-228). A systematic removal of the largest specimens has cascading effects for a given population of fishes, be these territorial or not (Wootton, 1990; Pitcher, 1993; Helfmann *et alii*, 1997). One recurrent effect of such a removal is an increased growth rate and earlier maturation of the remaining fishes, parameters that apparently change other life traits such as the migrating capabilities of certain species or their social behaviour, but that have not still been documented in detail. Until this is done, one needs to consider the possibility that some inferences made on how today's far smaller descendants are fished might not be at all applicable to their former, far larger ancestors.

Conclusions

Aside from taphonomic factors which influence the diversity of the remains recovered, the variety of fishing techniques, the modification of methods with the season of the year and life cycle factors combine to demonstrate how difficult it is to generalize about fishing technology on the basis of archaeological material. Having said this, it now seems appropriate to ask whether fish remains offer an adequate basis for inferring fishing gear in the archaeological record. There is really no straightforward answer to this question, since any single species, or even a size class within it, can be taken with a variety of tackle for which evidence will for the most part be absent from a particular excavation or even from the archaeological record at large.

Under such circumstances, the best the faunal analyst can do is to adopt a "programme of minima" that could at least exclude certain alternatives. Such "programme of minima", framed in the negative, is unlikely to bridge the gap between educated guesses and testable data. To proceed from mere speculation to fact, interdisciplinary collaboration will be required: no matter how defective or incomplete each set of data may be, combining all the pieces of the puzzle would produce a synergy allowing new lines of evidence to emerge and leading reseach down hitherto unsuspected, yet surely fruitful, pathways.

Acknowledgments

Guadalupe López Monteagudo (CSIC, Madrid) is gratefully acknowledged for providing the images from mosaic that appear in figure 1. Eufrasia Roselló (UAM) provided data and a review of the manuscript. Tønnes Bekker-Nielsen (USD, Kolding) is thankfully acknowledged for his reviewing of the final draft and for providing critical pieces of data.

Knot of a Roman fishing net (La Albufereta) (photo C. Alfaro).

2. Fishing Nets in the Ancient World: the Historical and Archaeological Evidence

CARMEN ALFARO GINER[1]

For obvious reasons, ancient objects made from organic materials are the great unknowns of the past. However, thanks to the techniques available to us today, we can now obtain more information from the few fragments that are preserved. This approach is more reliable than the traditional method based on written sources and iconography, which remain, nevertheless, very important as well. The scarcity of material information about artefacts such as fishing nets stands in curious contrast to the immense importance they had for everyday life in Antiquity. Neither freighters nor warships would have been possible in ancient times without fabrics (which were used for the sails and in the construction of the hull) or cordage (for the rigging, ropes and the extremely fine cord with which the strakes were sewn together).

In Antiquity, nets were usually referred to by various generic names: *lina* (Homer, *Il.* 5.487; *Anthol.Palat.* 6.12, 16, 27-29, 33)[2], *diktya* (Aristotle, *H.A.* 8.13, [598a]; 8.19 [602a]; Athenaios 7.284; Diodorus Siculus 17.43; *Anthol.Palat.* 6.11, 13-15, 24, 26, 30, 38), *halieutika, henodion* (Gr.), *rete, retis, plaga, cassis, jaculum, verriculum* (Lat.) etc. These terms, however, do not in any way describe the form of the nets in question.

---

1 This study was carried out with the economic support of R&D project reference number HUM2004-1984. I would like to express my thanks to the workshop organisers for all the practical experience we were able to have and which I personally found extremely helpful for my work.

2 It must be remembered, with Chantraine, that *linon/linum* does not always denote flax fibre *(Linum ussitatisimum)* or cloth made from linen or cotton. It can also mean string, fishing thread, sail, a thick yarn or rope, and even fishing net.

Nets were indispensable not only for making many types of bags or sacks (*retia ferre*), in which all kinds of objects were carried (Saglio, 1918, 855; Thomas, this volume, with excellent photos of nets from Myos Hormos used as sacks), but also as extremely useful items for other activities, such as holding small pots or containers. These nets were hung from the ceiling or other supports (Crowfoot & Crowfoot, 1961, 82, pl. XVIII.1; this particular net, dated to 132 AD, is a coarse-knotted net of Z-spun, S-ply threads identified as flax, with a coarse mesh (about 5cm between knots) and thick threads (0.3mm).

Nets were also used for making a wide range of military equipment, in gladiatorial combat (Berger-Joos, 1971, 25-28, 30-32), in the overland and maritime transport of goods (Beal, 1999, with incomplete bibliography; Wild, 2001), and even as clothing (hair nets, adornments on dresses, etc.).

Above all, however, nets in Antiquity were identified with hunting (Aimard, 1951; Guest-Papamanoli, 1996), and especially fishing (Bekker-Nielsen, 2002a, with bibliography). An epigram by Leonidas of Tarentum (*Anthol.Palat.* 6.13) shows the importance of different types of hunting and fishing nets as instruments worthy of being offered to the gods in their sanctuaries. To Pan three brothers have dedicated this equipment: *Damis a purse net for beasts that roam the mountains; Cleitor these nets for fish, and Pigres this unbreakable neck chain for the feathered birds of the air. For never did they return home with empty nets, one from the woods, one from the air, one from the sea.* These two activities are so closely related due to the instruments employed that one of the epithets used to refer to Artemis was *Diktynna*, "she who hunts with nets", and as such she was the protector of both terrestrial hunters and fishermen.

In the sphere of sea and river fishing, nets enabled their users to increase their catches enormously compared to those employing other techniques (Aelian, *N.A.* 12.43). This seems to be what can be deduced from the chapter which Aristotle dedicates to the migrating fish that swim in great shoals towards the Black Sea (*H.A.* 8.13 [598a-b]. Aelian follows him in almost all the details he gives (*N.A.* 15.5-6 and 9.42). The dominant idea we are left with is that nets provided abundant catches and enriched a large number of people who made their living from the sea.

Indeed, nets were regarded as a tool of the greatest importance within ancient commercial fishery. The naturalists knew that many fish fatten and breed in certain areas of the sea, but also in rivers (Aristotle, *H.A.* 6.14 [568a-b]; 8.19 [601b]) and ponds, where they thought the best fish were concentrated. It is not, however, possible so far to distinguish different types of nets for seas, rivers, lakes or lagoons. It may be useful to point out here that many of the best-preserved remains of nets, to which I shall be referring later, come from lakes and lagoons, which are more sheltered and more likely to accumulate mud which protects organic matter.

The size, thickness and quality of the material from which the nets were made (thread or yarn of varying thicknesses) always conditioned their usefulness. That is why we can say that there were enormously strong nets, such as one of those described quite precisely by Homer (*Il.* 5.487: *aphisi linou halonte panagrou*): a wide, knotted net that "catches everything". But there were also extremely fine-meshed little nets for catching very small species. Leaving aside the format of the net, which could render making it more complicated, what we might call "net-making technique" was relatively simple. Thick nets were used for catching large, strong animals and fish. One of the most beautiful scenes I know of is the hunting of wild bulls represented on one of the Minoan golden goblets found at Vaphio. Here the artist has simplified a great deal, leaving out the knots on the net and the repeated meshes, and replacing them with enormous ropes that were simply crossed over. The sensation of solidity in this case comes from the thickness of the rope (with the heavily pronounced detail of the S-twist). Tuna fishing always required strong nets, which Roman mosaics, especially, allow us to see clearly. At the other end of the scale, as examples of the finer nets used by the ancients, we might mention those hair-nets *(reticula)* made of extremely fine filaments, sometimes gold, that adorned the heads of women and girls of the well-off classes within Greek and Roman society (Crowfoot & Crowfoot, 1961, 82, n. 144; Alfaro, 1983-84, 77-81; 2001, 76-83; Bedini, Ferro & Rapinessi, 2004, 77-88).

It can therefore be said that the nets of the Greek, Roman and provincial world were instruments of immense utility due to their variety of shapes and uses, and the ease with which they could be made. In most cases no support or loom was needed to produce them; it was enough to fasten the initial structure, which defined the shape of the net, to a number of fixed points holding the net in its open position. The simplicity of net-making technology resides in the fact that it starts with a piece of rope that is gradually interwoven by means of simple or more complex knots. Some form of needle is usually employed for this purpose, although the knots can also be tied by hand. Occasionally techniques were employed that did require a small support (sprang technique). These were used to make certain extremely fine nets, but we do not know whether they were ever employed for fishing. We are going to look at the most common nets, i.e. those made with knots. The different types of net were developed by adding new types of knots, permitting a broad range of nets to be made.

The different types of nets used for fishing have a common property: they are translucent. This allows them to go unnoticed by animals. That is why net-makers strove for maximum efficiency, through their efforts to obtain as fine and as strong a cord or rope as possible, as we shall see below. We also have information to the effect that people tried to make nets of colours that would obscure the presence of the net or the line used in hook and line fishing. Aelian, *N.A.* 12.43 says, "of the dyed horsehairs, choose those of bluish colour and those of marine purple, as it appears all the others are bad"; naturally, because the blue hues are hidden by the waves when the sun is shining. The "bluish colour" must be light blue, in contrast to "marine purple", a bluish tone of real purple, which was employed when conditions were darker. Aristotle mentions net fishing in the early or late hours of the day (Aristotle, *H.A.* 8.19

[602a]). This tactic was obviously designed to make the fishing implements less visible to the animals. One might wonder whether, in certain cultural spheres of Antiquity, this reality and this practice of everyday fishing went beyond the barrier of *animism.* A study of the language taboos that still survive today in certain areas of South Asia revealed a tradition that considers nets almost as "(living) beings that make themselves invisible". Fish are supposed to be highly intelligent and the idea is to deceive them. One tradition that applies this idea to the very name of the instrument:

> Abbiamo parole che perseguono chiaramente l'unico scopo di agire sulla natura di un determinato spirito... La Weltanschauung animistica non conosce un netto confine tra gli esseri viventi e gli oggetti (dal nostro punto di vista 'inanimati'). E non si esclude la posibilita que nel nome sostitutivo... russo della rete da pesca *neved,* cioè 'non notata' (dei pesci), si abbia una espressione di questa stessa idea
> (Zelenin, 1989, 239. I am grateful to my colleague X. Ballester for this interesting piece of information).

Another phenomenon, perhaps more firmly grounded in reality, is the Lithuanian custom of protecting certain fishing-nets by padding them with strips of linden, to prevent them making any sound that might warn the fish of the presence of a net in motion (Rimantienë, 1995, 57). This is known through communications by fishermen in the early twentieth century who, curiously, used the same type of stone weights wrapped in vegetable material as those found at Neolithic sites in the same area, to which I shall return below.

The shapes of the nets

We have numerous written sources of information about the forms and applications of fishing nets in Antiquity, although they are not very descriptive. Especially significant are the hunting treatises that deal with the hunting of land and sea animals as a single subject; the techniques applied to catching fish are given the name *Halieutika.* Aelian mentions fishing nets as one of the four ancient types of fishing gear, the others being spear, creel and hook (*N.A.*, 12.43). Strangely, he lists all the items needed to make a net *(diktyeina)* and to fish with it: a rope *(sparton),* white and black linen *(linon leukon kai melan)* rush rope *(kyperoi),* cork *(phelloi),* lead *(molyboi),* pine wood *(pitoi),* straps *(imantei),* sumach *(rhouoι),* a stone *(lithos)* and a papyrus (*byblos*) but fails to explain how or for what purpose each item is used. Oppian is much more explicit regarding the forms of the nets, but he, too, fails to describe how, or from which materials they are made. This knowledge is taken for granted.

At all events, the sparse comments of these authors provide us with the quintessence of what in their time was judged to be the core knowledge necessary to obtain as precious a food as fish. Various other authors – Pliny the Elder, Diodorus Siculus, Philostratus, Ovid, Leonidas of Byzantium, etc. – complement our knowledge of the subject with more or less descriptive details. Oppian mentions a large variety of fishing nets, but this is undoubtedly only a summary of the numerous types that must have existed in the ancient Mediterranean. In its coastal waters, in the rivers that flowed into them and in the nearby lagoons, fishing was one of the most widespread economic activity and occupied a large part of the population (a good summary of these aspects is found in Bekker-Nielsen, 2005, 83-85).

The variety of knots used is one of the distinguishing features of fishing nets. There is another which is even more important, the raw material employed in making the yarn or rope that went into producing the net in question. As mentioned earlier, the carrying capacity of a given net depends on the thickness and quality of these items. In the analysis I offer here, I shall focus especially on the purely technical aspect and the materials employed. I shall describe the different ways of making nets and, in particular, I shall try to present, as far as possible, the results of the analyses of the very few material remains of the examples that have been preserved for us and been examined.

To help us in our analysis of the forms of the nets, we also have a wide range of images showing us fishermen in action with their nets. We can recognise extremely typical postures in the ways they work with the nets while fishing and see how many of these are the same as those still used today in some parts of the world (Brandt, 1984). Mosaics (Bekker-Nielsen, 2002a; López Monteagudo, in this volume) and paintings are particularly interesting. We also have scenes of people fishing with nets on Late Roman lamps (*lucernae*) and luxury tableware such as the famous silver plate with a scene depicting *erotes* fishing with a net from a boat, found at Augst (Switzerland) and kept in the site museum (Ginella & Koch, 2006, figures 61-71).

All these images offer us a picture in which the technical aspects of the business of fishing are combined with details of the clothing, the housing and the living conditions of the people engaged in it.

If the iconography is rich in representations of fishing nets in their context, the enormous spread of the motif of the net as a decorative pattern on many ceramic vessels in ancient art in general is a more curious phenomenon. This type of pattern (a repeated rhombus) became an extremely common ornamental element. It is found, for example, in certain pre-historic ceramics such as those of the Mesolithic period in the area of Lithuania (a time from which we also have some remains of actual nets, see Rimantienė 1995[3]), on Minoan ceramics and even on some Athenian and Rhodian geometric amphorae of the seventh and sixth centuries BC (De-

3 I am grateful to Margarita Gleba for the translation of a part of this book.

los Museum, some geometric bowls in Corinth Museum and the Athens Agora Museum, among many others). Its presence naturally inclines one to think of the net motif as a sign of the cultural identity of predominantly fishing peoples.

Among the sources for the study of the fishing gear of the ancients, there is a series of objects that were especially close to the hands of their users and conserve their direct imprints. These are the fishing instruments that have survived to this day and are the focus of our interest at this time. I am referring to the nets, the needles for making and mending them, the lead, stone and ceramic weights, the wooden and cork floats, etc. All of these items tell us something about the types of nets to which they belonged.

I have said earlier that both ships' sails and nets are objects, whose chances of survival are very poor, as they are highly perishable. Nonetheless, we do possess some important examples – more numerous in the case of nets – allowing us to gain direct knowledge of their fundamental technical characteristics: the raw materials employed, the type of manufacture of the yarn, types of knots and the openings in the mesh structure constituting the net. Very seldom is it possible to go beyond an intuitive guess at their overall shape, since the material remains are extremely small in size, making it hard to identify specific net types. However, the ancient written sources provide us with information that can be filled in by looking at the forms still in use in many areas of the Mediterranean. I am, however, not going to develop this theme here. It has already been extensively dealt with (Corcoran, 1957; Brandt, 1984; Powell, 1996; Bekker-Nielsen, 2005) and, as I have already said, there are no precise descriptions of the forms and their different uses in the ancient sources, nor are there any remains of whole nets. All I intend to do here is to recall briefly the information furnished by the literary sources.

It was commonly believed that the most efficient way of fishing as regards the quantity of the catches is net fishing (Bekker-Nielsen, 2005, 84-86). In the *Halieutika*, Oppian described, without going into excessively technical details, various types of nets which he attempts to associate with particular species of fish or animals, but does not go into the matter in any great depth. The extremely beautiful mosaic in the "Maison de la Cascade" in *Utica*, in Tunisia, and other representations in various mosaics in the Bardo Museum are typical of some of the different types of nets described by Oppian (*Hal.* 3.80-84). The following kinds can serve as examples within a broad range of types:

- *Amphíblêstron, or casting net*, with two variants: one called *sphairôn*, or round cast-net, that is thrown over the fish by the fisherman and the fish are then taken out through one end (probably the same as the type of net referred to in Spanish as *esparavel*), and the *skolion panagron*, a broad, curved, *all-enveloping* net that can catch all kinds of prey and was already mentioned by Homer (*Il.* 5.487).
- *Trawl nets*, including the following: the *griphos*, a draw-net. The *gangamon* (Gr. *gaggamon*), which seems to be a small net for catching oysters (similar to those

called *lamberos* in Spanish). Also belonging to this group were the *péza*, or small ground net, as its name suggests, as it was pulled along the ground by the feet; the *periegês hypochaí* or "rounded bag nets" which may have been used for fishing for purple, and may be identified with what is in modern Greek called *apochê* (with the meaning of distance). It is like a pocket of very fine mesh with a narrow opening. One of the most commonly used may have been the *sagênê*, a large driftnet (the same term was used for the big hunting nets). And the last one in this group is the *kálymma*, a cover-net (Bekker-Nielsen, 2005, 90-93).

Nets and fabrics

The oldest surviving examples of fabrics are preserved as simple impressions and are really closer to basketry than fabrics proper. These first fabrics were made with the aid of supports to keep the vertical threads of what we might call the warp in place (Alfaro, 2002a, 19 and n. 23). These are manual weaving techniques and the result is (wrongly) called "corded fabric". Corded fabrics were made with the aid of a structure that may be confused with a loom, but is not one. The difference lies in the fact that the weft is threaded through by hand, step by step in the former case, but with the aid of heddles that make the job easier and speed up the work in the case of the loom. They are typical of the Eastern pre-ceramic Neolithic: Nahal Hemar, Israel, Tell Halulla, Syria (Alfaro, 2002a), Wetzikon-Robenhausen (Altorfer, 1999; Altorfer-Médard, 2002), etc.

Netting is a very special case. As I said, we cannot properly refer to a net as woven, as the surface of the netting emerges out of the intertwining of a single thread with itself. It is strange how the ways of working we shall be looking at seem to be completely universal, and how cultures that were never in contact with each other arrived separately at identical solutions. A long period of time must have elapsed between the moment when humanity first had the wish to imitate animals (spiders) and when they achieved a practical technology with which to do so (Brandt, 1984, 204-5). I shall examine some of the net-making techniques, not all of which were used exclusively for fishing nets:[4]

a) The so-called Sprang technique (figure 1A; Hald, 1980, 256). This was widely used in ancient Egypt. There are a few examples from the Neolithic in the north of Europe and Switzerland, but they are not very common (Rimantienë, 1995,

4 These ways of making nets were described many years ago by authors such as S. Müller (1897), E. Vogt (1937), K. Schlabow (1950), M. Hald (1950), A. Seiler-Baldinger (1973), Jenkins & Diferí Williams (1985) etc., and their findings are reproduced, very often without citing the source – in these times of free acquisition of data that are held prisoner in another type of net, the Internet.

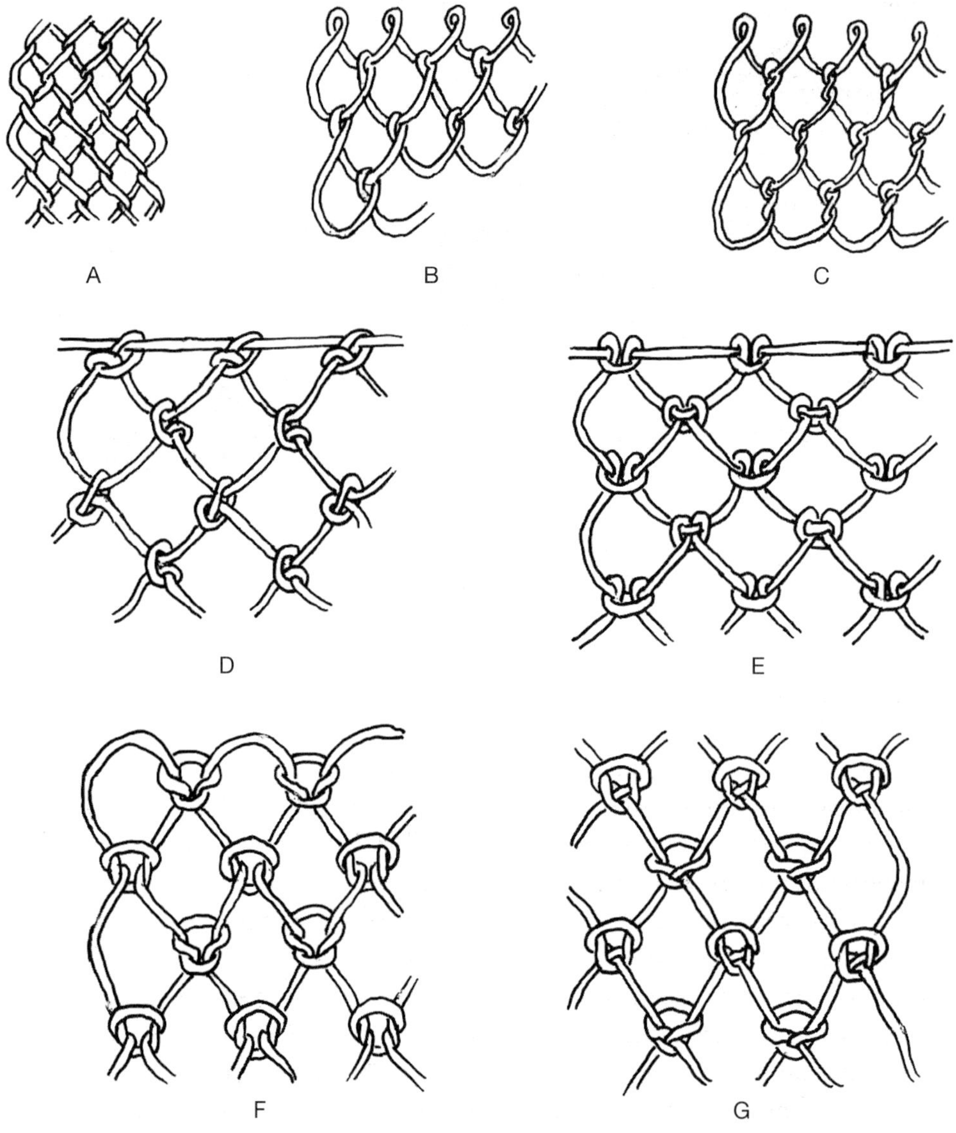

Figure 1. Fishing knots: A: Sprang technique, B and C: Primitive knotless netting (after Seiler-Baldinger), D: Palaphitic net, E: "cow hitch knot", F: reef-knot; G: weaver-knot.

55). This technique employs a lightweight stand, which may be formed simply of two cross pieces of wood, the higher one held in place by cords or string, the lower one held in tension by weights. A continuous piece of thread is placed between and around these two supports, creating what may we might call a false warp. The manual intertwinings of these threads are temporarily held in place by four simple bars of thin wood until the surface of the net has been created. When the piece has been completed, the place occupied by the last bar, held in

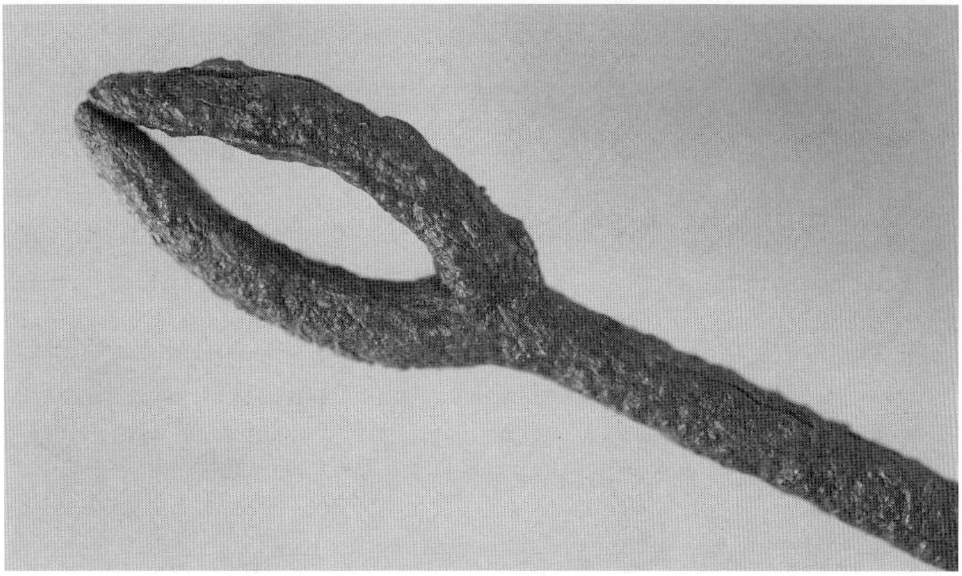

Figure 2. Net needle: Mediterranean filet type. La Albufereta, fourth century BC (photo C. Alfaro).

the centre, is taken by a small cord fixing in position all the nodes, what we might call "false knots" (Hald, 1980, 256). As the technology evolved, this system gradually became more complicated: mobile frames came into use, allowing the two wooden bars to move towards each other as the work progressed (Hald, 1980, 252-255; Jenkins & Williams, 1985). Many kinds of small nets were made with this type of technique (sacks, fine riverine and lacustrine fishing nets, etc.). This type of net has a drawback for fishing, since, if the thread forming it breaks, the whole net unravels and the catch may be lost.

b) Primitive looped knotless netting. Netting made with this type of technique, using a needle or a small manual shuttle, presents a number of variants (figure 1B and 1C), depending on the material employed. It can be used to make nets for fishing in rivers or lakes. It too has the drawback that if a breakage occurs in one place, all the knotless joints are also likely to open up (Hald, 1980, 284-310; Seiler-Baldinger, 2003, 56). However, its advantage lies in how simple it is to make, which was an important factor given the short service life of nets made from many of the vegetable fibres we shall go on to look at.

c) Knotted netting (figures 1D-1G). The technique for making this type of netting is, as the name implies, based on the continuous tying of knots on a thread that is, in its turn, attached to a thicker cord. This base is as long as the maker wants this part of the net to be. A special type of needle is used for this, varying from one country and one culture to another (Brandt, 1984, figure 394). The amount of thread that could be loaded into the so-called "Mediterranean needle" with two forks opening up slightly at the ends of a slim rod (14cm long; figure

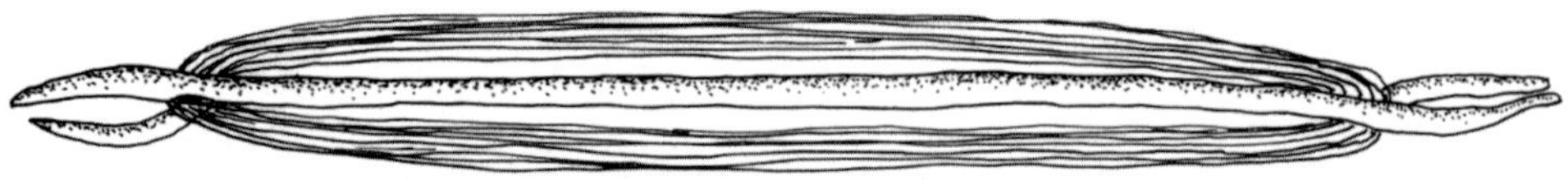

Figure 3. Net needle with thread (C. Alfaro).

2), enabled the work to advance on successive passes. The greater length of the needle meant that more thread could be loaded, while the closed shape of the forks (figure 3) enabled the needle to pass through the, frequently small, openings of the net (defined by the length of the sides of the apertures in the mesh). Ancient needles, even in Roman times, were generally of bronze. There were needles of bone as well (Højte, 2005, 137, figure 4) and, in all probability of wood, but very few examples of needles made from these materials have survived.

Work on rectangular netting always begins at one of the edges with a fastening cord or *linea* (Crowfoot & Crowfoot, 1961, figure 117). Starting at the first knot (on the left-hand side of the cord, for example) the *linoplôkos* (the net-making specialist) gradually makes the knots of the first row. When the net-maker reaches the final length of the net, he changes the direction of working from right to left. The result of this changeover is that the knots made in one direction are different from those in the other (figures 1D-1G). When the netting is finished, alternate rows have knots that will appear to be different.

A different system is used to make circular nets (such as the Spanish *esparavel).* The first rows are made as shown, starting from a fastening cord (or *linea*). Then this cord is knotted and hung from a hook in the ceiling, and the work continues, always in the same circular direction. Here the problem consists in making the net grow bigger so that it reaches the desired size and becomes a flat circle. To achieve this, new knots are added radially (figure 4) as often as necessary to give the net the desired shape.[5]

---

5 In the splendid small-scale model made by the Cádiz Brotherhood of Fishermen as a gift for those attending the conference, one can clearly see the mathematical frequency of the knots as the net grows bigger. The increments are gradually introduced from the second row onwards, leaving five holes in the mesh between the first ones. The increments are made every two rows in the form of 14 radiuses. By the end of the five increments that are introduced, the distance between them is 14 holes. After two final rows the net is closed and fastened to a thread knotted onto the last cord. The lead weights are placed between the knots. Naturally, the size of the net and this final cord determine the size of the lead weights. The *esparavel* would probably always carry long, cylindrical weights, similar to those found in the sixth century BC fishing village of Oropos (Mazarakis, 2002), at the fourth century BC site in La Albufereta, Alicante (figure 14, weights) and in many other places.

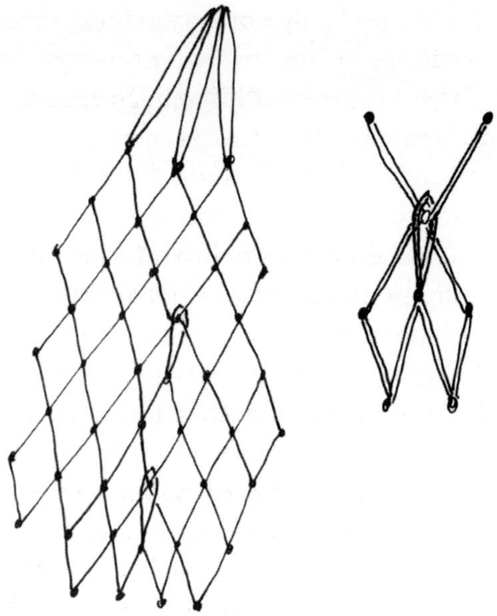

Figure 4. Fishing net knots: growing system (C. Alfaro).

The archaeological evidence. The scant remains of nets and accessories: weights, floats and needles

The raw materials employed in making nets

Generally speaking, we can say that throughout history, nets have been made of fibres from different plants and tree bark. The types of fibres used in ancient nets seem to have come from the phloem or bast (a membrane between the external bark and the wood) of certain trees. Two main kinds of tree fibres seem to have been used since the Mesolithic (8000 BC) and the Neolithic for textile production in general (Rimantienë, 1995; Medard, 2003, 81-82). The fibres of Gramineae and other types of plants were also used.

a) *Linden bark.* In the Mesolithic and Neolithic, linden fibres appear to have been the most frequently used. This is a fibre that gives very good results in regard to the useful service life of the net. It has a long tradition in northern Europe, from prehistoric times right up to the nineteenth century. Once the phloem layer of the linden has been extracted, it is torn into as fine strips as possible which are spun into simple yarn. Then two of these yarns are twisted together to obtain a double yarn, which is what normally forms part of the structure of the fishing nets.
b) *Willow bark* can be treated in the same way. Willow fibre has been identified in a net found in Korpilahti, Finland (Clark, 1936, 227).

c) *Flax*, probably not very refined. In any case, flax or linen nets withstand the passage of time badly (lasting two or three months at the most, according to the experience of Lithuanian fisherfolk, as recorded by Rimantienë, 1995, 55).
d) *Hemp* appears to have been much used from the Iron Age onwards. This is an ideal material for nets, as it is stronger than flax. The Iberian net from La Albufereta (Spain), which we shall later see in more detail, is made of this material.
e) *Other fibres* such as broom, aloe, mulberry and palm can also be used to make nets (Ginella & Koch, 2006, 116).

Remains of nets and their accessories in their archaeological context

In our study, the first general study (as far as we know) of fishing nets preserved from Antiquity, we had to start from the basic fact of the small number and size of the fragments that have survived from that period. Naturally this makes it difficult to draw conclusions about the evolution of the net, the types of net that existed, the extent to which different types of fibres were used, or mesh sizes. Therefore I have to state in advance that I do not claim my list as exhaustive: there is certainly still a lot of work to be done. I must also say that the material is extremely sparse and I have been able to analyse only the three Hispanic samples I shall present later on. The rest were found through the existing literature or simply by studying the display cabinets of some museums.

It is worth remembering that it is the from oldest periods that the greatest variety of specimens, and those in the best condition, originate. Obviously, environmental conditions play a part here. For us, these remains from such a remote time are magnificent precedents that it is essential to discuss briefly here. The oldest remains of nets I know of come from the Mesolithic (eighth millennium BC) and were found in Korpilahti, Finland. The material used was, as mentioned earlier, willow fibre. The remains of several nets were also preserved in the Karelia region of Russia, especially one example in Kamenogorsk. It turned up inside a wet well mixed with bones (which probably acted as the conserving agent), stones and various instruments, as well as three pine floats. The size of the net fragment (roughly 20cm long) was enough to ascertain that the mesh aperture was 6cm wide, which suggests that the net was intended for catching large fish. Another net was found at a site in the northern Estonia, near the town of Narva. It was made from the roots of a plant that has yet to be determined. Some knots from a net have been discovered near Syktivkar, in Russia (Rimantienë, 1995, 53). In total there are 37 known specimens.

Small remains of net from the Maglemose period (Early Northern Mesolithic, 8000-5000 BC) have been found in fairly good condition, for instance the so-called Antrea net from the Karelian Cape in Finland. The context suggests that all these specimens were carrying nets, made from an indeterminate vegetable fibre (Bender-Jørgensen, 1990, 2-3). A similar item was found in Friesach near Potsdam, Germany (Pälsi, 1920; Gramsch, 1987, 89, plate 25).

In Lithuania, a large number of nets made of linden have been found on the Neolithic site of Sventoi (Rimantienë, 1995, 55). The combination, in these finds of the nineteen-sixties, of weights, floats made out of pine bark, fairly large net fragments and pieces of wood identified as sides for holding long, rectangular nets for lagoon fishing, provide us with a comprehensive and extremely instructive overview. A net from this period was also found in Latvia (Sarnate), but subsequently disappeared (Rimantienë, 1995, 54).

With the experience built up over the years, archaeologists began to learn to detect this type of material in the mud of lakes and to collect all the organic remains found with them. It was then that the remains of nets began to appear in larger numbers.

Remains of ropes, basketwork and nets from the Neolithic period, conserved due to the dryness of the environment, have been found in various places in the Near East. There is always a doubt as to the original function of such small remains as the fragments of nets usually found at archaeological sites, especially in places not near a sea or lake. In the case of Ohalo II (on the Sea of Galilee: Nadel *et alii*, 1994, 455-57) the presence of fish remains makes one think that the site was associated with fishing or the storage and transport of fish.

The Neolithic period has provided material of great interest from across Europe. Many remains of nets made from beautifully spun yarns with very long fibres that at first looked like flax, although subsequent analyses showed them to actually be linden, were found at Sventoi, in Lithuania. The linden is double-spun, with a fairly strong S twist, then Z twist (Rimantienë, 1995).

The Swiss lakes have furnished a great deal of very interesting material. For example, the finds of the excavations at the palaphitic site in Wetzikon-Robenhausen, Zürich (Switzerland), which were made as far back as 1863, but whose composition is unknown, as they have not been analysed. More recent studies (1999) have found further specimens of linden bast fibre nets made from 3mm-diameter Z-ply double yarns (Altorfer & Médard, 2000, 57-58, figures 21; 22; 35). Several examples of nets – perhaps not all for fishing – belonging to the cultures of Cortaillod (4200-4000 BC), Pfyn (Middle Neolithic, 4000-3500 BC) and Horgen (Recent Neolithic, 3800-2700 BC) have been found (Bazzanella *et alii*, 2003). The excavations in the Mozartstrasse in Zürich have provided a large number of net fragments of great interest (Rast-Eicher, 1992, 10-11; 14 and figure 5).[6] At Hornstaad and Wangen, two Neolithic sites on Lake Constance, some nets were found in a fairly good state of preservation (some carbonised, others not). Treating the fi-

6 I would like to thank my colleagues Antoinette Rast-Eicher and Regine Fellmann, of the Bern Museum, for all the information they gave me during the preparation of this article.

bres with ammonium hydroxide made them more elastic and made it possible to see their structure clearly: the material is flax (*Linum usitatissimum* L.). The yarn used is double, made from simple yarns 0.5-0.6mm in diameter which were in a carbonised state. The diameter of the fibres was 6-12 micrometres and remained the same for more than a millimetre (Körber-Grohne, 1990, 11-12, figure 2.1). It is interesting to note the variation in the size of the meshes in this net, which suggests that it was not made with any great care.

The Sutz Rütte site at Sutz-Lattrigen, in the Canton of Bern, has provided a considerable number of fragments of a large flaxen fishing-net (Rast-Eicher, 2003, 237), of which the largest is 40cm long and 16cm wide. They were found during an underwater excavation in a lake and came from a fire horizon (2704 BC, Final Neolithic) of the Auvernier culture (2700-2400 BC). The yarn is a double Z-ply made from two simple S-spun yarns 1mm in diameter. It should be recalled that the fibres from the secondary wall of hemp and jute have dextrogyral, or clockwise, helixes (lending themselves naturally to Z-spinning), whereas flax and nettle fibres have levogyral, or anti-clockwise, helixes (that lend themselves better to S-spinning) (Strasburger, 1994, 108).

Two examples of nets made from tree fibres were also found in Swiss lakes. The first is fairly large (26cm × 20cm) and is carbonised. It is made of double Z-ply yarn made in turn from two simple S-spun yarns. It comes from the lake site of Felmeilen-Vorderfeld, Meilen, Zürich Canton, within the so-called Horgen culture (3300-2800 BC) and is kept in the Swiss National Museum in Zürich (Rast-Eichler, 2003, 221). The second, from the same site and kept in the same museum, measures 17cm × 12cm, is also made from an undetermined tree fibre, has not been carbonised and its yarns are 2mm-diameter double Z-ply made from simple S-spun yarns. The technique employed is the unknotted simple mesh (figure 1B) (Rast-Eichler, 2003, 222), which might indicate that it was used as a bag rather than a fishing net.

A net in a fairly good state of conservation was discovered in the palaphitic settlement of Hornstaad. It had double flax yarn, 2cm mesh and knots as shown in figure 1D (Feldtkeller & Schlichtherle, 1998, 26).

The Museo Tridentino di Scienze Naturali has an interesting 16cm × 12cm piece of linen net from an Early Bronze Age stratum at Molina di Ledro, Trento. The piece, which is completely carbonised, is made of S-spun yarn with a diameter of 0.6mm. It was made according to the pattern shown in figure 3A (Bazzanella-Mayr, 2003, 172).

A fine, rolled-up net was discovered among the ashes at Santorini (Thera) in Greece. A report of the find is to be published shortly. Its state of preservation is excellent. According to Spantidaki, it may be a triple net of different meshes. In the photographs I was able to see at a recent round table in Athens (October 2007) it was clear that the smallest of them, with a mesh size of approximately 1cm, is laid over another with a slightly bigger mesh size of about 3cm. The yarns are quite tightly twisted

Figure 5. Fisherman with net, Museum of Corinth (photo Museum of Corinth).

(Z for the simple yarn, S for the double yarn). This suggests that the raw material may have been hemp, since this fibre is more easily Z-spun, as noted before. In spite of their different mesh sizes, I do not think this is a composite net of the *trammelnet* type (Brandt, 1984, figure 709). These triple nets, arranged in parallel, form a sandwich made up of a fine net between two others with a much larger mesh. Moreover, there is a bigger difference between their respective mesh sizes – a ratio of approximately 1:12 – than in the case of the Thera piece, in which the ratio is 1:3. It is not known for certain whether there is actually another net inside, i.e. whether this is really a triple net, but in any case, the fine net would have to be between the two coarser ones and not outside them. The technique of these triple nets consists in getting the fish to go through the first and third barriers (the ones with the larger meshes) so that they become caught up in the finer net and are trapped in it when they come out again through the larger outer net. It is the inner net that traps them in a sort of pocket from which it is impossible to escape.

There are numerous remains of fishing weights and other items that can help us to delimit or reconstruct to some extent the form of the net with which they were used. Precisely because there are so many remains, it would take too long to mention them all. Although a systematic study of these artefacts has yet to be made, this is not the place to do so. Here I shall refer only to some of them, for their special interest. Those

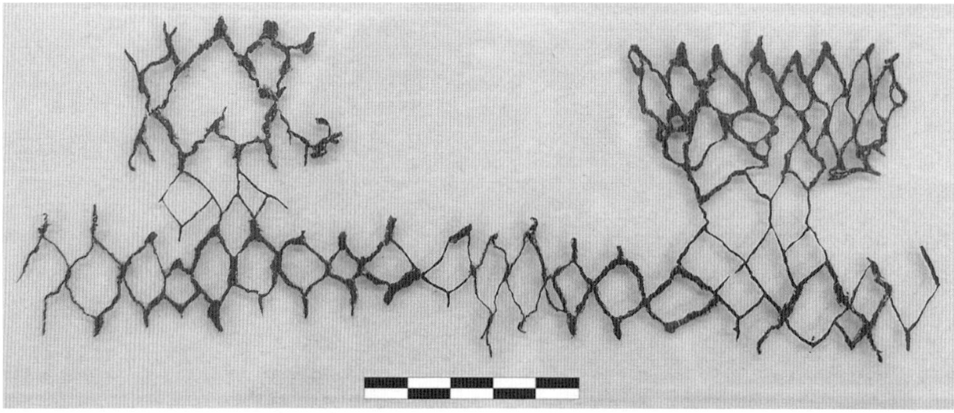

Figure 6. Iberian fishing net (net 1), La Albufereta Necropolis (fourth century BC) (photo MARQ).

found at Oropos, near the modern Skala Oropou, on the border between Attica and Boeotia, opposite Eretria (Mazarakis, 2002, 157; I am grateful to Manuel Arjona for this information) fall into this category. The excavations carried out there in the 1980s revealed some interesting fishing implements from the eighth to sixth century BC, including 22 lead fishing net weights. In an industrial neighbourhood, a very special disc was discovered that has been associated with a fishing net. It is made of ceramic material and its interest lies in the fact that it bears an inscription (*sinistrorsum*), with the owner's name in the genitive case: PEIZALIMO. Its morphological features suggest that the language is a dialect of Eritrean, so this area may have been settled by people from the island in this period of the Early Iron Age.

The image of a fisherman from the Archaic Period in Greece on the internal face of a plate in Corinth Museum (figure 5) may help us. The fisherman is shown about to cast his net. In his left hand he is holding a coiled-up rope, on which can be seen a number of tightly attached weights. In his right hand, he is holding the end of the rope coming out of the middle of the net that is used to haul it in. In spite of the poor quality of the photo the scene depicted allows us to see that the distance between the weights is quite small. It might be an *amphiblêstron (sfairon).*

Few fishing nets or fragments of fishing nets from Roman times have been conserved. The most complete may be those found at *Herculaneum.* I have been unable to trace one of them, which is apparently kept in the Museo Archeologico Nazionale in Naples (Deiss, 1995, 58; I would like to thank T. Bekker-Nielsen for this reference). However I have seen, though not studied in detail, the thirteen fragments of a very fine carbonised net kept in the small Boscoreale Museum (Stefani, 1990, 15; a large number of fishing nets have apparently been found along the coast of Pompeii, especially at Bottaro). These fragments are also from *Herculaneum* and their state of conservation makes it possible to guess at a double yarn structure with a very small mesh size.

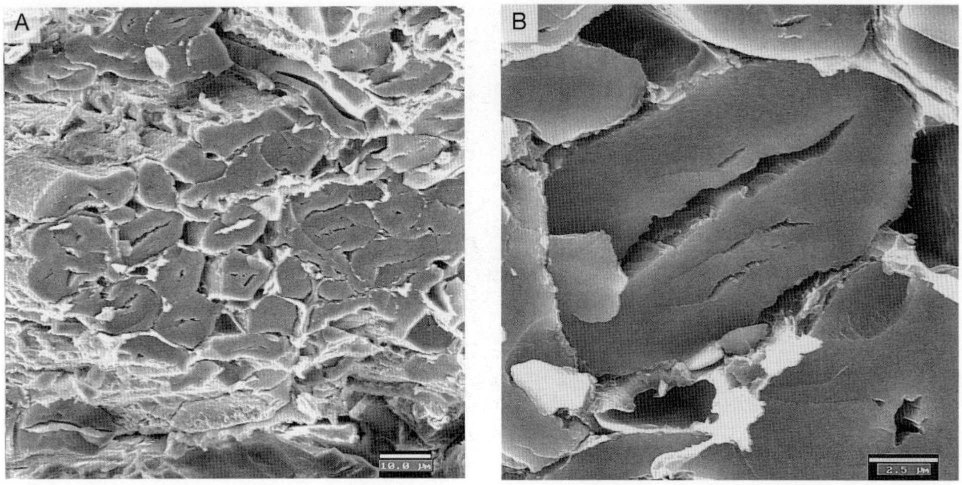

Figure 7 A-B. Photos SEM: hemp of the Iberian fishing net of La Albufereta (photo C. Alfaro).

A net in a very good state of conservation was found among materials of the second century AD in the so-called Cave of the Letters (Israel). Fully extended it measured 10 × 6.5m. It was rolled up and tied with its own rope, which had then been threaded through the mesh. However, it may have been intended for catching birds (Yadin, 1962, 233).

By coincidence, three of the four fishing nets found in Spain come from the same place: the Iberian-Roman site at La Albufereta in the province of Alicante. They belong to different periods, however. The first (figure 6) was found in one of the tombs of the Iberian necropolis on this site and can be dated to the fourth century BC. It suffered the effects of the cremation of the body it accompanied and is now kept in the Alicante Museum.[7] It was difficult to analyse this piece with the scanning electron microscope due to the extent of the degradation of the yarns of which it was formed. When it was discovered, the fragments were covered with an organic substance which, in the course of time, had taken on a yellow colour and was replaced a few years ago by another protective layer that also hardened what today remains of this piece. Despite this, it proved possible to ascertain that the raw material is hemp (figure 7A-B). Its technical characteristics are as follows: simple Z-spun yarn (the most common for hemp due to its natural tendency, mentioned above) and two-ply S-spun yarn. The simple yarn is 0.2mm thick and the double yarn 0.4mm thick. The sides of the mesh openings are about 1cm long, although there are considerable differences between them, ranging from 0.7 to 1.2cm.

7 I would like to thank the director of the MARQ of Alicante, Manuel Olcina, and his collaborators who gave me every facility at all times during my study of this and other pieces.

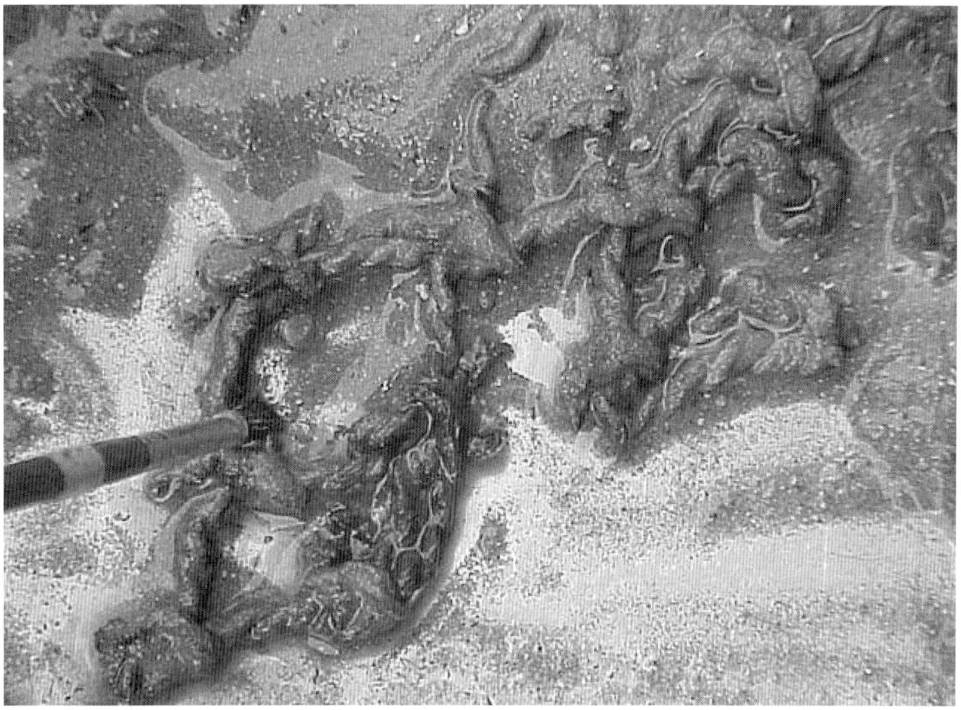

Figure 8. Roman fishing net from La Albufereta (photo A. Fernández, CASCV).

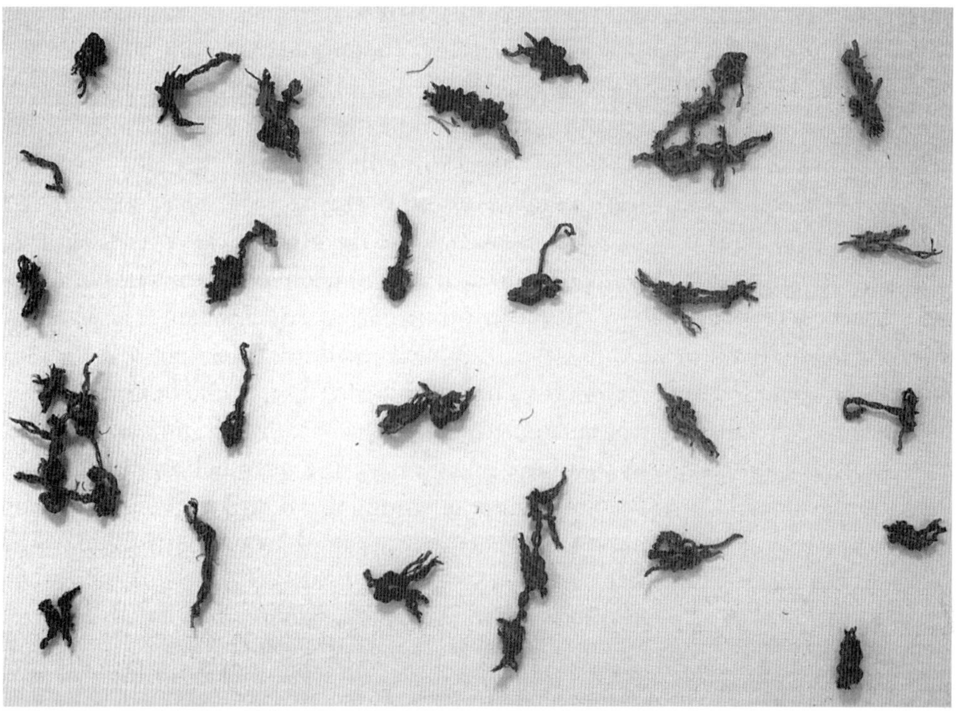

Figure 9. Roman fishing net from La Albufereta (photo C. Alfaro).

Together with these fragments were found remains of double S-spun yarns (also the product of twisting together two simple Z-twist yarns) that still had some knots in them. They were most likely part of a somewhat larger net. In this case, the double yarn is 1mm thick. Not enough has survived to be able to tell the exact mesh size, but the sides of the mesh apertures were more than 2cm long. Remains of a slightly thicker (2mm) Z-spun yarn made from three simple, also probably Z-spun, yarns, may be related to the first net (Alfaro 1984, 150, plates XXXV-XXXVI).

The third net from La Albufereta (figure 8) is of considerable interest, in spite of the greater difficulty involved in analysing the fibre, which has so far made it impossible to determine the raw material exactly, though it is probably hemp or flax. It was found some years ago in a Roman shipwreck from the Flavian period near the settlement of La Albufereta. All the items from this important find are kept in the Centro de Arqueología Subacuática (Underwater Archaeology Centre, CASCAV) in Burriana (Castellón).[8] In addition to the net, a large number of ropes of different thicknesses, remains of baskets and a bilge pump that still had part of the cord for operating it, were found. All that remains today are various undone knots, although there was more of the net intact at the time it was found (figure 9). The netting was made in the so-called "cow hitch" system (figures 1E and 10-11). The yarn is double and follows the S2z pattern, that is, two simple yarns with z-twist giving the double yarn a S-twist. The simple yarn is 2mm thick and the double yarn 2-3mm thick. The distance between knots is 1.2cm. As mentioned above, the little fibres from the secondary wall of hemp and jute have dextrogyral (Z) helixes, whereas those of flax and nettle have levogyral (S) helixes which suggests that in this case, the raw material here is hemp.

To conclude this no doubt incomplete list of ancient nets that have come down to us from such remote times, I shall describe an extremely interesting piece found some years ago in the river port of *Caesaraugusta* (Zaragoza) (Royo & Acín, 1991). It is a net from the Roman period, but has not been dated specifically for want of a precise stratification (figure 12). It is clearly a conical river fishing net of a type that is still in use today, although there is no mention of such a type in the classical literary sources. It is most likely a tubular trap made of netting that was fastened to a metal, cane or wood arc that kept it open when it was held against the current in areas where the current was concentrated. According to Brandt (1984, 166) "tubular traps are funnel-shaped gear, mostly closed at the smaller end and without any non-return device". The piece still has an apex that is in fairly good condition and is 28cm long in its current position, but it is likely to have been a metre

8 In this case I am grateful to Asunción Fernández, head of the Underwater Archaeology Centre of the Autonomous Community of Valencia and the person responsible for excavating this piece of wreckage.

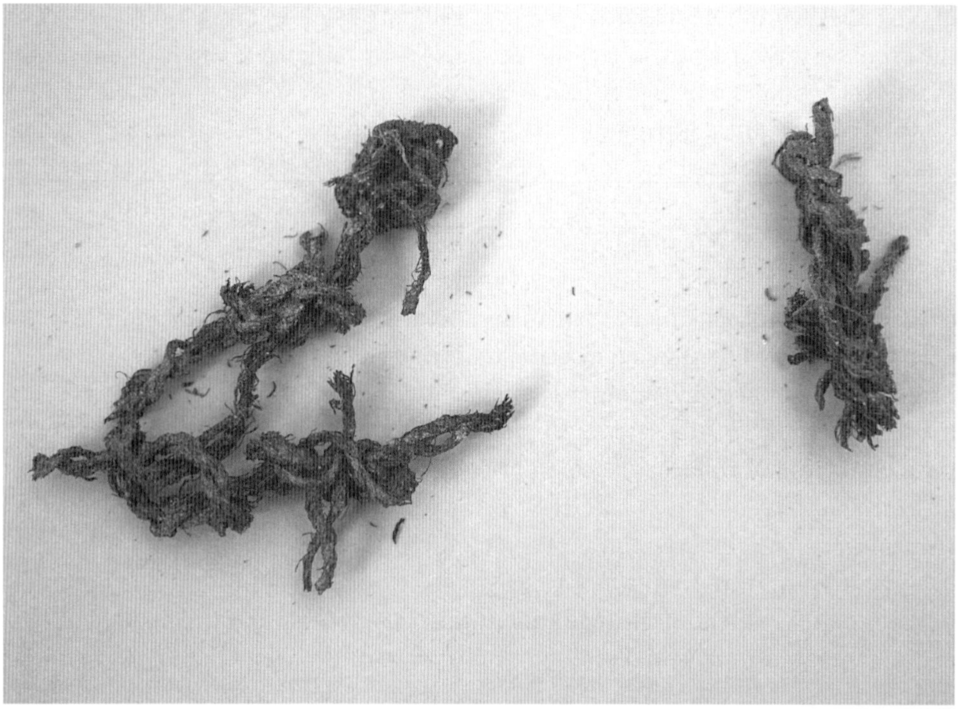

Figure 10. Roman fishing net from La Albufereta (photo C. Alfaro).

Figure 11. Knot of the Roman fishing net from La Albufereta (photo C. Alfaro).

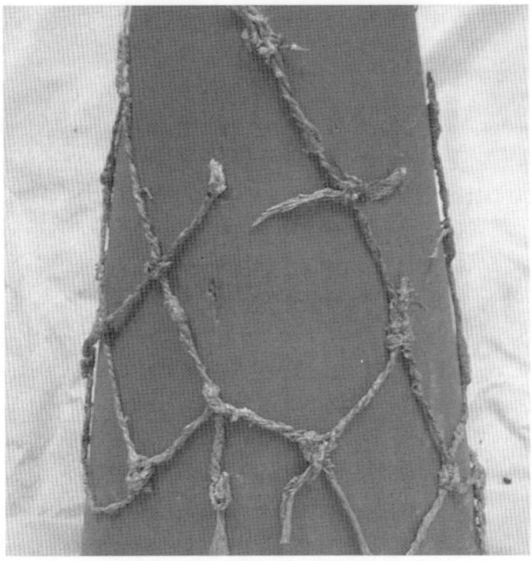

Figure 12. Roman fishing net from Caesaraugusta (photo P. Conesa).

Figure 13. Roman fishing net from Caesaraugusta, detail (photo P. Conesa).

long or more originally.[9] The diameter of the lower part is 12cm. I should say that such instruments are employed today in shallow river areas and so are used by men, but also by women and children. From the technical point of view, it is formed by double S-ply yarns made from single z-spun strands. The knots used are reef-knots (figures 1F and 13); the mesh size is 3cm. The net has some repairs which seem to be original. The fibres have yet to be analysed.

Net floats

Here we are interested in these items only as accessories for nets. A comprehensive study of this type of materials has yet to be made and so the data cited here derive from a variety of bibliographical sources. A few examples will serve to illustrate that they were made from various types of tree bark (all the varieties of *Pinus*, as well as *Quercus suber*), from the wood of other trees with lightweight wood, such as the black poplar, and even of plants that tend to float for a certain time, such as papyrus. Some of these models, such as the pine bark floats, were used in different cultures and over a considerable period of time.

9 I would like to thank my students, Pablo Conesa Saez and Asunción Ferrer Calderer, for informing me of the existence of this piece in the Zaragoza Provincial Museum and for the photographs they have kindly lent me for this article.

Some pine bark net floats were found in the mud at the bottom of the lake at Sventoi (Lithuania; Mesolithic). The bark had been treated in a special way: small rectangular pieces had been dried slowly near a fire and then highly polished (and perhaps oiled?). Before this, a hole had been drilled in each so that they could be hung from one of the ropes – the one on the inside – of the draw-nets. Bark prepared in this way does not absorb water, thus does not swell, nor does it break easily. That is why it floats so well, and its polished appearance makes it look almost like amber. At Nida, also in Lithuania, large numbers (between 60 and 100 samples) of lagoon net weights of this type were found. The tradition of using pine bark floats dates from very late Byzantine times. Some pieces of pine bark were found in the Serçe Limani shipwreck (eleventh century AD) and identified as fishing floats (Powell, 1996, 105). Such floats have also been found in some ports and shipwrecks of the Roman period. I shall mention a few them, by way of example only. In the famous find of the port of Pisa (Camilli, De Laurenzi & Setari, 2006, n. 28-30), some pine-bark floats were found that had been conserved almost whole. They are between 2.70 and 3.47cm long and between 1.39 and 2.27cm wide. Another very similar kind of float was used at the Roman site of Untereschenz on Lake Constance (Switzerland), dating from the early first century AD, as shown by this interesting example of black poplar wood (Ginella & Koch, 2006, 116-17).

The Minoans, who lived in a volcanic region, used floats of pumice stone (Powell, 1996, 113). Papyrus floats are known to have been used in Egypt in Roman times. These are quite different from the usual, more compact float types. They consist of a tight roll of papyrus folded into three segments. The ends are joined to form an isosceles triangle through which the upper cord of a vertical-type river net is threaded (Powell, 1996, 105). In antiquity, cork was not native only to Portugal, as some have held, but also grew in extensive areas of Spain (especially in Extremadura, but also in Catalonia and Valencia). It must have been widely used as floats for fishing, but there is no actual archaeological evidence of this. Some have spoken of small gourds or little ceramic pots, but this is extremely hypothetical and based only on ethnography or the doubtful interpretation of tiny designs on Minoan seals.

Net weights: lead and other types of objects used

The variety of net weights in antiquity was enormous. Some were metal (mostly lead) with a large range of shapes, others were stone and yet others were ceramic, but virtually anything small of a certain weight could be used. This is not the right place to examine the matter in detail, but I shall try to illustrate some of these different types.

Stone fishing net weights from the Natufian period were found at the Mallaha site in Israel (Nadel *et alii*, 1994, 456). Two different types of net weights were discovered together with a splendid collection of Neolithic articles at the Sventoi site in Lithuania. Some were made of stone, occasionally quite large, and attached to the net only by linden cords, while others were made of stones wrapped in thin linden bark tied up with cords of the same material, which served to attach them to the net. Now that the wrapping has disappeared, the instrument is no longer recognisable as a net weight. Wrapped stones of this kind were used in this area of Lithuania until fairly recently, as I mentioned before. However, the stone weights sometimes had notches cut into them so that the cords fixing them to the net could get a good grip on their surface. Examples of this type were found at Neolithic or Greek Bronze Age sites such as Saliapos (Cyclades), Poliochnis (Lemnos), Troy VII, etc., in obvious fishing contexts (Powell, 1996, 115). The danger in identifying objects only by their shape is that, for example, these pieces are very reminiscent of those used in mining by the Romans in Riotinto (Huelva), so in order to distinguish them, various factors such as context, size, weight, etc. must also be taken into account.

Fishing net weights made of pottery were also common, their shapes varying from one area to another. With this type of material, the same problem arises as I described earlier. It is sometimes difficult to determine the actual function of such a simple object as a not very heavy (about 60-200gr) truncated pyramidal or discoid piece of fired clay with one or two holes for attachment. Here, I shall describe just two instances showing the difficulty of interpreting such artefacts. Some pyramidal terracotta weights have been identified, in view of the context, as net weights instead of weights for a warp-weighted-loom (Højte, 2005, 135). In other cases, the dividing line separating the two different uses (loom weights or net weights) is more difficult to determine. The typical flat discs with one or two holes in them, which are very abundant (particularly at Thera) and always interpreted as loom weights, could be mistaken for fishing net weights. On the face of it, the larger size of the holes could be of some help, but the context is probably the best guide to identification. As is still done today in certain parts of Greece, these ceramic discs could be hung from the inside of a stationary net in which the fish were trapped by their fins or gills. If one looks carefully at some Minoan seals, it seems as though the intention was to show this type of weights as being attached to what might be a net rope. A fish with its fins clearly marked can be seen at the top of the space of the seal (Powell, 1996, figure 58; Cheval, 2008).

In the Graeco-Roman period, however, the most common fishing net weights were metal, made of lead in different shapes. The type most frequently employed was a long, cylindrical weight of varying thickness, depending on its use. As I have said, these are ideal for casting nets, although until not very long ago they were used, or are still even used today, for drag-nets (at least in the Balearic Islands). They were already known in the Late Greek Bronze Age, when metal became more easily accessible, and in slightly later periods, being very often found in shipwrecks. In the Uluburun wreck, 107 pieces of different formats, and therefore used for dif-

Figure 14. Fishing weights for net. MARQ, Alicante (photo C. Alfaro).

ferent nets, were found. Eighteen such weights were discovered at Cape Gelidonia. There are numerous other examples (Powell, 1996, 118-20). They were also abundant in the archaic period, a good example being those found in the previously mentioned fishing village of Oropos in Greece (Mazarakis, 2002). Here 22 specimens were collected, of different length and weight, which suggests that there were several different nets together on the site before their organic components decomposed. The fishing-gear sinkers found at Carmel (Israel) are of great interest due to their variety, including lead sinkers in the form of tubes, rings and folded rectangles (Galili, Rosen & Sharvit, 2002). The team from the Swedish Institute at Athens, led by B. Wells, asked me to study a small collection of folded rectangular lead fishing net sinkers discovered in the Poseidon Sanctuary of Kalaureia (Poros, Greece) which are very similar to those found at Oropos and Carmel. In this moment, I am studying the fibre remains conserved inside them. It is new way to identify the raw materials of the ancient nets.

This type of long, slender, folded lead weight was also widespread in the Western Mediterranean between the sixth and fourth century BC. In *Emporiae* (Ampurias in the Spanish province of Girona), for example, a large collection was found. In some cases, the weights are homogeneous and carefully formed; the 28 examples found at La Albufereta, in the province of Alicante (fourth century BC) are the most interesting in this respect (figure 14).

Figure 15. Fishing weights for nets, Cala Olivera, Ibiza (photo B. Costa).

This type of weight was firmly attached to the net cord. This was done by fastening it to the lower cord, perhaps while the lead was still warm; it was given its final shape by hammering the half-open forms of the lead. The pressure was so strong that it prevented the weight from moving and in many cases the weights still have the imprint of the cord or rope inside them. They may have been used, as I have said, on round casting nets. This would have required them to be of uniform weight and regularly spaced so as not to unbalance the net during the cast.

Seven lead weights (figure 15) were recently found at the *muricidae* shell deposit at Cala Olivera in June 2005 (Alfaro & Costa, 2008). These lead pieces are considerably worn, but only on one side – the one that was dragged along the sea floor. The weights were discovered in a not very precisely dated unit (between the third century BC and the second century AD). They are 3.2-5.1cm long and 2.6-4.1cm wide and weigh between 72.43 and 187.71 grams. The inner diameter is 1.54-2.30cm, showing that the space occupied by the cord varied depending on how tightly it was joined to it. We can also imagine, based on the wear-and-tear marks on some of the pieces, that they would have been used for small nets, such as trawl nets on sandy sea beds.

But lead was also employed in making other types of net weights. In the river port of the Ancient *Moguntiacum*, Mainz, in Upper Germany, other types of weights were found together with some cylindrical leads such as those I have just described. These have a hole in them, are fairly flat and rectangular, but of very irregular widths and shapes, measuring between 4.5cm and 9cm (Ginella & Koch, 2006, figure 56).

Conservation of the nets

As indispensable tools of the fisherman's trade, nets were carefully looked after. Although our sources of information on this point are few and far between, we do know that certain precautions were taken to lengthen their service life. First, we should mention the daily repairs required by the use of the nets to mend breakages and tears as they occurred. It was because of frequent breakages that nets needed to be knotted, rather than made by the weaker systems I described at the beginning. The use of two-ply yarns made the overall surface more elastic and, therefore, stronger. After each use, it helped to wash the nets thoroughly. In this connection, it is worth recalling the episode of the *purification* of the nets recounted by Aristotle in a somewhat obscure passage (*H.A.* 8.13 [598a-b]). Here he is talking about migratory animals and says they normally reach the *Pontos* (Black Sea), pass through it and then come out, suggesting that this was an excellent area for net fishing. But in the case of sardines, he comments that an exceptional phenomenon occurs, as they do not follow the general trend. He tells us they leave the Black Sea by strange river routes that run into the Adriatic and introduces an interesting piece of information: some of these sardines became disoriented and swam back into the Mediterranean through the straits of the *Pontos*, and if they fell into the nets of any of the fishermen, this forced the fishermen to purify their nets (*hoi halieis ta diktya perikatharousi dia to mê heiôthenai ekplein*). This text also contains technical information on the general rule for treating nets (they should be cleaned frequently) together with some ideas that may have more to do with superstitions or beliefs. He seems to be suggesting that the fishermen had caught these sardines accidentally, and this had a harmful effect on the nets, being an abnormal event. Of course the text gives us an insight into the care that was put into maintaining the big nets. The New Testament also provides some information on this point (Luke 5.2). The scene occurs after a whole night's fishing from boats and from the shore, following which the nets had to be left out in the air to dry: "…the fishermen were gone out of them (the "boats"), and were washing their nets" (*hoi de haleeis ap'autôn apobantes epiplynon ta diktya*). The description of the scene, on the shore of Lake Genezareth, shows that the nets were being cleaned of seaweed or other matter stuck to the nets.

After use, the fibres were full of salt water or dirt, and the nets had to be hung out to dry before they were put away. For the ancients, the salt in seawater was a preservative, but the continual soaking and drying of the fibre had extremely harmful effects. Fibres such as flax and hemp, which were used to make nets for sea fishing, required special care to make them last longer. Nevertheless, it has to be taken into account that nets made of both these materials, as well as other fibres I have mentioned, needed to be constantly repaired and new ones had to be frequently made. This was not the case with esparto, which held up well in seawater, but we do not have any remains of esparto nets. Although some colleagues told me that an esparto net had been discovered in the Algarve, in Portugal, I have so far not been able to find out any more about it. Pliny (*N.H.* 19.26) says that esparto must be retted with seawater and it was generally believed that the esparto ropes used in ships lasted longer, the more they were in contact with salt water. However, the difficulty of spinning fine yarn from esparto may be the reason why there are no reports of its being used for fishing.

Nets made of hemp and other fibres before the advent of nylon were not very resistant. Because of this, traditionally and until relatively recently, natural fibre nets were treated with a certain type of dye, or rather mordant, that impregnated them and to some extent slowed down their natural rapid deterioration. There are reports of the Ibiza fishermen's custom of using ground pine bark (a magnificent mordant and protector of natural fibres) to make a vat of dye in which to immerse the net for a while before casting it into the sea for the first time. The colour of the net then turned dark brown.[10]

10 This operation was repeated on other occasions with the same net. I owe this information to some fishermen from Ibiza and my colleague B. Costa.

Semis from the mint of Carteia (40 BC-15 AD), in the National Archaeological Museum, Madrid (detail).

3. Fishing Tackle in *Hispania*: Reflections, Proposals and First Results

DARÍO BERNAL CASASOLA

The identification of fishing tackle in the archaeological record

The present paper deals with ancient fishing equipment, a topic that has largely been overlooked by archaeology in Atlantic and Mediterranean contexts – that was one of the reasons behind the organization of the international workshop of which this volume is the outcome. The common denominator for the entire *Mare Nostrum* is a shortage of specialized research publications, barely ten monographic works in all. In Spain, only three titles have been published, concerning the possible ancient origin of fishing enclosures or "corrales" (Moreno & Abad, 1971), the typology of fishing equipment (Gracia, 1981-1982) and the general issue of fishing techniques, especially the *almadraba* (Martínez Maganto, 1992). Only in recent years have these shortcomings begun to be addressed, with some monographic works focusing on specific locations, e.g., *Emporiae*/Ampurias (Castanyer, 2006) or the Strait of Gibraltar (Bernal, 2009a), although neither systematically nor with sufficient empirical support.

The situation is similar elsewhere in the Roman world, as shown by the works of Bekker-Nielsen (2002a and 2005), including those focused on fishing techniques in the Black Sea (Bekker-Nielsen, 2009). As mentioned, only recently have these deficiencies begun to be tackled, as shown by a recent work of synthesis, as yet unpublished (Ayodeji, 2004), or by A. Trakadas' doctoral thesis, in which a full chapter is dedicated to the fishing equipment used in *Mauretania Tingitana* (Trakadas, 2009).

Among the reasons for this hardly justifiable research reticence is what we have termed the "*garum* factor", especially apparent in the area around the Strait of Gibraltar. The study of different aspects of the fish processing industry sector – the factories themselves, the foodstuffs processed there (by means of archaeozoological or residue analysis), the amphorae types used to transport the finished product – have dominated the research field from the work of Ponsich and Tarradell in 1965 (synthesised and updated by Ponsich in 1988) and down to the International Congress CETARIAE (Lagóstena, Bernal & Arévalo, 2007). In this situation, the study of other topics related to ancient fishing-related topic have been relegated to a subordinate position on the research agenda.

A further factor working against the archaeological characterization of the fishing equipment has been a general belief that the iconographic record was a sufficient source. Thus, Roman fishing has been illustrated by images taken from mosaics and, to a lesser degree, wall paintings. Since 95% of the evidence comes from *Africa Proconsularis* or from Pompeii/*Herculaneum*, to what degree is it permissible to extrapolate these data as sources for other parts of the *Mare Nostrum*? The well known Tunisian mosaic of the Tomb of Hermes, found in *Hadrumetum* and now in Sousse museum (inv. no. 10.455), dating to the late second century and showing several fishing scenes (angling, netting, trapping, etc.), has been used over and over again, under the assumption that the techniques shown were common to all coastal Mediterranean provinces. As shown in the paper by López Monteagudo (this volume), the iconographic representation of fishing activities in *Hispania* is limited to six mosaics, with a further two in the neighbouring province of *Tingitana* (the House of Desultor and the House of the Ephebe, in *Volubilis*), while none have been found in the Black Sea region (Bekker-Nielsen, 2009, 296). Were the fishing techniques described by Oppian in his *Halieutika* (3.72-91) from rod-fishing to the different net types, employed in the same manner throughout the Mediterranean? We think not, arguing that the Tunisian and the Pompeian/Herculanean representations have been erroneously taken as a basis for repeated extrapolation.

For progress to be made in this direction, a new source of empirical evidence for fishing techniques must be introduced, focusing on material remains from fishing sites. This approach, as already developed in European (Cleyet-Merle, 1990; Camps, 1998) and even Iberian (Aura & Pérez, 1998) Prehistory (with good support from ethnoarchaeology) can serve as a point of reference for Classical studies. So far, none of the evidence for fishing tackle found in the fish-processing installations around the Strait of Gibraltar – of Phoenician-Late Phoenician (such as Las Redes, Puerto 19 or Plaza de Asdrúbal) or Roman dates (such as the well known fisheries of *Baelo Claudia, Carteia, Cotta* or *Lixus*) – has been published in detail.

Normally, at least in the western Mediterranean, the published data fall into two groups:

1. Isolated references to items of fishing tackle, within the general category of metal artefacts and without any further elaboration.
2. Equally isolated references to such implements, normally hooks, to substantiate claims about the presence of fishing activities in the associated contexts (Castanyer, 2007; Fernández Pérez, 2002).

In general, these approaches are linked with a tacit assumption that the historical and technological conclusions to be obtained from the detailed study of this sort of archaeological evidence can only be very limited. Therefore, the most pressing task for the near future will be the systematic study of fishing equipment on archaeological sites, allowing the elaboration of regional *corpora*. This task is already being undertaken in the province of *Baetica*, by the SAGENA project (Bernal, Bustamante, Díaz, García, Hernando, Lagóstena, Ramos, Sáez, Soriguez & Zabala, 2009a and b), in *Tingitana*, recently by A. Trakadas' doctoral thesis (2009), and in *Campania*, by an Italian-Spanish project (Bernal, Cottica & Zaccaria, 2009b).

The original purpose of this paper was to present a preliminary analysis of the Roman and Late Roman evidence for fishing techniques on the Strait of Gibraltar, based on the ongoing study of material from the city and bay of Cádiz, the fish factory in *Baelo Claudia*, the site of *Traducta* on the Bay of Gibraltar and the city of *Septem Fratres* (Ceuta) on the African shore of the Strait. Considering the lack of a general archaeological background, as already mentioned, it was felt that the development of a broader (Atlantic and Mediterranean) typology for Roman fishing implements would be of more use, drawing on finds from a variety of locations, which naturally will include the Spanish examples currently being studied. The proposed typology will approach the issue diachronically by establishing, as far as this is possible, starting and ending dates for some of these types. In order to make that possible, and also to permit the investigation of potential technology transfers, a wide chronological and cultural framework will be adopted, including comparisons between Roman evidence and that from the Phoenician world, and the continuity of the record into the medieval period, both in the Muslim and the Christian worlds.

As a first draft, this typology is presented for discussion, refutation and, obviously, the incorporation of new evidence. The criteria followed for the typology are based on the morphological features of the recorded evidence. Special care has been taken to use only using items with a solid archaeological contextualisation for which the connection with fishing activity is beyond any doubt.

The evidence has been divided into hooks, weights and other tackle (figure 1). It is important to note that fishermen's tools such as needles or shuttles ("navettes"), frequently found in fishing contexts, but considered only as complementary to the fishing activities, are not included in the typology. Regarding the fishing equipment in itself, much more attention has been paid to the first two categories – hooks and weights – as being the most common in the archaeological record and also the most problematical, grouping the rest, for which presence in the record is minimal, in a single category organized alphabetically. In the final section of the paper some conclusions will be drawn concerning possible research strategies for the study of ancient fishing techniques based on the available evidence, even if it is not easy, at the present state of knowledge, to relate every archeological item clearly and directly to an ancient fishing technique, as we will see in this paper.

Fishing tackle		Name				
FISHING GEAR						
Hooks		AI	Simple			
		AII	Chained			
		AIII	Double/Multiple			
Weights	Clay	CI	Cylindrical			
		CII1	Disc-shaped	Simple		
		CII2		Grooved		
		CIII	Spherical/globular			
		CIV	Spindle-shaped			
		CV1	Trimmed	With central hole		
		CV2		With lateral notches		
		CVI	Truncated pyramid-shaped			
	Correspondance with Galili *et alii* 2002 (in brackets)					
	Stone	PI1	Perforated	Angular hole	Simple (S 1.1.2)	
		PI2			Double (S 1.1.3)	
		PI3		Straight hole	Irregular (S 1.2.1)	
		PI4			Shaped and with one hole (S 1.2.2)	Pear shaped
		PI5				Circular/Doughnut
		PI6			Shaped with two holes (S 1.2.3)	
		PI7			Shaped with three holes (S. 1.2.4)	
		PII1	Grooved	With lateral grooves		Crossed (S 2.1.1)
		PII2				Simple (S 2.1.2)
		PII3		With notches		Crossed (S 2.2.1)
		PII4				Simple (S 2.2.2)
		PIII1	Simple	Pebbles (circular)		
		PIII2		Irregular		
	Lead	PLI1	Ring shaped	With lateral perforated appendix		
		PLI2		Whole section (L 1.3.2)		
		PLI3		Partial section (L 2.1)		
		PLI4		Flat-convex section (L 1.3.1)		
		PLII1	Cylindrical	Hollow/Tubular		Whole section (L 1.2)
		PLII2				Partial section (L 2.2)
		PLII3		Solid		Rolled (L 3.1)
		PLII4				With grooves
		PLIII1	Cone/truncated cone		With distal appendix (L 1.1.2)	
		PLIII2			With groove	
		PLIII3			Perforated	
		PLIV1	Cubical		With groove	
		PLIV2			Perforated	
		PLV	Clamps-rods			
		PLVI	Off-centre/Crescent-shaped			
		PLVII1	Spherical		With groove	
		PLVII2			Perforated	
		PLVIII1	Sphenoid		With groove	
		PLVIII2			Perforated	
		PLIX1	Rolled plate		Square shaped	
		PLIX2			Rectangular shaped (L 2.3)	
		PLX1	Pyramidal/Truncated pyramid		With horizontal groove (L 1.1.1)	
		PLX2			With metal appendix (L 1.1.4)	
		PLX3			With vertical groove	
		PLX4			Perforated (L 1.1.3)	
		PLXI1	Pear-shaped		With groove	
		PLXI2			Perforated	
		PLXII	Triangular perforated			
		PLXIII	Tubular			
Devices	Harpoons					
	"Arcaduces/Cadufos" or clay pots for octopus fishing					
	Hand nets or "Camaroneras/Esquileros/Zalabares"					
	Fishing Rods					
	Coral fishing devices					
	Creels or baskets, either simple or line-linked					
	Night fishing with torch-light					
	Wooden Traps/Stone Traps or "Corrales"					
	Tridents					

Figure 1. Table with the suggested typology for archaeological evidence on fishing equipment. Annotations between brackets refer to Galili et alii *(2002).*

Hooks, the clearest archaeological find

Hooks are the finds most easily linked to fishing in the archaeological record for the Roman period, because their similarity in shape with modern examples makes their identification easier. Found at numerous locations, they have been published by the hundreds, but in almost every case the inferences drawn are minimal, normally going no further than that fishing was practiced at the site.

Typologically, hooks can be divided into traditional shapes and other shapes, worked in bone or wood, of which a multitude of varieties exist (fish-bone shaped, straight, etc.); common characteristics are the pointed end, the attached line, and the working principle with the potential catch having to swallow the bait (figure 2A and B). In Spain, there is evidence for this kind of bone artefact, called "double micro-points", from the Upper Palaeolithic, for example in Cueva de Nerja, on the coast of Málaga (Aura & Pérez, 1998). For the Roman period, several artefacts, worked in bone or ivory and showing a double perforation around the middle (figure 2C), found at the factory in *Cotta*, have been interpreted, among other things, as hooks (Ponsich, 1988, 85, figure 32, 5); this is also the case with some recent finds from Thamusida (Cerri, 2004-2005, 41, figure 27). The use of these artefacts as hooks has not been proven, and, as Ponsich pointed out, their function could indeed be difficult to determine, since they could have also been used to pull nets or to carry fish. Further progress in this direction depends on a systematic study of the evidence.

The prevalence of bronze hooks of traditional shapes is, however, clear from the Phoenician period, at least around the Strait of Gibraltar. The oldest recorded examples for the Iberian Peninsula come from the site of Morro de Mezquitilla in Málaga, in Phoenician contexts dating to the eighth and seventh centuries BC (Schubart, 2006). Other similar finds, also dating to the archaic Phoenician period, such as those in Calle Ancha, in the city of Cádiz itself (excavations of F. Sibón), or in eighth century contexts in Torre de Doña Blanca in the Bay of Cádiz (Ruiz Mata, Ruiz Gil & López Amador, 2006, 291) remain unpublished. The preliminary data indicate their absence from late prehistoric contexts, which may suggest the introduction of the traditional shapes, among many other things, to the West during the Phoenician colonization of the eighth and seventh centuries BC. This would indirectly explain the absence of bronze hooks in the area around the Strait of Gibraltar prior to the arrival of the Phoenicians. But most of the evidence comes from Late Phoenician sites such as Cerro del Prado, in The Bay of Gibraltar, or "Factoría P-19" in Puerto de Santa María, both from the fourth century BC (Arévalo, Bernal & Torremocha, 2004, 126 and 140); the introduction of new metal technologies would have displaced – or at least, reduced – former fishing techniques in which bone, ivory or wood hooks were used.

For the Roman period, hundreds of hooks have been found throughout the coastline of Spain. All are made of bronze, apparently the only metal used for this purpose, against the testimony of some ancient authors, e.g., Aelian, who also mentions

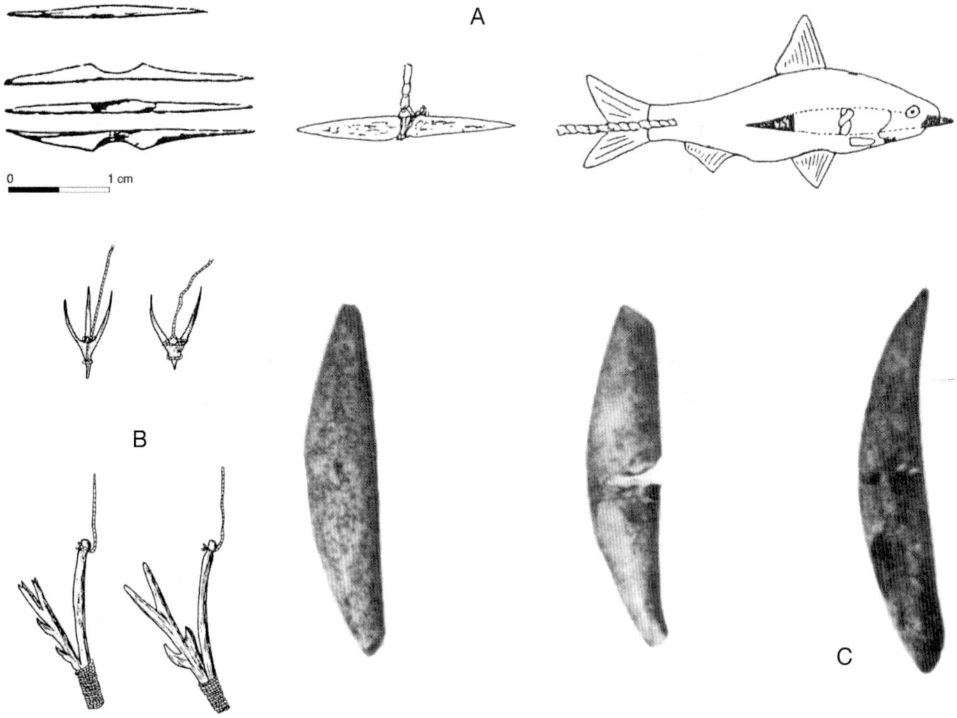

Figure 2. Straight hooks with double point and central groove, from European Upper Palaeolithic contexts – Dordogne – with a sketch showing their use (A), and assembly suggestion for fish-bone-shaped hooks (B), following Breuil (Cleyet-Merle, 1990, 86-87 and 78); and Roman bone hooks with central perforation found in Cotta (according to Ponsich, 1988, 85, figure 32, 5).

the use of iron. Bronze seems to remain the exclusive – or almost exclusive – raw material for hooks into Late Antiquity, as shown by the finds from the factories in Calle San Nicolás in Algeciras, dating around 500 AD (Arévalo, Bernal & Torremocha, 2004, 110-111), from *Carthago Spartaria* (Vizcaíno, 2005, 63) or, outside Spain, from *Castrum Perti* on the Ligurian coast (De Vingo & Fossati, 2001, tav. 95, 2-7), among many other examples also dating to the sixth and seventh centuries AD (some examples made of bone, such as those found in the Canary Islands – Tenerife Archaeological Museum – come from chronologically uncertain contexts (González Antón, 2004, 301) and, in any case, do not correspond with the sort of fishing equipment used in the Mediterranean or the Atlantic in the Roman period).

The evidence mentioned so far suggests that raw materials can give us at least three chronological milestones regarding hooks: 1) the use of bone, ivory shells or wood during Prehistory; 2) the substitution of these materials for bronze at least from the eighth century BC, following a Phoenician technological innovation; 3) the reversion to perishable materials at the end of the Roman period, suggested by the disappearance of bronze hooks from the archaeological record from the eighth century AD, contrasting with the important economic role that the iconographic, literary and zooarchaeological records seem to claim for fishing in medieval times, both in the Muslim and the western Christian worlds. The manufacture of metal hooks does not resume until the Modern period, although is yet unclear whether this innovation is a generalized Mediterranean pattern. We know of some isolated examples of iron hooks of Medieval date, for example in Rhodes (Zaphiropoulou, 2005, ed., 27, no. 16) or elsewhere in the Aegean, dating between the 11th and the 14th centuries AD (Καθημεριν" ζω" στο Βυζάντιο, 2002, 156-157, no. 174-175).

Typologically, and in spite of some pioneering efforts made in Spain (Gracia, 1981-1982), no chronological conclusions have been drawn from morphological features so far. This fruitless trend, also followed in other works (Déchelette, 1910; Galliazzo, 1979; Kuniholm, 1982), has led many scholars to assume that no relevant historical or technological conclusions can be obtained from the study of hooks. Although the current state of our knowledge, marked by the lack of stratified records and a coverage on a truly Mediterranean scale, does not as yet allow for the elaboration of a typological *corpus*, the experience from our own region and the work done within the SAGENA project allows me to offer some suggestions.

First, the metrical criteria appear to be the potentially most useful. A numerical classification according to size is required, either considering the distance between the point and the shank or the total length, which latter is the feature considered here. In this way, hooks can be classified as very small (< 2.5cm total length), used for recreational fishing or very small catches; small (2.5-4cm); medium (4-8cm); and large (> 8cm). In each region, the different sizes can be related to the available fish species, which offers a chance to verify this metrical classification.

For each group, the following trends can be identified (figure 3):

- *Very small hooks:* Scarcely represented, they seem to develop especially during Late Antiquity, as seen from the examples found in San Antonino di Perti in Liguria (De Vingo & Fossati, 2001, 659, tav. 95, no. 3-7). Some unpublished examples are known from the sanctuary at La Algaida, dating to the Republican period, and some Phoenician examples from Cerro del Prado (4th century BC), between 2 and 3cm in length (Arévalo, Bernal & Torremocha, 2004, 126). The most logical interpretation links them with coastal recreational fishing, because of the small size of the potential catches. Reduction in fish size due to overexploitation could explain a reduction in hook size, but the state of our knowledge does yet permit such an interpretation.
- *Small and medium hooks:* The most common in the archaeological record, they are found together on many sites. They suggest angling, or maybe longline fishing, for very lucrative species (Sparidae, Serranidae, etc.) that are well documented in the ichthyological record.

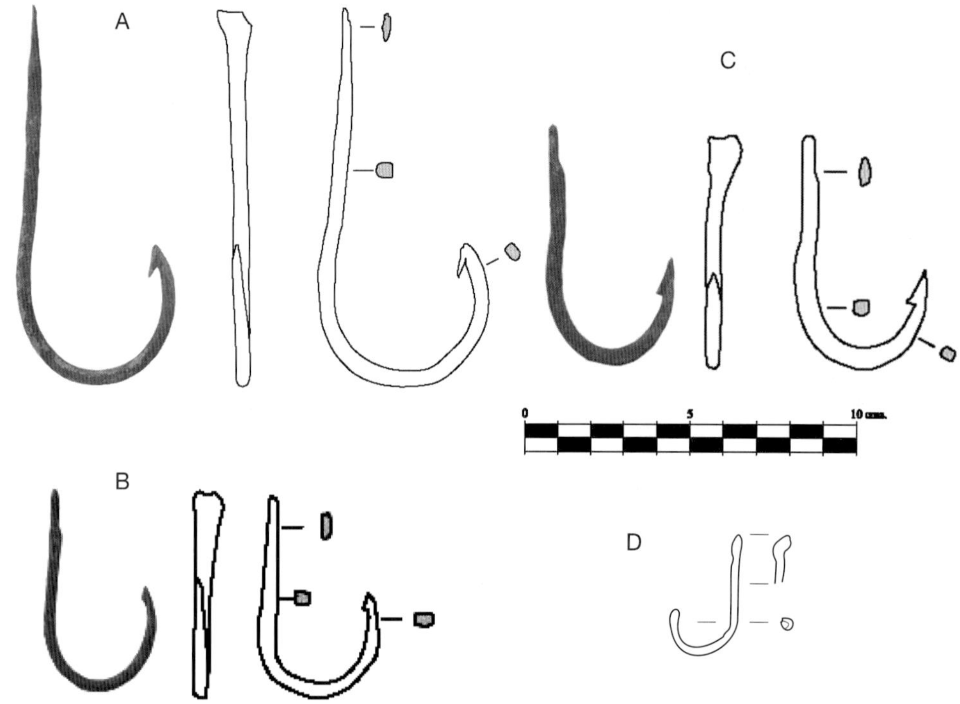

Figure 3. Hook size classification: Large (more than 8cm), medium (4-8cm) and small (2.5-4cm) from Baelo Claudia, and very small (less than 2.5cm) from Castrum Perti (De Vingo y Fossati, 2001, 659, tav. 95, no. 7).

- *Big hooks*: These are exceptional, and very few whole specimens have been found, such as an example from *Baelo Claudia*, 11cm in length (figure 3A). They seem to cluster around the Early Empire, being used for catching large pelagic species – Scombridae – from boats. In numerical terms, they represent a minimal proportion of the finds; in *Baelo Claudia*, they amount to 1.5% of the total number – 1 out of 79 – on a preliminary assessment (Bernal, 2009a, 185, figure 1). Their scarcity shows that the hand capture of big species was seldom practiced, net fishing being much more profitable.

Regarding typology, some initial observations can be made. First, the limited formal evolution of these functional artefacts. Over time, they have remained almost unaltered since the first definition of the shape (eye, shank, bend, point and barb) and for one and a half millennia. Thus, modern classifications – Limerick, Aberdeen, Kirby, etc. (López, 2003, 255-256) – are inadequate for ancient hooks. The only evolution may be a possible shortening of the shank, unduly elongated in some Phoenician examples, which would not facilitate the fish taking in the bait (Arévalo, Bernal & Torremocha, 2004, 141), although it is also possible that such shapes are oriented to the capture of large species (flat fish as nowadays?).

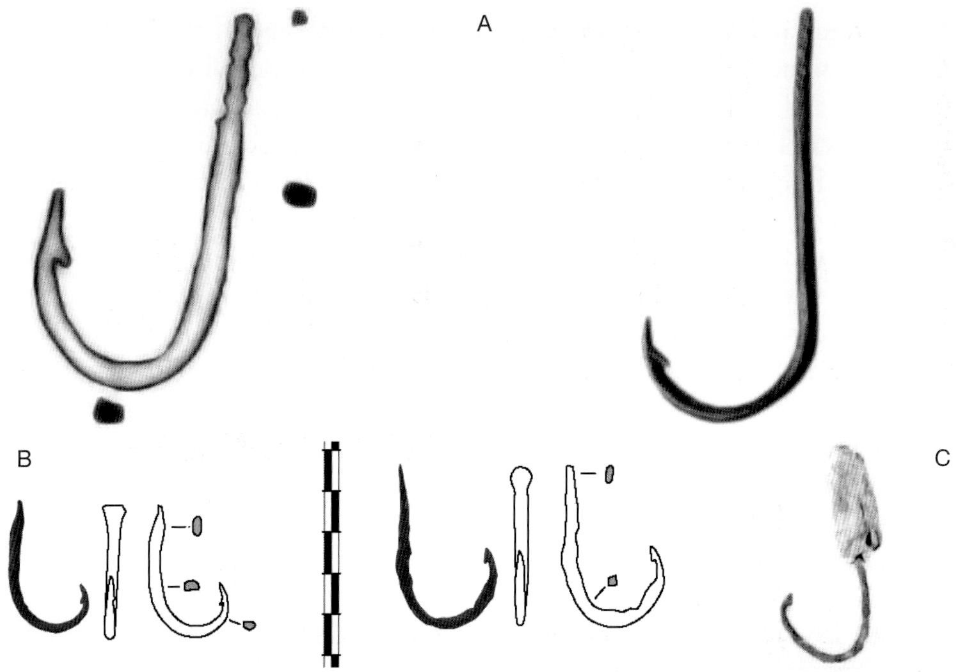

Figure 4. Line attachment systems in Roman hooks: A. Grooved, from Iberian (Oliver, 2006, 38) and Punic (Arévalo, Bernal & Torremocha, 2004, 140) contexts. B. Hammered (from Baelo Claudia, unpublished). C. With lead fitting (from Baelo Claudia, Arévalo, Bernal & Torremocha, 2004, 108).

The biggest changes concern the attachement of the hook to the line (figure 4). Grooved hooks, with several grooves running along the shank from the eye to avoid the loss of the line (figure 4A), seems to be the oldest system, being documented in Spain both in Iberian – two examples in Puig de la Nao (Castellón) dating to the fifth century BC (Oliver, 1994, 38 y 54) – and in Phoenician contexts – for example from the fish-processing site at "Factoría P-19" (El Puerto de Santa María) or in Cerro del Prado (San Roque), at least from the 4th century BC (Arévalo, Bernal & Torremocha, 2004, 126; 140). We do not know yet if this type overlaps with the "plain shank" type, because the state of preservation makes it difficult to identify the grooves, as is the case in many of the Phoenician sites considered. Although very rare, grooved hooks are still used in the Roman period, as shown by an extremely short-shanked example found in a *taberna* in the *insula* 30 of *Emporiae*, dated to the second-third centuries AD (Castanyer, 2006, 20).

Throughout the Roman period, the hammered end, with eyes presenting a triangular or an oval section, resulting from the hammering of the metal rod (figure 4B) appears to be the standard, as seen from numerous examples from the Roman imperial period in the western Mediterranean, and also been documented in the East (Καθημεριν˝ ζω˝ στο Βυζάντιο, 2002, 156) and even the Black Sea (Bekker-Nielsen, 2009, 299, figure 7).

The third known system uses lead fittings at the eye end of the hook (figure 3C), as on some examples from *Baelo Claudia* (Arévalo, Bernal & Torremocha, 2004, 109) and *Tingitana* (Ponsich, 1988, 86, figure 32, 1), although the scarcity of examples in the record suggests that this was an emergency solution restricted to the early Empire; another possibility is their being related to more complex fishing techniques only known fragmentarily. Perforated-eye hooks are so far absent from the record for the Roman period – with the exception of linked hooks, as we shall see – so overall, this technique should not be considered as typically Roman.

The most useful typological difference, based on the available examples from the Iberian Peninsula, distinguishes between simple hooks and the combination of several hooks in the same tackle (double or multiple hooks); "chained hooks" are distinct enough to be considered as a third category in their own right.

Simple hooks (AI)

These are the most common type, being ubiquitous on our Atlantic and Mediterranean sites, as opposed to the other two categories, which are only represented by isolated examples.

Double or multiple hooks (AIII)

Very few examples are known. So far, these have been found in five Mediterranean locations (figure 5). In Pompeii (*Regio* VI, *insula* 16, *entrance* 8), a small example made of a single metal piece bent over itself with a lead-reinforced joint (figure 5A) and interpreted as a double hook (Stefani, 1990, 14, inv. no. 3100b). Unfortunately, none of the ends are preserved, so the interpretation cannot be confirmed or discarded, because as we shall see below, double hooks are normally tied with tin strips. It could also be a specifically Campanian type, although the absence of further examples in Pompeii, or indeed of any other fishing equipment in the specific context of this example, calls for caution (it is not rare to mistake double barbless metal hooks, used for furniture support, for fishing hooks). The find of a complete double hook from the Pisa wrecks, however, dating to the Flavian period (Camilli, De Laurenzi & Setari, 2006, ed., 53) (figure 5B), confirms the use of this type of hooks in the Mediterranean basin.

In one example found in the Comacchio shipwreck, the hooks join at the back of the unusually well developed shanks (figure 5C). Since some of the points are barbed and, the shipwreck was also rich in other fishing equipment, including numerous simple hooks and weights, the interpretation of this example is beyond any doubt. The old collection of Fréjus museum includes another example: 5 hooks joined by a lead fitting with a perforated head, of uncertain date (Sciallano, 1997, 12) (figure 5D).

In the Iberian Peninsula, two exceptionally interesting examples, now kept in the Archaeological Museum of Alicante – MARQ –[1] were found in the necropolis of La Albufereta. The larger of the two (inv. no. CS 5800) – the original size would have been around 15cm (figure 5E1) – was found in the so called "Fisherman's Tomb" (L-17) during J. Lafuente Vidal's excavations (1931-1933), among bronze hooks and charred net fragments (Carmen Alfaro Giner, this volume) which appear to confirm its function. The tomb is dated to the first half of the second century BC by analogy with the nearby tomb L-15, which also contained hooks, along with a black glazed shallow bowl of the Lamboglia 36 type. The other example (figure 5E2; CS. 5254) lacks any archaeological contextualisation, although it was probably found during the same excavation campaign by J. Lafuente Vidal.[2] Their shape is very similar, with a central rod, especially visible in the section of the fragmented piece (figure 5E1), around which the "legs" of the hooks were welded, their shanks fused and laced into a single piece; the whole example only lacks the bends and points of the hooks, whose curved shape can nevertheless be guessed at. Of the other example, only its fragmented upper part (figure 5E2) is preserved: the upper end of the rods and the bronze horizontal thread holding them together. Comparisons with the example from the Comacchio shipwreck (figure 5C) leave little room for doubt concerning their interpretation as multiple fishing hooks by local scholars.

This type of multiple hook is traditionally connected with octopus and squid fishing, as some scholars have already pointed out in relation to the Comacchio example (Rossi, 1990, 271; Gianfrotta, 1999, 26). They offer an important archaeological corroboration for the ethnoarchaeological evidence for the specialized capture of Cephalopods. Although there are but a few examples, their distribution, eastern coast of the Iberian Peninsula (Albufereta), in the French Mediterranean coast (Fréjus museum) and the Tyrrhenian (Pisa in Tuscany and Pompeii in Campania) and Adriatic (Comacchio) Italian coasts, seems to suggest a widespread practice in the western Mediterranean. Due to the inherent identification problems, especially acute for the fragmentary examples, interesting surprises can be expected for the future.

Similarly, the evidence allows for the division of multiple hooks into: a) double hooks, manufactured bending a single metal rod, with an open upper eye, either reinforced with lead (Fréjus) or not (Pisa); b) long-rod multiple hooks, with elongated shanks welded together (illustrated by the examples from Comacchio and La Albufereta) and; c) lead-joint hooks, in which several single hooks are joined in a central lead fitting.

---

1 I am grateful to Mr Manuel Olcina Doménech, Director of the MARQ in Alicante, for his kindness and help for the study and publication of these pieces.

2 I am grateful to Mr Enric Verdú Parra, from the Gabinete de Colecciones e Investigadores, MARQ, for his kindness in providing contextual references for both pieces, as well as for the illustrations.

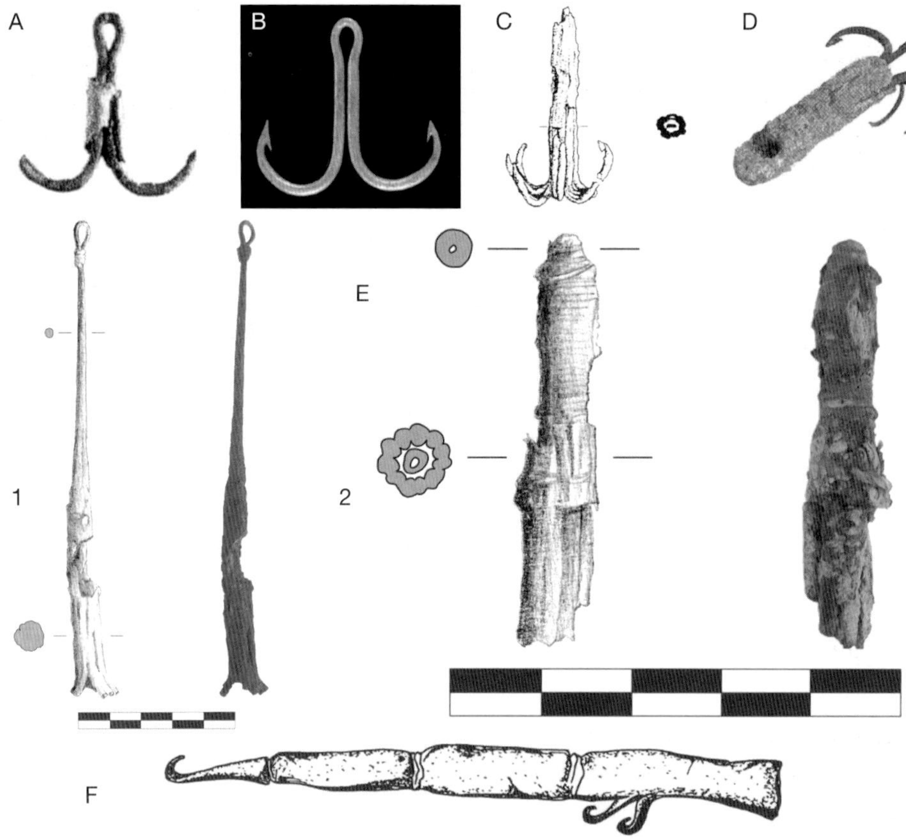

Figure 5. Double or multiple hooks from Campania (A. Stefani, 1990, 14), the Pisa shipwrecks (B. Camili, De Laurenzi & Setari, 2006, 53), the Comacchio wreck (C. Berti, 1990, 271), Antibes (D. Sciallano, 1997, 12) and Albufereta, on the coast of Alicante (E. unpublished; drawings by E. Verdú Parra), along with an item of complex interpretation (Joncheray, 1975, figure 46), maybe used for fishing with live bait (F).

Finally, an artefact of uncertain type found in the La Chrétienne C shipwreck, sunk on the French Mediterranean coast in the third or second century BC with its cargo of Italian wine in Greco-Italic and Dr. 1 A amphorae (Joncheray, 1975, 79). This is a lead sheet, wrapped around its own axis, with at least three perpendicular grooves dividing its overall length of 15cm, into several segments. One end carries a hook-shaped appendix, matched by two other similar implements set closely together at the opposite end (figure 5F). The interpretation of this object is uncertain; a raw material as soft as lead seems to rule out its use as a hook. It has been included here, however, as it might be part of a multiple hook used for fishing with livebait, as is common practice when fishing for groupers (*Epinephelinae*) or even some Scombridae such as tuna. Squids or other small fish, trapped in the three hooks, would have served as a lure for bigger catches, a traditional fishing technique that is known throughout the Mediterranean but which is not documented in Antiquity.

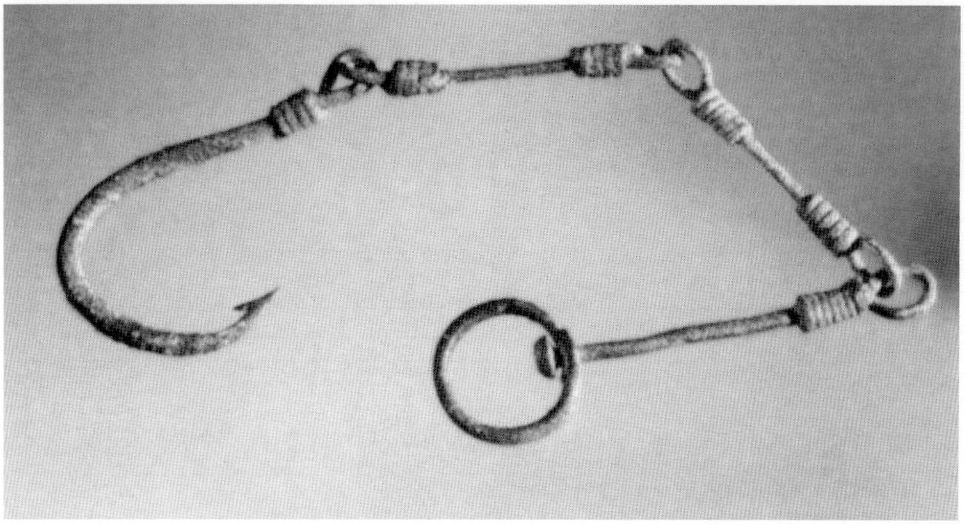

Figure 6. Bronze chained hook from Asciutta, a suburban villa south of Pompeii (Stefani, 1990, 14).

Chained hooks (AII)

Medium-large size hooks attached directly to a metal chain are known, as shown by an example found in the suburban villa of Asciutta in Campania (Stefani, 1990, 14). Its is wrapped around itself and attached to three tied bronze pieces and a circular ring (figure 6). The full rotation of the ring would prevent the tackle from tangling and the bronze links strengthen the line at the point of maximum tension where it was attached to the hook, *cf.* Oppian's description in the second century AD in his *Halieutika* (5.139-146):

> A coiled chain is cast about the butt of the dark hook – a stout chain of beaten bronze to withstand the deadly violence of his teeth and the spears of his mount. On the midst of the chain are set round wheels close together, to stay his wild struggles and prevent him from straightway breaking the iron in his bloody agony, as he tosses in deadly pain, but let him roll and wheel in his fitful course (translation by A.W. Mair).

Although this specific text refers to whaling (Bernal, 2009b, 259-265), the technical specifications apply to the Pompeian example.

This sort of tackle was obviously intended for large species, probably Scombridae. So far, only this Italian example has been found, in a basket along with other fishing implements (40 hooks, a knife and weights for nets and lines), which may suggest that it is typical of the area in the first century AD, although the scarcity of finds calls for caution.

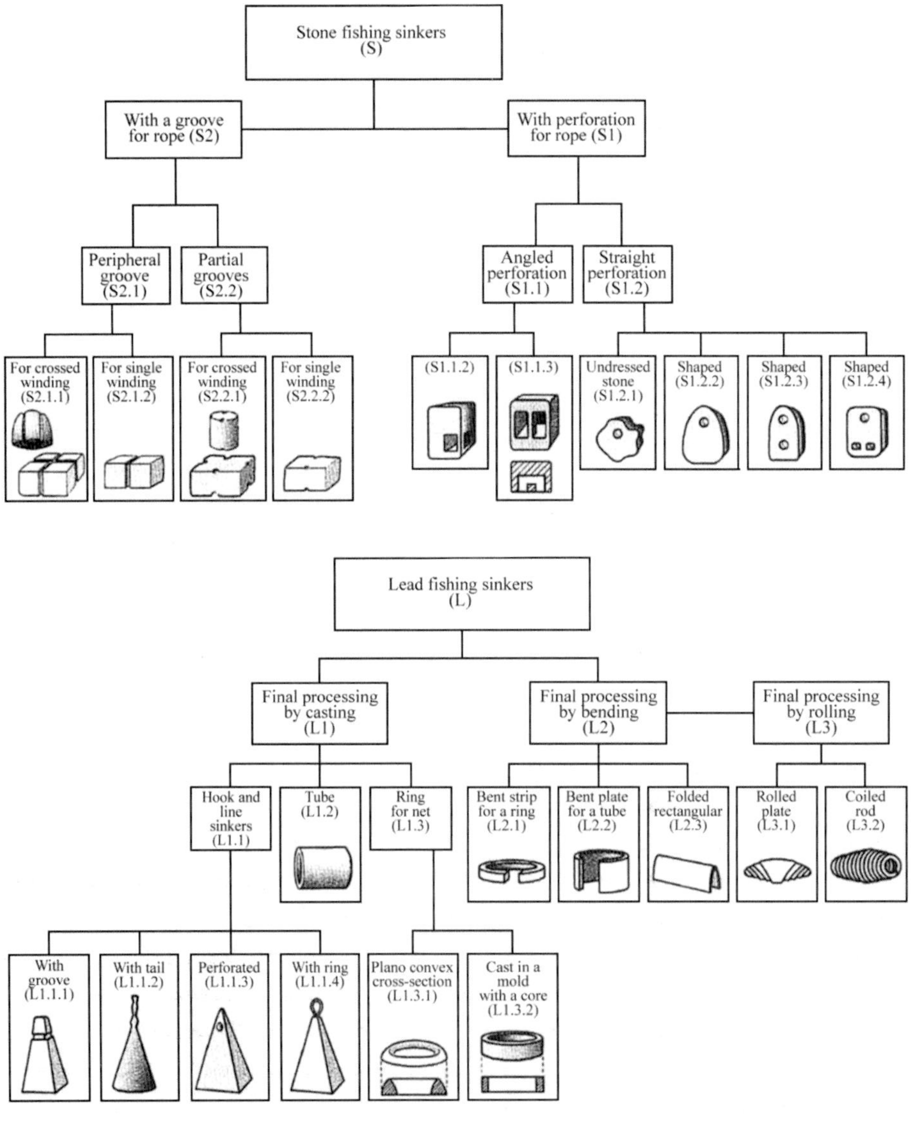

Figure 7. Typological classification of stone and lead weights from Israel (Galili, Rosen & Sharvit, 2002, 183, figure 2).

Weights (sinkers) for nets and lines: the most common find

The study of fishing weights is still in an embryonic stage of development, as shown by the lack of consensus on typological categories, which only serves to prove the complexity of the task. Generally, typologies have relied on finds from underwater contexts; one of the best illustrated and contextualized collections comes from the Yassi Ada shipwreck, which allowed for the identification of 7 types – conical, pear-shaped, spherical, sphenoid, crescent-shaped, triangular, and folded over (Kuniholm, 1982).

An interesting and original typological suggestion, based on underwater material from Israel (Galili, Rosen & Sharvit, 2002), combines raw material, shape and manufacturing technique criteria. More recently, new typological categories have been suggested: axe-shaped, truncated cone, flat disc, parallelepiped/cylindrical and bell-shaped, all based on Tuscan evidence (Ciampoltrini & Andreotti, 2003, 214), to add to the ten categories defined in a doctoral thesis (Ayodeji, 2004, followed by Trakadas, 2009), and a first draft of a typology of weight shapes in the Roman *Baetica*, based on finds from Andalusia and neighbouring areas and following the raw material criterion, which includes ten types (Bernal, 2009a).

Confronted with this complex scenario, I propose a combined approach based on that followed by the Israeli colleagues for stone and lead weights (Galili, Rosen & Sharvit, 2002, 184, figure 2), shown in figure 7, with two modifications: addition of new types (simple pebbles for stone weights and spheroid shapes for metal weights) and not focused on the manufacturing techniques (smelting, bending and rolling); because duplicities ensue, e.g., smelted tubular shapes (Galili L1.2 type) and bent plate for tubular shapes (Galili L.2.2), burdening clarity, because manufacturing techniques have no uni-directional effect on function, and because many of our finds will not fit the pre-defined categories.

Regarding *stone weights*, Galili's typology stands unaltered, with only the addition of simple weights (PIII) – pebbles or irregular stones – used as sinkers, known through ethnographical accounts (figure 8). Obviously, it is possible to subdivide each of these categories on the basis of stone shape, but that would make the classification unnecessarily complex.

13 categories of *metal weights* have been defined, considering only types fully represented in the archaeological record and merging variations such as the "ad accetta" lead weights (Ciampoltrini & Andreotti, 2003, 213, figure 5, 1-3) – which are considered as variations of the truncated cone with perforated appendix type – into the main categories. Some perforated, disc-shaped examples, probably reused (Beltrán, 2007, 63) and other lead artefacts, whose relation to fishing is not certain and may have been used for weighing (Ponsich, 1988, 86, figure 22, 3) have also been left out.

For *clay weights* the recently published typology (Bernal, 2009a) is retained with the sole addition of the perforated truncated cone type, also typical of loom weights, but present also in the underwater record. The double truncated cone with longitudinal perforation found in some shipwrecks, for example, Comacchio (Rossi, 1990) is also omitted since although it may be related to fishing, the similarity with spindle whorls makes it unclear whether these were in fact used for fishing or for spinning and repairing lines on board.

Our proposed typology, which is of course open to revision and future additions, is shown in figure 1, along with a proposal for an alphanumeric nomenclature defined by three features: raw material (C: clay; P: stone; Pl: lead); basic shape (expressed in Roman numerals); and variations within the shape (Arabic nume-

Figure 8. Modern stone weights to sink longlines, including pebbles (PIII1) and masonry (PIII2), in use at Sagres, Portugal (August, 2006).

rals). 22 categories have been defined (50 if variations are taken into account), of which 6 (and 8 variations) correspond with clay weights, 3 (and 13) in the case of stone weights, and 13 (and 29) in the case of metal (lead in all instances) weights.

While it is not possible to define each group in detail here, a general presentation of each will be given below, with some chronological references as to their identification, their frequency in the Atlantic and Mediterranean coasts and, as far as possible, their relation with specific sorts of fishing tackle. Those types not included in Galili's classification (figure 7), are illustrated with Spanish or, in their absence, western Mediterranean examples.

Clay weights

This is the least well-defined category to date, and thus less easily identified in underwater archaeological fieldwork. To remedy that situation, a first typological and chronological draft has been put forward (Bernal, 2009a, 188, figure 3), which, with some modifications, will be followed here (figure 9). According to this preliminary study, clay weights do not appear in major quantities (25% of the total according to the evidence from *Baelo Claudia*), although they are attested without interruption at least from the fifth century BC to the Early Byzantine period (sixth-seventh centuries AD), at least in *Hispania* (Bernal, 2009a, 204).

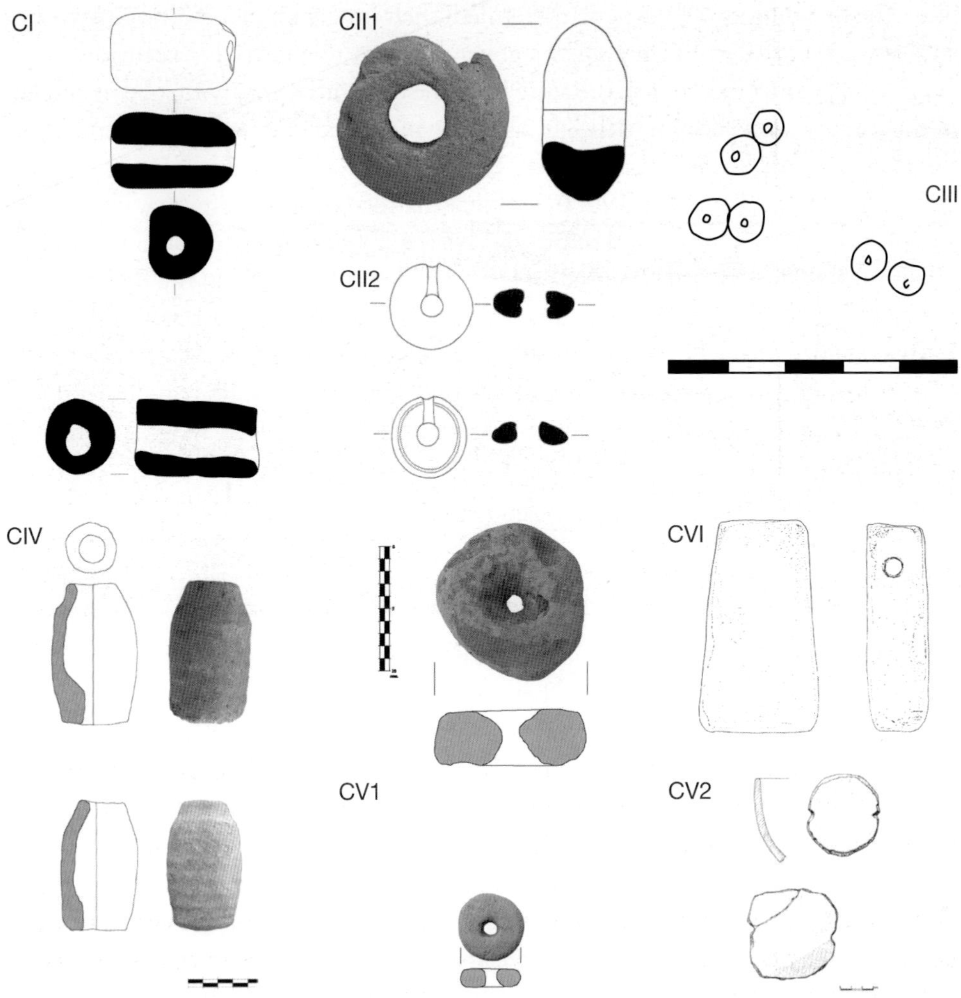

Figure 9. Typology of clay weights; cylindrical (CI), plain disc (CII1), grooved disc (CII2), spherical/ovoid (CIII), spindle-whorl (CIV), trimmed with central perforation (CV1) or with lateral notches (CV2) and truncated pyramid (CVI), from a number of sites (CI. SIIIC, Ramon et alii, *2007, figure 102, no. 271 and Sáez, 2008, 661, figure 42; CII1. Ceuta, PP, unpublished; CII2. Pinheiro, Mayet & Silva, 1998, figure 41, 61-65; CIII. Tavira, following Maia, 2006; CIV. Ceuta, Bernal, 2009a, 199, figure 10; CV1. Baelo, unpublished; CV2. Valencian wreck (Aparicio & Climent, 1985); CVI. Comacchio (Rossi, 1990).*

Cylindrical weights (CI)

Easily recognisable due to their cylindrical shape, their length is normally twice the maximum inner diameter, although some examples are shorter (in the analysed examples, 4-8cm total length, against 3-4cm external diameter). Length seems to increase over time from their first appearance in the Punic period, although this remains unconfirmed due to the lack of complete series, especially for the early Empire. The finishing is not particularly refined; their longitudinal perforations are not always well centred, and they show a general lack of regularity in execution.

They appear with complete certainty from the early fifth century BC, being present in the pottery workshop of SIIIC, in San Fernando, showing up in big numbers in the Roman levels, although those may be discards (Ramón, Sáez, Sáez & Muñoz, 2007, 21-39; Bernal, 2009a, 190-191). Their manufacture in Torre Alta, in the third and the main part of the second century BC (Sáez, 2008, 660 and 661) confirms their use in the late Phoenician/Republican periods. Their continuity in the Bay of Cádiz after the first century BC is uncertain, but their presence in the fish salting factories at *Barcino* in the third century AD and at Cullera, around 500 AD, confirms the continuity of the type well into the Late Empire (for references, Bernal, 2009a, 191).

The size of the holes, 1-1.5cm in diameter, indicate that these weights were used with large nets.

Disc-shaped weights (CII)

Easily recognisable due to their circular shape, these are between 6 and 8cm in diameter, showing a central perforation between 1.5 and 2.5cm wide. Some have a pseudo-circular shape. Generally, they are concave-convex in section, with an internal inflexion, although in some cases the surfaces are straight, with rounded edges due to weathering. They divide into simple discs (CII1), and discs carved with a straight groove (CII2), the purpose of which is obvious (figure 9).

Although some pieces dating to the Punic period (fifth-fourth centuries BC) have a similar shape (Ramon, Sáez, Sáez & Muñoz, 2007, 89), their attribution is uncertain, because of the presence of decoration and the small size of the hole in relation to that of the artefact; they might be loom weights. This is one of the possible sources of confusion within this group; in particular, the small size of the orifice would make rotation difficult. This type is attested with certainty from the second and first centuries BC, in the factories at the Plaza de Asdrúbal site in Cádiz (unpublished); many examples have been found in *Mauretania Tingitana*, dating to the second century AD and later (*Lixus, Septem* or Metrouna). Their use continues through the third century, as shown by the Barcelona example; some examples, from the sixth and seventh centuries have been found in *Carthago Spartaria* (detailed references, in Bernal, 2009a, 195).

The grooved examples (CII2) seem exclusive of the Lusitanian Atlantic coast, as their presence is only attested on sites around *Olisipo*, such as the pottery workshop in Abul, from the reigns of Augustus or Tiberius to almost the third century; they are 7-7.5cm in diameter and the orifice is 1.5-1.8cm wide, showing a flat surface and a rounded one (Mayet & Silva, 2002, 253-255). They also appear in Pinheiro (Portugal) from the late second century to the first half of the fourth; their diameter is 6-7.5cm and their orifice is 1.8-2.4cm wide (Mayet & Silva, 1998, 61-65).

Such weights must have been used with medium/large nets, judging from the size of the hole. The grooved ones may have been attached to the net's terminal lines or to nets/tackles made up with single lines. The holes are too large for the weights to have been used for hook and line fishing.

Spherical/globular weights (CIII)

This is a general category encompassing all spherical or pseudo-spherical shapes with a more or less central perforation. Generally, they are large and show poor workmanship or, at any rate, a non-serial mode of production. The most relevant site is that of Tavira, in the Portuguese Algarve, where dozens have been recorded; they are crudely made but their faces seem to be dressed; they are heavy (750g) and large (13-13.5cm in diameter). These weights come from a Punic context (fifth to third centuries BC) clearly related to fishing practices (Maia, 2006, 455-465). The type may be the typological predecessor of the disc-shaped type, until the latter became standard. No examples are known from the Roman period.

Spindle-shaped weights (CIV)

Cylindrical in shape, narrowing at the ends, especially in the upper half. Typically, the material is thickest towards the lower end, creating a concave profile on the inside. Their morphology is similar to that of small *tubuli*, but their presence in fishing environments, such as the fish-processing factories at Tahadart, active from the Augustan period to the fifth century AD (Ponsich, 1988, 149-150, figure 32, 4) or in *Septem*, of uncertain date (Bernal, 2009a, 197) suggest a function as net sinkers. Their dimensions are also suitable for that purpose (around 10cm long with perforations of 1.5-2cm). So far, their distribution is limited to *Tingitana*, so they seem to be related to middle- or large-sized nets used along the Maghreb.

Trimmed weights (CV)

Weights not made for the purpose, but adapted from miscellaneous pottery fragments which, if adequately trimmed, can be reused as fishing sinkers. The first variant, trimmed weights with central perforation (CV1) is very common, consisting of common pottery sherds, normally from amphorae, trimmed to a circular or pseudo-circular shape and drilled in the centre for the insertion of the line (figure 9). They show remarkable variation in size, from 2-3 to more than 10cm in diameter; the size of the central hole also varies. The edges can be more or less weathered de-

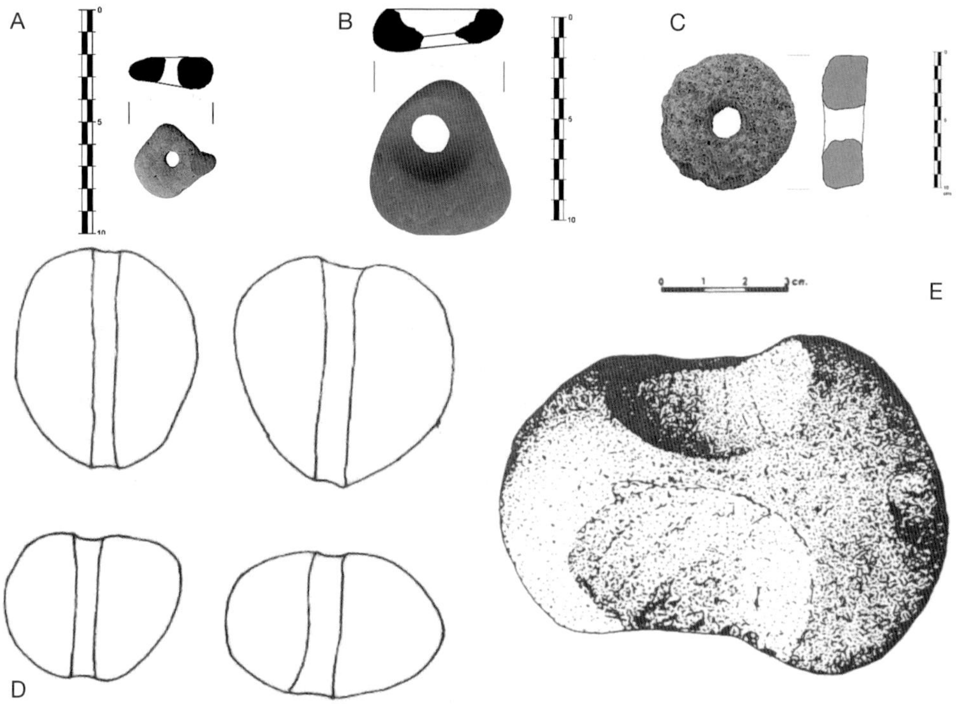

Figure 10. Main stone weight types documented in the Iberian Peninsula. A. PI3 from Baelo; B. PI4 from Baelo; C. PI5 from Ceuta; D. PII 2, from CDB (Ruiz, Ruiz & López, 2006, 293, figure 8); E. PII4, from Valencia (Aparicio & Climent, 1985, 15).

pending on use and care of execution. Their interpretation is obscured by the fact that very similar artefacts can be put to other uses, such as spindle whorls, weights for measuring etc., thus the context of the find is crucial for interpretation.

This variant is known at least since Protohistory, being common in Iberian contexts (Playà, 1998, 399). In the Roman period, they are attested all throughout the Empire. In Spain we have at least some examples from different contexts within *Baelo Claudia*, of imperial date, made out of amphorae or plain pottery sherds (Bernal, 2009a, 197). Continuity into the Late Empire is confirmed by the finds in the *cetariae* of *Barcino*, dating to the third century AD (Beltrán, 2007, 279) and Pinheiro in Portugal (Mayet & Silva, 1998, 189-220) of the fourth century AD.

The poor workmanship, with no regularity in size and an overall diversity (in *Baelo*, no two examples are similar) suggest episodic use. Most likely, they were used for line fishing.

The other attested variety (CV2), also pottery sherds trimmed into a pseudo-circular shape and around 6-7cm in diameter, show two V-shaped lateral notches; they are attested in fishing sites in the Valencian coast during the Bronze Age (Aparicio & Climent, 1985, 11-15). They could have been used as net weights (with lines tied around the notches) or as reels to haul the line, although the shallowness of the notches makes the latter possibility less likely. Although they are not attested in the Roman period, it is also possible that they may have gone unnoticed by archaeology. Most likely these were also used for line fishing.

Truncated pyramid-shaped weights (CVI)

In this category, we include the well known clay loom weights, with the shape of a truncated pyramid and a perforation (figure 9), commonly used as *pondera* in the weaving industry. Their frequent appearance in shipwrecks confirms their use as fishing implements (Beltrame, 2002, 68) from the sixth century BC onwards, for example in the Gela shipwreck (Panvini, 2001, 62) to the Roman period, as the Comacchio examples confirm (Berti, 1990, 271, figure 256). Generally, they appear in isolation or in small numbers, so their interpretation as sinkers for angle or line fishing seems the most likely.

Because they are closely similar to loom weights, a detailed knowledge of their context is crucial to verify their use as fishing tackle; multiple functionality and reuse also seem very likely.

Stone weights

The use of stone for fishing weights in *Hispania* is very limited. In *Baelo Claudia*, a site systematically studied, they barely amount to 15% of the total, a good example of their rarity.

The recent proposal of Galili and others is valid and well structured, so we will not elaborate further on it (Galili, Rosen & Sharvit, 2002, 183, figure 2). They divide the weights into perforated and grooved types (S1 and S2, our PI and PII respectively), further subdivided according to the position and the morphology of the attachment (figure 7). We will only make two small additions. 1) The inclusion in this category of perforated weights of flat circular stones with a central orifice (PI5), sometimes referred to in the literature as "doughnut" weights, for example, one example in the Ceuta museum (figure 10C). And 2) an additional category called simple stone weights (PIII), encompassing pebbles (PIII1) or undressed masonry (PIII2), used as net sinkers, wrapped into a mesh or simply tied to the nets (figure 8). Some Egyptian models – around 2000 BC – illustrate this system (El Shahawy, 2005), traditional in contexts in which metal was not easy to come by (Sáñez Reguart, 1791-1795).

Stone weights are attested in the Iberian Peninsula at least since the Bronze Age (Aparicio & Climent, 1985, 15) and more clearly from the eighth century, as seen at the Phoenician site of Castillo de Doña Blanca (CDB) in the Bay of Cádiz (Ruiz Mata, Ruiz Gil & López Amador, 2006, 291), along with many earlier European examples (Cleyet-Merle, 1990, 140-147). The cited examples are the oldest weights that can be linked with fishing nets with total certainty, due to their weight and size, found to date in Spain. Some grooved examples now in the Tenerife museum (González & Chaves, 2004, 304), of uncertain chronology, could also belong to the same group. Their sequence may suggest that grooved weights are the typological predecessors of the perforated ones (PII4 and PII2). So far, no well contextualised examples of the grooved series have been found in *Hispania*, with the exception of some items currently in the A Coruña-Bahía de Coruña, Casa Martelo (Bernal, 2009a, 188) and Irún museums. Regarding simple stone weights, no example has yet been clearly associated with fishing activities, although many potential weights – especially small and medium sized pebbles – are known from Spanish coastal sites, eventually discarded due to the problems of identification.

In contrast, perforated weights are well represented; they can be divided into undressed masonry (PI3) and most commonly, when a defined shape is followed, pear-shaped (PI4) or, less frequently, circular (PI5) (figure 10A and B). All of these show straight perforations.

A first bibliographical survey suggests the almost total absence of stone weights from Roman contexts, due not only to their actual scarcity but also to the lack of research interest, especially in the case of simple weights. It is likely that a prejudice towards considering stone weights as typical of prehistoric contexts has tacitly contributed to their underestimation.

The known examples can be associated with diverse fishing practices. Those from CDB were clearly intended to ne used with nets – they are medium size and were mass produced – similarly to the simple (PIII) and perforated weights (PI5). In contrast, the pear-shaped (PI4) or irregularly-shaped (PI3) ones with straight single perforation must, due to their size – less than 10cm in all cases – have been used as sinkers in angling.

Due to the recurrence of these makeshift systems in traditional fishing environments when ready-made weights were scarce, it is difficult to evaluate the chronological span for each of these types.

Finally, we must bear in mind that weights, especially those with two or three holes (PI6 and PI7) are similar in shape to traditional anchors, so identification must rely on size. To date, no examples for stone sinkers so large as to make identification doubtful have been found, at least in our own region, although examples found in Israel, almost 30cm long (Galili, Rosen & Sharvit, 2002, 187, figure 3) confirm their use, at least in the eastern Mediterranean. It is, however, pointless to try to distinguish between anchors and sinkers for large nets, because the same artefacts could have been used for either purpose.

Metal weights – lead

These are the most common pieces of fishing equipment, amounting to more than the 60% of the total number of weights in *Baelo Claudia*. Lead seems to be the only metal in use, both because of its weight and of its resistance to corrosion when in contact with water. In the West, lead weights are well attested from the Punic period, in the fish salting factories on the Bay of Cádiz, they appear with certainty from the fourth BC, and probably much earlier; in other areas, they date from as early as 1320-1295 BC, as shown by the Uluburun shipwreck off Anatolia (Pulak, 1998, 210 and 214). Their continuity into the end of the Late Antiquity was confirmed years ago by the wreck of Yassi Ada, dating to the seventh century AD (Kuniholm, 1982), later confirmed by further finds. A future Atlantic-Mediterranean contextualisation of the finds will allow for further elaboration.

The classification of these artefacts is complex, once the triple division based on manufacturing technique – smelting, bending and rolling – is discarded to avoid ambiguities. Our draft, which draws exclusively on typological criteria, has defined thirteen categories, explained below. The complexity of the group is in itself sufficient evidence of the popularity of these weights in the Roman period and of the number of technological innovations introduced.

Ring-shaped weights (PLI)

These show a circular and flat shape, similar to a ring. They can form a full, made by smelting (PLI2-Galili L132) or a partial circle, made by the bending of a lead piece without fully converging edges (PLI3-Galili L21). Sometimes, the cast shapes show a very characteristic flat-convex section (PLI4-Galili L131). They are common in the eastern Mediterranean, especially Israel (Galili, Rosen & Sharvit, 2002, 186, table 1), but much less so in the western Mediterranean, due to the lack of systematization, among them one example from *Sexi*, on the eastern coast of Andalusia (Molina Fajardo, 2000, 178). This group includes the so-called "lead rings", traditionally interpreted as brail rings for sails, but for which a possible use as net sinkers has been proposed, based on the abnormally high concentrations found in some shipwrecks – such as *Caesarea* and Giardini Naxos, Cap Lardier 4 – (for a synthesis, see Beltrame, 2002, 67-69; this volume 234-238). This idea has received some recent support (figure 11). There are at least three advantages of these weights against the tubular or lineal (Galili, Rosen & Sharvit, 2002, 198-199): they do not get entangled in the net; they can be attached and detached without dissembling the net or changing the lines (figure 11A); they can also be used to close and haul the nets, as it is common today with ringed net systems in use in Russia and Turkey, as well as in other Mediterranean contexts such as central Tyrrhenian Italy (figure 11B).

In *Hispania*, among many unpublished ones, there is an interesting example from the Culip IV shipwreck, which sank in the last quarter of the first century AD to the northwest of the peninsula, and which has been tentatively related to some other examples from the Greek shipwreck of El Sec in the Balearic Islands (Nieto

Figure 11. Attachment system for lead ringed weights (A), following Galili, Rosen & Sharvit (2002, 199, figure 18), along with examples of ringed nets from a vessel in Cetara (Campania, Italy) (B, June 2009).

& Pujol, 1989, 235, figure 153, 1). Many more examples have been found in *Herculaneum*, in a region where ringed nets – for example, in Cetara – are still in use, equipped with large iron rings.

Not included here are the rings with a lateral perforated appendix (PLI1), which many authors, following Foerster (1985), interpret as tools to liberate hooks and anchors stuck at the sea bottom, attached to a line tied to the lateral piece. More recent interpretations see them as brail rings used to hold the sail's bowstrings (Beltrame, 2002, 67-69, figure 105), a hypothesis apparently confirmed by practical reconstructions (Galili, Rosen & Sharvit, 2002, 198-199).

Cylindrical weights (PLII)

In this category, we include all weights of cylindrical shape; divided into hollow or solid. The hollow ones, tubular in shape, are divided into those which show a full circumference, made by casting (PLII1-Galili L12), and those which only show a partial section, by rolling up a plate (PLII2-Galili L22). The solid ones can be the result of rolling a plate (PLII3-Galili L31) or can show lateral notches (PLII4).

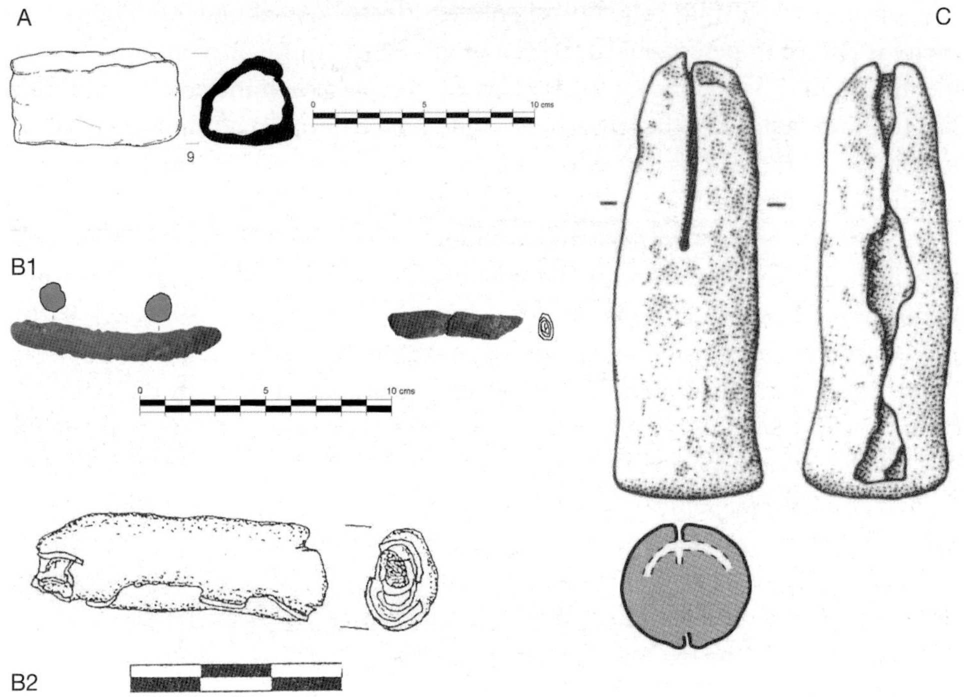

Figure 12. Cylindrical lead weights: hollow and partial (PLII2), from Baelo Claudia (A) (Bernal et alii*, 2007, 476), rolled and solid (PLII3; unpublished, from Ceuta (B1) and Castrum Perti, (B2) following De Vingo & Fossati, 2001, 659, figure 95, 16) and solid with grooves (C) (PLII4; from Cala Culip, Arévalo, Bernal & Torremocha, 2004, 153).*

Both versions of the tubular type (PLII1 and PLII2) were used on large nets. The first type is well attested in the eastern Mediterranean (Galili, Rosen & Sharvit, 2002, 186, table 1), but no well contextualised examples have been found to date in the west, where, on the other hand, many examples of the PLII2 group have been recorded. Their use on large nets can be inferred from their remarkable circumference: for example, the lead sinkers found in *Traducta* or *Baelo* are between 2 and 2.5-3cm in inner diameter (Bernal, 2009a, 200); normally they also show several bulbous protrusions on the inner face, showing that they were hammered against the multiple lower lines of the nets (figure 12). On the two sites mentioned, they are dated from the early imperial period to *c.* 500 AD; no earlier examples are known, although lead plating and rolling were widespread activities in the Phoenician period (Bernal, 2009a). Due to the lack of a complete series, it is not possible to draw inferences as to the types of nets they were used on, apart from the large size; probably almadraba-type nets used this sort of weight. Their length is normally around 2½ times the maximum outer diameter (which is between 2 and 4cm). Typically, the smaller examples show a pseudo-triangular or squared section, belonging to the rolled plate type (PLIX).

Conversely, the solid rolled plate type (PLII3) is very abundant. The central cavity is almost imperceptible in section, especially because of corrosion, making identification difficult (figure 12B). These pieces were hammered around a thin line (1-2mm) rendering the lead, which could not be unrolled, unsuitable for reuse, in contrast with other types, such as the partial section type (PLII2). Their use for line fishing seems clear.

All this goes to explain the scant attention paid to these leads in the bibliography, although many examples have been found; for example, in the cargo of the Chrétienne M shipwreck, dating to the first century BC (Joncheray, 1975, 93, 9-10), the fish processing factory in *Septem Fratres* on the African shore of the Strait (figure 12B1), of the Middle Imperial period, and *Castrum Perti*, to the sixth and seventh centuries AD – figure 12B2 – (De Vingo & Fossati, 2001, 659, figure 95, 16), showing the longevity of the type.

The last variety includes fully cylindrical weights with lateral perforations (figure 12C), as shown by one example in the Culip VIII shipwreck, dated from 10 BC to 10 AD (Carreras *et alii*, 2004, 156-157, figure 94, CP VIII-03 0-0-30). This type does not appear to be very common. The line is attached through insertion, in a manner similar to the pyramidal weights (PLX3).

Cone/truncated cone weights (PLIII)

Very common in the archaeological record, with the exception of the first variety, which shows a distal appendix (PLIII1-Galili L112), so far only recorded in Israel. Their body is that of a cone or a truncated cone (figure 13), and sometimes they are called "bell-shaped" (Ciampoltrini & Andreotti, 2003, 215, figure 6 *infra*). The only difference between the varieties is the groove (PLIII2) or perforation through which the line is attached (PLIII3). Generally, they are of small size (below 5cm) and poor workmanship resulting in irregular shapes. The perforations were usually made after casting, fixing the line in the weight by hammering its walls, producing the previously mentioned deformities. Chronologically, they are attested from the Republican period in southern France (Joncheray, 2002, no. 2-5) and from the Augustan period in the north-eastern coast of *Tarraconensis*, for example in the Culip VIII shipwreck (Carreras, 2004, 153), to the seventh century AD, as shown by the Yassi Ada shipwreck, where they are one of the dominant types (Kuniholm, 1982, LW1, 2, 6, 7, 10, 12 and 13). They are distributed all across the Mediterranean. The perforated variety (PLIII3) is also widely distributed and shows a similar continuity, for example from La Chrétienne M, from the first century BC (Joncheray, 2002, 93, no. 8) and Anatolia, in the seventh century AD (Kuniholm, 1983, LW 4, 8 and 14), overlapping with the previous variety, although no clear indications of an evolutionary sequence can be drawn yet. Their use for line fishing is widely agreed upon.

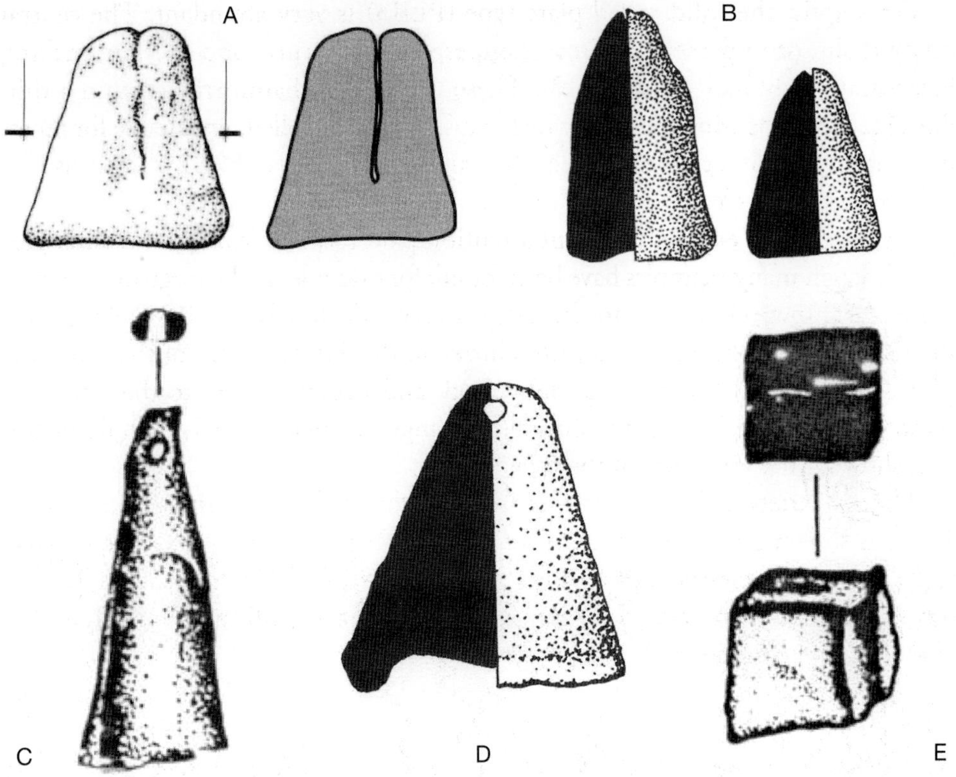

Figure 13. Lead weights: Grooved conical, type PLIII2 (A. from Culip VIII, Carreras et alii, *2004, 153 and B. Yassi Ada, Kuniholm, 1982, LW 1 and 2), perforated cone (C. PLIII3, Joncheray, 2002, 93; D. Kuniholm, 1982, LW14); and grooved cubical from La Chrétienne M (E. Joncheray, 2002, 93, no. 1).*

Cubical weights (PLIV)

Similar in all but shape to the previous variety (figure 13E). The method of attachment is identical, through grooves (PLIV1) or perforations (PLIV2). They also seem to have been used on individual tackle (rod and line fishing).

The grooved variety has been found in La Chrétienne M, first century BC (Joncheray, 2002, 93, no. 1) and the perforated variety in Lucca, with an uncertain date somewhere between the second century BC and the third century AD (Ciampoltrini & Andreotti, 2003, 213, no. 10 and 215). Both seem to be totally absent from Late Roman sites (Yassi Ada, *Castrum Perti* or Dor shipwreck) and from Israeli waters (Galili, Rosen & Sharvit, 2002). Future work should confirm the restriction of their distribution to the western Mediterranean and the Republican and Early Imperial periods. At any rate, they are not too common, and they are often reused fragments of other weights. Also, their angular form and low weight would render them functionally defective.

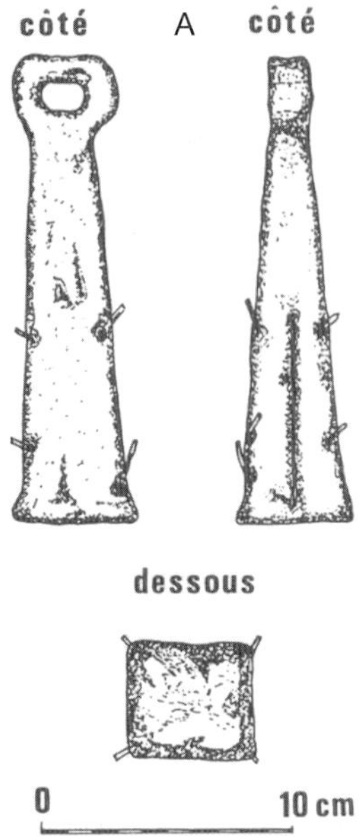

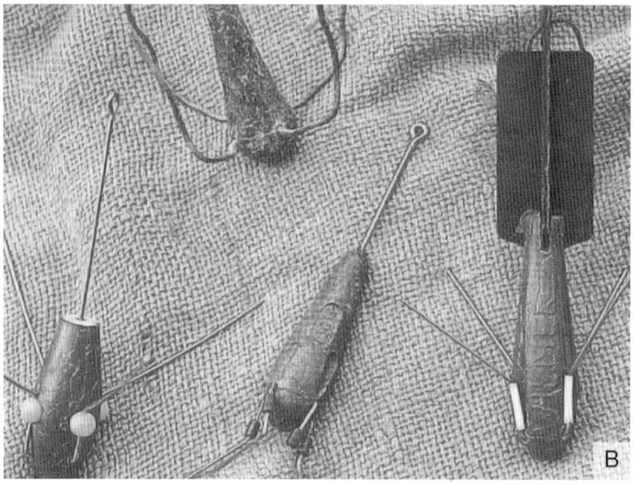

Figure 14. Clamp weight (PLV) from La Chrétienne C (Joncheray, 1975, 97, figure 46, no. 3), and modern examples (López, 2003, 253).

Clamp-rod weights (PLV)

This variety has not previously been identified, which increases its interest. We only know of one example, from the republican shipwreck of La Chrétienne C (second-third centuries BC), originally identified as a sinker for a sounding line (Joncheray, 1975, 95). It has an irregular shape, worked by hammering, is 8cm long and 7.8cm wide at the base, and weights 2026g (figure 14A). This piece, with the shape of a truncated pyramid, was topped by a large fitting appendage with a series of metal rods on the sides (probably eight, on two parallel lines) facing up, resembling the so called "clamp weights" or "rod weights" used in sandy sea beds; being movable, the rods would "anchor" into the sand, rotating when the line was pulled (López, 2003, 253; figure 14B). Rod weights are ideal for rough waters or heavy currents, because the rod's "anchorage" prevents them from rolling over, entangling the line. Curiously, they are found in French waters, but not in the Atlantic. They show a remarkable level of technical achievement, having remained almost unaltered to our days.

Its use in line-fishing seems obvious, for its weight rules it out for rod-fishing. Maybe this type of lead weight was used on multiple lines or sinking large hooks for big catches, although its use on longline ringed nets on sandy sea beds cannot be discarded.

Off-centre weights (PLVI)

This category includes those weights previously referred to as crescent-shaped (Kuniholm, 1982, 301, LW 20 and 21). Only two examples are known, from Yassi

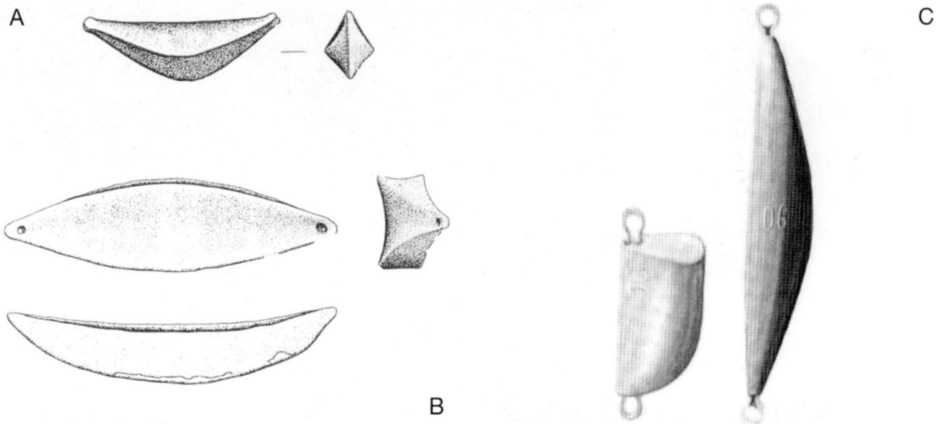

Figure 15. Off-centre leads (PLVI), from Yassi Ada (A., B. Kuniholm, 1982, 301, LW20 and 21), and modern examples (C. López, 2003, 254).

Ada, seventh century AD, of 107 and 479g in weight, and 8 and 13.5cm in length respectively. As shown (figure 15A) they have a curved shape, in one case with perforations at the ends, not preserved on the other example (15B).

This sort of weight, normally called off-centre, is intended, by means of the perforations at the ends, to prevent the lines from tangling (López, 2003, 254). Generally they are big in size, and used for trolling. Nowadays there are several models, boat-or banana shaped (figure 15C), all of them with an off-centre mass from the projected bisection of the ends.

It is interesting to note the remarkable technological achievement represented by these weights. To date, they are only attested in the Yassi Ada shipwreck, Turkey, dating to the Late Antiquity, interestingly being absent from Israeli waters (Galili, Rosen & Sharvit, 2002), the western Mediterranean and the Atlantic, although the scarcity of finds should be pointed out.

Spherical weights (PLVII)

They are an uncommon item in the record: one of the few examples known – some are yet unpublished – from the Yassi Ada shipwreck, of late Roman date, has only been identified as fishing tackle thanks to its being found in a halieutic context (Kuniholm, 1982, LW 16). The item is fully spherical (figure 16A) and it seems that in the past it had a ringed attachment for inserting the line; this was eventually lost, so the weight itself was drilled. Thus one example illustrates both types (PLVII 1 and 2).

It is, as said, very uncommon, and never shows perforations running all through the body, as in contemporary varieties. It may be a variant from the conical, pyramidal or pear-shaped types, all scarcely used in Antiquity.

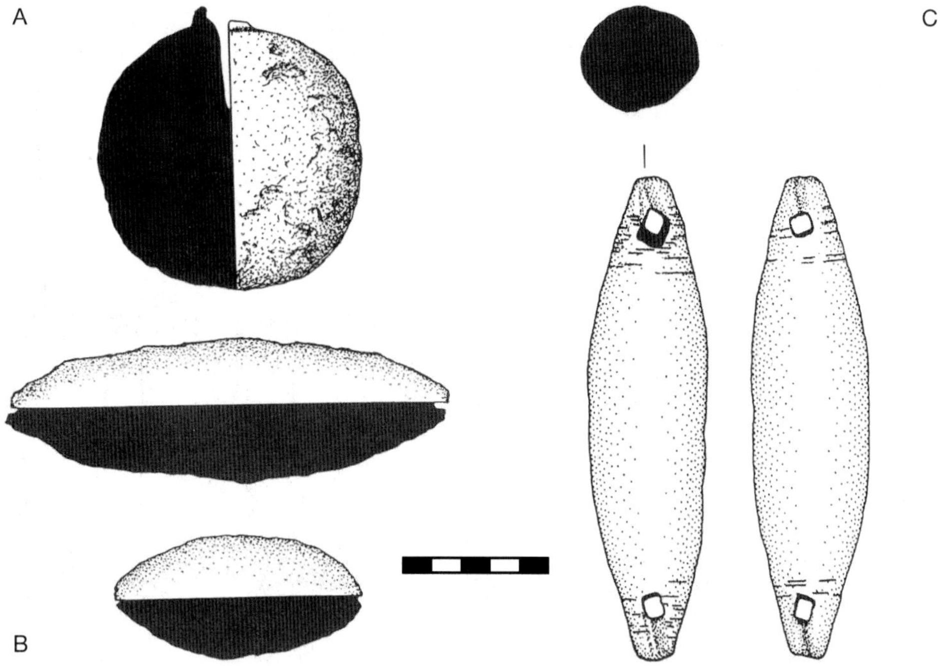

Figure 16. Spheric (A. PLVII) and sphenoid leads, grooved (B. PLVIII1) and perforated (C. PLVIII2) from Yassi Ada (Kuniholm, 1982, LW 16, 17, 18 and 19).

Sphenoid weights (PLVIII)

So called in the study of Yassi Ada, they have either two grooves (PLVIII1; figure 16B) or two holes (PLVIII2; figure 16C) at each end for inserting the line. They have only been found in this late Roman shipwreck. The grooved ones had the advantage of being attachable to the line at any time, by rotation of the weight. The perforated type, of which only one example is known, could have functioned as a running weight, although the friction marks on the only example known (figure 16C) are perpendicular to the axis, suggesting attachment to both the free line and the line carrying the hook. However, the isolation of this example should prevent us from taking the interpretation any further.

Rolled plate weights (PLIX)

The most common type in Spain and probably over most of the Graeco-Roman Mediterranean. These are manufactured by simply bending a plate, normally rectangular in shape, over itself, resulting in a U or V-shaped section which also tells about the width of the line with which it was used (figure 17). Many varieties could be drawn from the differences in size, but it is more convenient to divide only between those made

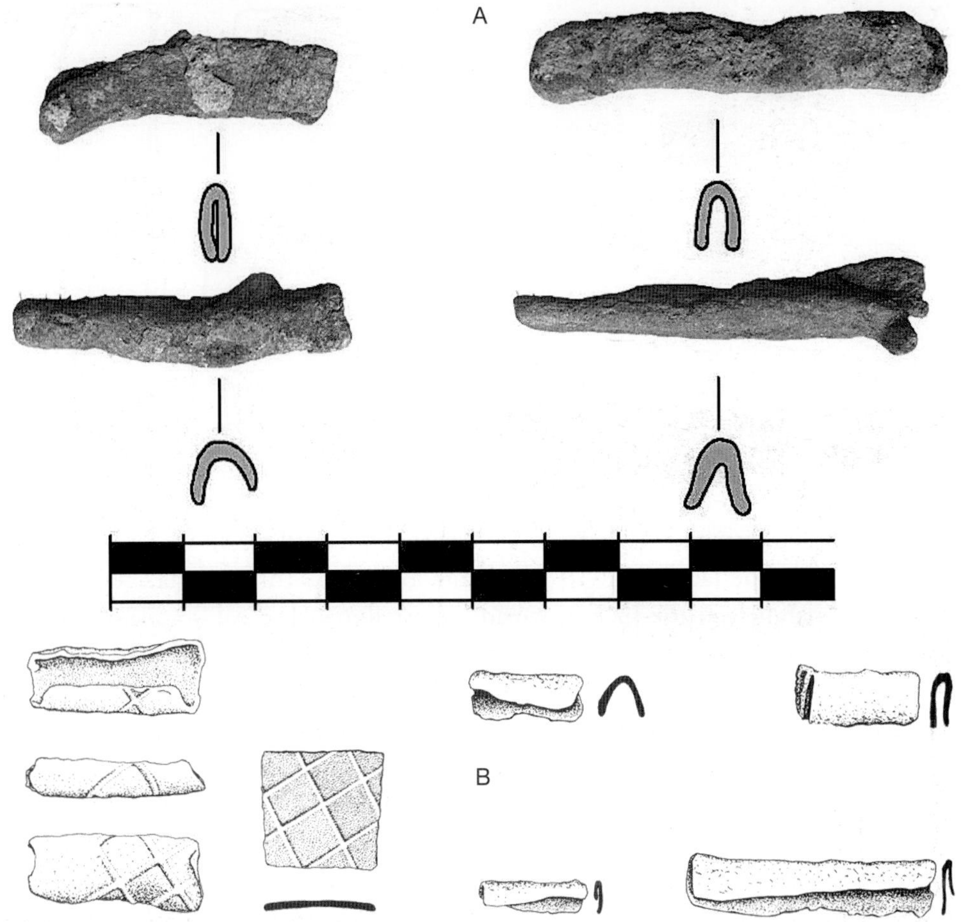

Figure 17. Rolled rectangular plate weights, from Baelo Claudia (A. Bernal, 2009a, 203, figure 12, 5-8) and Yassi Ada (B. Kuniholm, 1982, LW 31-36).

from square shaped plates (PLIX1) and from rectangular ones (PLIX2-Galili L23). Formally, they are similar to the smaller varieties of the sub-type PLII2, although their length is normally more than 2.5 times the diameter, often as much as five times. Some have suggested a division according to their width – not length – into large (more than 37mm), medium (20-37mm) and small (15-20mm), with central cavities between 7 and 9mm in diameter for the bigger ones to 0.8-1.4mm for the smaller sizes. For the moment, and until wider series for the western Mediterranean are known, we will not apply this classification (Galili, Rosen & Sharvit, 2002, 187, table 1).

The oldest square shaped plate weights (PLIXI) in the Iberian Peninsula, where they are an uncommon type, have been recorded in the Phoenician factory "P-19" at the Bay of Cádiz, a clear halieutic context, dating to the fourth century BC (Arévalo, Bernal & Torremocha, 2004, 2000, 138). The rectangular shaped ones (PLIX2), on the other hand, are very frequent, in spite of having been misinterpreted as ri-

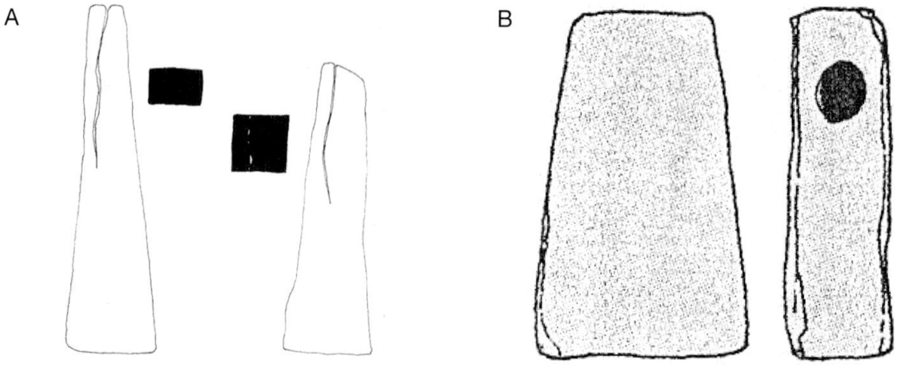

Figure 18. Truncated pyramid weights (A. PLX3), grooved, from Culip IV (Nieto et alii*, 1989, 214, figure 153, 4) and perforated (B. PLX4) from Lucca (Ciampoltrini & Andreotti, 2003, 215, LR 46).*

vets or as undetermined lead artefacts by archaeology until recent years. For example, those from the factory "P-19" (Arévalo, Bernal & Torremocha, 2004, 138) have many parallels in the Iberian world – Castellones de Ceal – and in Phoenician and Late Phoenician contexts yet unpublished. They are distributed all throughout the Iberian Peninsula, from *Emporiae* (Castanyer, 2006, 21) to the Strait of Gibraltar, including *Portum Sucrone,* in Cullera (Valencia) in the second half of the fifth, or the sixth century. Outside, they are attested at Yassi Ada for the seventh century (Kuniholm, 1982), in many locations on the Israeli coast, also dating from Late Antiquity (Galili, Rosen & Sharvit, 2002), and many other sites and shipwrecks, resulting in a full list well beyond the scope of this paper.

It should be noted, however, that eastern examples display decorated motifs, as shown by the examples found at the Turkish shipwreck of Yassi Ada (Kuniholm, 1982; figure 17B) and more recently the Dor shipwreck, in Israeli waters (Galili & Rosen, 2008). Both wrecks are early Byzantine, dating to the seventh century AD. Recent studies have analysed the variety of designs, lineal, diagonal, geometric, or even inscribed, among others (Galili, Rosen & Sharvit, 2002, 188-190, figure 8). This may help identify the presence of eastern fishermen in the western Mediterranean, where no equivalent practices have been attested to date.

Scholars fully agree on the relation between these weights and casting nets, small individual nets used from aboard vessels – thus, their abundance in shipwrecks – or from the coastline.

Pyramidal/truncated pyramid weights (PLX)

Closely related to the conical/truncated cone type, their only difference being the flat and square shaped base. They only differ in the way the line is attached: through

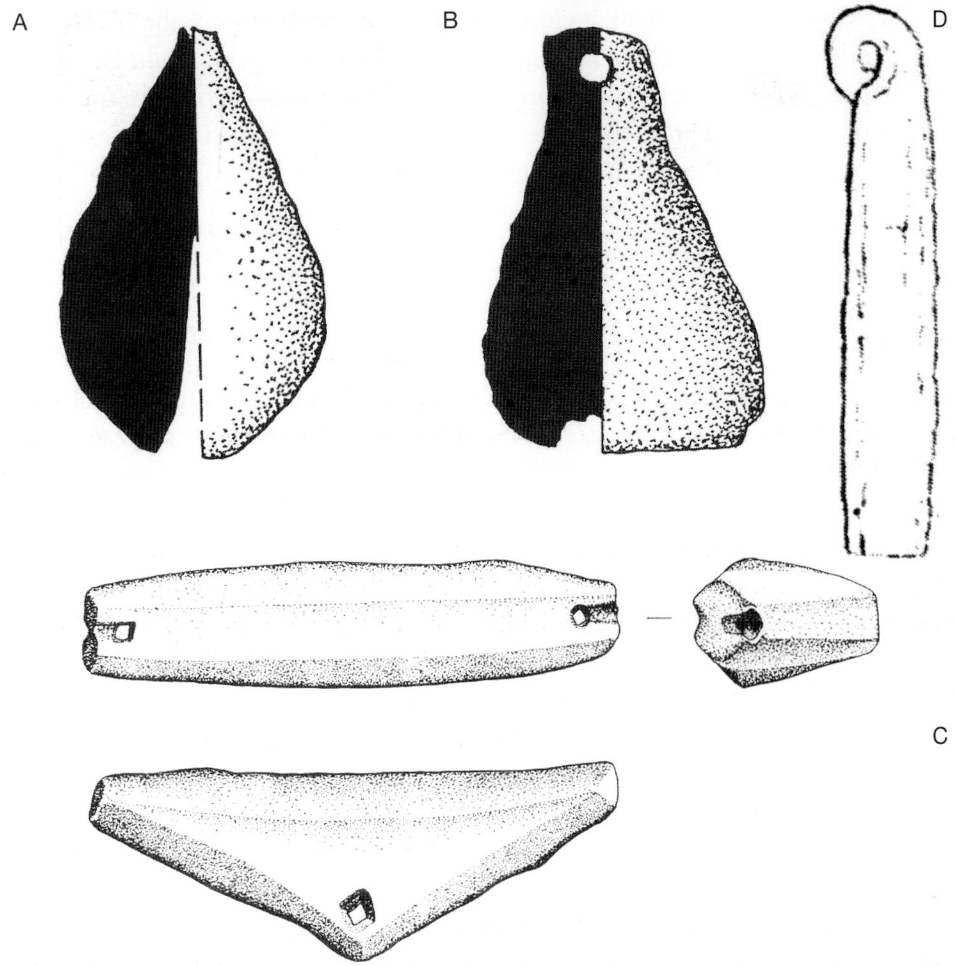

Figure 19. Pear shaped weights, grooved (A. PLXI1) and perforated (B. PLXI2), along with a triangular perforated example (C. PLXII) from Yassi Ada (Kuniholm, 1982, LW 11, 8 and 22), and a perforated tubular example (D. PLXIII) from Villaricos (Siret, 1906, no. 47).

a horizontal groove (PLX1-Galili L111), a metal piece (PLX2-Galili L114), a groove (PLX3) or a hole (PLX4-Galili L113).

The perforated (PLX4) and vertical grooved ones are the most common types in the west (figure 18), due to their easy identification, as shown by the examples found in Lucca (Ciampoltrini & Andreotti, 2003, 215, LR 18, 23, 37 and 46) for the earlier and the collection of nearly 40 sinkers from the Escombreras 1-4 shipwrecks, at *Carthago Nova*, for the later (Pinedo, 2004, 172); the type is, however, very uncommon on the Strait of Gibraltar. The other two varieties are attested in Israel (Galili, Rosen & Sharvit, 2002, 183). It would seem that these types, maybe evolved from the traditional loom weights, were progressively replaced by the truncated cone type, very common in Late Antiquity.

These sinkers were used for angling, from aboard vessels or from the coast. Their suspicious absence from Phoenician and late Phoenician contexts around the Strait of Gibraltar suggests that they may originate in Italy, arriving in the west during the colonisation process started in the second century BC, of which the Escombreras shipwreck is a good exponent, although this hypothesis will have to be confirmed with further well contextualised examples.

Pear-shaped weights (PLXI)

Only different from the conical/truncated cone type (PLIII) in body shape. The line can be attached through a groove (PLXI1) or through a ringed appendage (PLXI2), as shown in figure 19A and B. Some conical examples may have become pear-shaped due to corrosion. Used for angling, they are uncommon, being attested in imprecise contexts on the Tyrrhenian coast of Italy from the Republican and the early Imperial periods (Ciampoltrini & Andreotti, 2003, 216, LR-34 and 38) to the seventh century AD (Kuniholm, 1982, 297, LW 3, 5, 8 and 11).

Triangular perforated weights (PLXII)

Only one example is known, once again from Yassi Ada, in the seventh century. It is quite heavy – 918g – and 13.2cm long (Kuniholm, 1982, 301, LW22). Figure 19C shows perforations on all vertexes, and a groove linking the two on each end on the flat side, suggesting a line running between both. This shape may indicate the insertion of a line carrying a hook into one vertex, with the general line running through the other two, although the groove makes it more likely that the weight functioned on a slip line. Its use was obviously related to angling, but its isolation does not allow for further inferences.

Tubular weights (PLXIII)

This group includes an example from Villaricos, in eastern Andalusia, found along with hooks and identified as a late Roman sinker (Siret, 1906, 467, XXVIII, no. 47); it has been recently republished (Menasanch, 2007, 147-148, figure 9). It shows poor workmanship; a tubular piece of lead rolled on itself, creating a cylindrical cavity (figure 19D). This category aims at the inclusion of a miscellanea of elongated lead pieces, so common on fishing sites, crudely worked and used in angling. They only appear on late Roman sites, and the lack of more systematic series prevents the development of further inferences.

Finally, it should be emphasized that this typology does not pretend to be exhaustive, but only to include those categories of fishing tackle found in *Hispania* and the rest of the Mediterranean, complementing Galili's proposed typology. Some types have been left out, however, such as the ringed spring-shaped type (Galili L32) or some varieties within the group PLX2 which sport a ringed metal appendage, for example in Culip VIII (Carreras *et alii*, 2004, 153), so this draft is open to future expansion.

It is very interesting, however, how this first classification already eases the task of evaluating ancient traditional techniques and later developments. Some categories of leads in use in our days are not represented in the archaeological record, such as "pearl" or "star" weights, while others are: for example, the pear-shaped truncated pyramid type (López, 2003, 253). Slip or running leads or grilles, aimed at preventing the line from getting entangled, are also absent, suggesting their introduction in later times.

From archaeology to fishing: a long path ahead

The complexity revealed by the study of nets perfectly illustrates the caution needed in this branch of research. Oppian's *Halieutika* and other sources inform us about the existence of several types of net, the subject of some recent studies (especially, Bekker-Nielsen, 2002a; 2005). From our perspective, and considering the available archaeological record, it is as yet risky to draw a clear set of relationships between weight types and net types. We must wait to have more publications on the matter, for, as we have seen, the evidence is scattered, scarce and complex to interpret. The elaboration – as the SAGENA project aims to do in Andalusia – of complete metrical series of each type would be a huge step forward. Some conclusions have already been drawn, such as the increment in weight diameter (and thus, line diameter), between the Phoenician and the early Imperial periods in the area around the Strait of Gibraltar (Bernal, 2009a, 206-207), or the validity of the inner diameter in some categories (CI, CII, CIII, CIV, PLI, PLII, PLIII or PLIX) as a metrical proxy to infer the size of the lines. Determining the fibres used – through analysis of the inner face of weights – is another issue to pursue. Nor can more inferences be made on the basis of net mesh size, since, and although according to the ethnographical evidence the dimensions of the navette head is equal to mesh size, their apparent lack of evolution over time and the use of needles complicate the situation, made almost unapproachable by the limited range of the record, which includes less than ten pieces of ancient physical evidence (see Carmen Alfaro Giner, in this volume). The recent find in *Oiasso*, northern Spain, of a wooden reel interpreted on the basis of the ethnographic evidence (figure 20) as a tool to haul the nets, fully illustrates the complexity of these reconstructions.

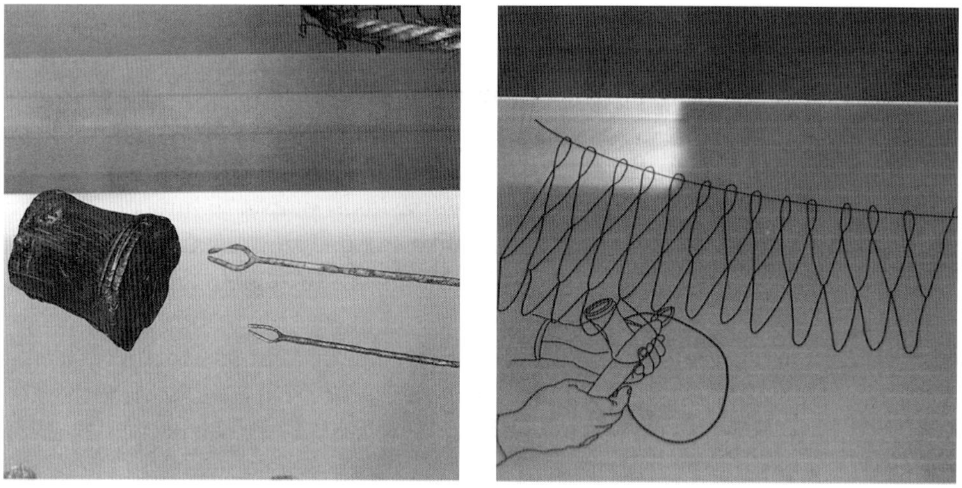

Figure 20. Wooden reel from the first century AD in the Oiasso museum (Irún, Spain), and suggestion for reconstruction and use (after M. Urteaga).

The conclusions about specific net types is complicated even further if we consider the overlapping presence of several types and raw materials in the same contexts, as recently shown in the case of *Baetica* (Bernal, 2009a, 204), or in the paradigmatic shipwrecks of La Chrétienne M (Joncheray, 2002), Yassi Ada (Kuniholm, 1982) or Dor, recently published (Galili & Rosen, 2008), among other examples.

Sadly, the poor preservation of buoys and the exceptional nature of the finds outside specific areas such as Egypt or the Red Sea, for example in *Herculaneum* or the recent excavations in Pisa (Camilli, De Laurenzi & Setari eds., 2006, 54), also play against the characterisation of net types.

It seems that there were clear regional differences, but we only are beginning to glimpse them. That is the case of the rolled plate weights (PLIX2), which in the east were decorated, whereas in the west were completely plain; in another example, clay weights seem to be absent from Israel, according to a reliable recent study, based on a sample of 1200 pieces (Galili, Rosen & Sharvit, 2002, 183-184). Future progress will depend on the development of a general Atlantic-Mediterranean *corpus.*

Aiming at a clear understanding of the issue, our conclusions are divided into two categories of evidence: direct evidence for a particular type of technique, and reliable proxies.

Proxies

Archaeological evidence which in isolation cannot be totally conclusive, but which in association with other artefacts allows for the reconstruction of the tackle it was used with. Within each category, we suggest the kind of evidence to be looked for.

Casting nets

The weights used in these nets were of the lead rolled plate type, normally rectangular in shape (PLIX2-Galili L23), at least it seems so in Spain. Their reduced size seems to suggest exclusive use in this sort of tackle, although the diversity of these practices calls for caution. Confirmation relies on multiple finds, normally more than five, as in *Emporiae* with at least 6 examples (Castanyer, 2006, 21), *Baelo Claudia*, 8 examples (Bernal, 2009a, 203, figure 12), Yassi Ada, with 16 (Kuniholm, 1982, 303-304, figure 13-5), Tossal de la Cala (Benidorm, Alicante) still unpublished, with around 20 (Bernal, 2009a, 202) or Dor, with 153 (Galili & Rosen, 2008, 69-70). The finds indicate their use from boats and from land. The possibility of these nets using other sort of weights in some regions should not be discarded (Cottica & Divari, present volume).

From medium size nets to "almadrabas"

As previously mentioned, the association of a type of weight with a specific type of net is a difficult task. With a few exceptions, such as the circular nets known as casting nets, our current knowledge does not allow us to draw any connections between raw materials, type of weight and fishing technique. An approach based on metric parameters will, however, permit some progress in the middle term, because, although we still lack the necessary complete series of types, some colleagues have already started to work in that direction (Galili, Rosen & Sharvit, 2002, 185-186). There appears to be good arguments for two net categories: medium and large.

A good example of the *large* nets are the, so called "almadrabas" in modern fishing in the Strait of Gibraltar with large, hollow cylindrical weights (PLII1 and 2), the diameter of which is often of more than 1.5-2cm. Other proxies related to similar nets are whole or fragmentary anchors: ethnographical fishing practices in Cádiz area tells us that "almadrabas" were fixed with anchors (Ponsich, 1988, 34-36, figure 12, among others), the dispersion of which – or of their lead stocks – can be used to indicate the location of the nets, as shown by Ponsich in *Baelo Claudia* and other authors in Akrotiri (Gianfrotta, 1999; García Vargas, 2004, 224) or, more recently, in northern Morocco, around Cape Spartel (Erbati & Trakadas, 2008, 65-67). Mosaics showing people carrying an anchor, or the well-known beating scene so often reproduced by Ponsich (1988, 37, figure 13), seem to be depicting *almadraba* scenes.

Ringed nets

A recent suggestion (Galili, Rosen & Sharvit, 2002, 198-199) links ring-shaped leads of diverse typologies (PLI) to ringed nets, where they were attached to the lower

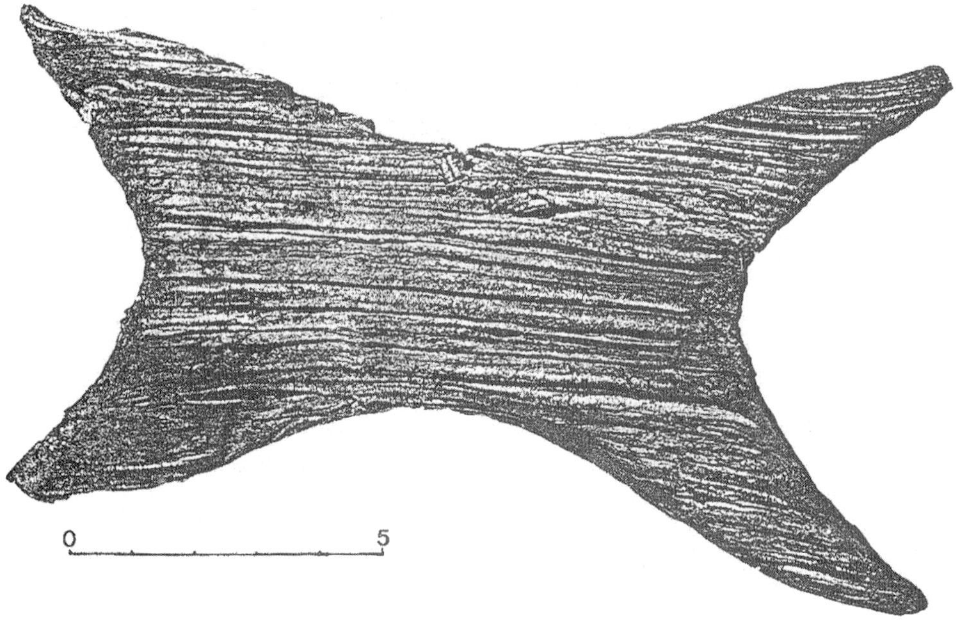

Figure 21. Wooden reel from Saguntum (Aranegui, 1982, pl. XXVI).

lines, as shown in figure 11. This system would facilitate the hauling of the nets on board, as well as their storage; a similar system is found as a traditional technique in some Mediterranean areas, such as central Tyrrhenian Italy. There are, to our knowledge, no references to this technique in the written sources. The wide distribution of these weights, from Israel to Spain, makes their study particularly relevant. At any rate, the problematic archaeological identification of these artefacts as weights by archaeologists in the field drives us to expect some future news.

Line fishing, a highly specialized technique in the Roman period

The automatic association of weights and nets is very common in the bibliography; this should be avoided in the future (Bernal, 2009a, 186). Both the written and the pictorial sources indicate the importance of fishing with rods or with lines, from the coast or from aboard vessels (Yacoub, 1995, 237-239, figures 120-121), but the practice is archaeologically elusive, with a few exceptions, as when remains of buoys or an "ox hide" shaped reel to roll the line can be found, for example in *Saguntum* (figure 21), fifth century BC (Aranegui, 1982, pl. XXVI), and maybe some of the trimmed potsherds with lateral grooves (CV2).

Some categories seem to be linked with line fishing, especially, among the metal types, e.g., the conical/truncated cone (PLIII), pyramid/truncated pyramid (PLX) and pear-shaped (PLXI) types, very similar shapes only differing in tiny for-

Figure 22. Longline from Herculaneum, preserved inside a basket. The bronze hooks and the folded lines are still visible (Bernal et alii*, 2009c, figure 9).*

mal details, indeed more relevant to us, easing their identification, than to the actual operation of the weight. Other, less common, series are the cubic (PLIV), rods (PLV), spherical (PLVII), sphenoid (PLVIII), perforated triangle (PLXII) or tubular (PLXIII) shapes. Also trimmed potsherds with central perforation (CV1) and stone pear-shaped perforated weights (PI4), especially the smaller examples, whose poor workmanship and irregularity of suggest individual efforts to create weights for angling from the coast, probably for recreational purposes.

Especially interesting, as shown in previous pages, is the remarkable number of categories of weights specifically designed for rod and line fishing, as many as 10 categories just for lead weights (PLIII to VIII and X to XIII), suggesting this to be a rather specialised activity.

First, the use of multiple hooks (AIII), suggests, according to ethnographic parallels, the capture of Cephalopods, largely squids. Second, the off-centre sinkers (PLVI), suggest the practice of trolling with live baits or lures from boats – as confirmed by the finds in Yassi Ada wreck. The multiple hook from La Chrétienne C could also have been used for fishing with live bait, as already mentioned (figure 5F).

Finally, we must bear in mind that it is impossible, without more evidence than the weights on their own, to determine whether the weights were used for rod-fishing or simply for line fishing, since both techniques are practiced from both aboard vessels and the coast, with the exception of particularly heavy weights – around 500g – which rule out the possibility for rod-fishing.

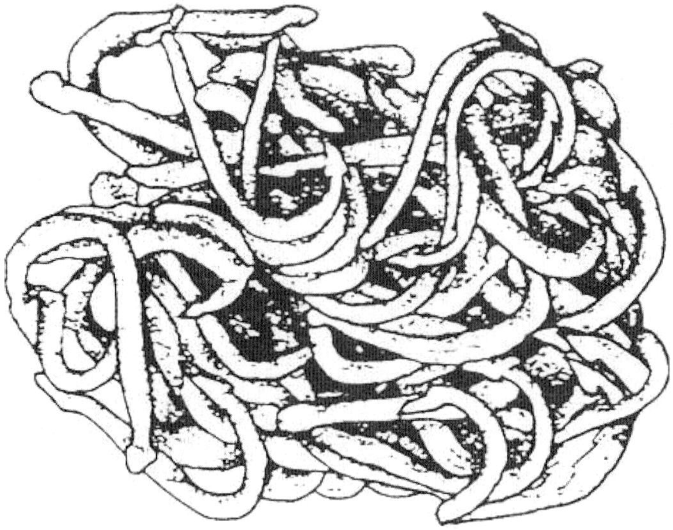

Figure 23. Hooks from Comacchio wreck (Rossi, 1990; Gianfrotta, 1999, 27, figure 17).

Multiple-line fishing

The use of this sort of tackle is confirmed by Oppian, when he mentions that the fishermen "rejoice in many-hooked fishing lines" (*Halieutika*, 3.78). The archaeological identification of the tackle is not easy, since the preservation of physical remains of the lines to which the hooks were attached is exceptional; the only known example in the Mediterranean, in *Herculaneum* (Pappalardo, 1990, 202, figure 5A), is currently being studied within the framework of a Spanish-Italian project (figure 22) (Bernal, Cottica, Zaccaria, Acqua, Arévalo, Bermejo, Bustamante, Cappelletto, Díaz, Jiménez-Camino, Lagóstena, Lara, Lorenzo, Sáez, Vargas & Villada, 2009c, figure 9). They are much more common in anaerobic conditions, such as those found in the Red Sea (see Thomas, this volume).

The clearest indication for the identification of multiple lines is the find of many hooks. That is the case with the Comacchio shipwreck, where several dozens of linked hooks were recorded (Rossi, 1990; Gianfrotta, 1999, 27, figure 17), as shown in figure 23. Although in this case the hooks were of a similar size, that need not always be so, for the tackle can combine diverse hooks and baits aimed at different catches. The vessel could also be equipped with a large lead or stone to anchor the line to the sea bed, as the one recorded in Cala Culip (Carreras *et alii*, 2004, 153, CPVIII-02 2-12B-25), although its isolated find is not sufficient indication.

Direct indicators for tackle. "Talking" fishing equipment

In this section, we refer to archaeological evidence directly pointing at the sort of fishing tackle involved, giving also geographical indications.

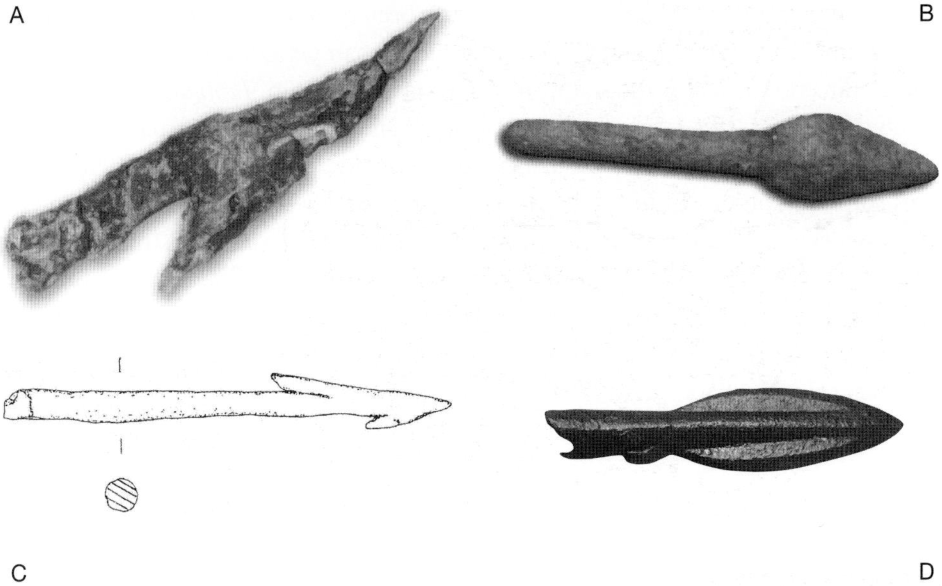

Figure 24. Bronce harpoon point from Emporiae, dating to the sixth century BC (A. Castanyer, 2006, 22), late Roman bone harpoons from Traducta (B. Bernal, 2009b, 275, figure 7B) and Castrum Perti (C. De Vingo & Fossati, 2001, 659, figure 95,1), and D. Macalón arrow head, from the fourth century BC, from the Punic preserves factory "P-19" (Arévalo, Bernal & Torremocha, 2004, 140).

Harpoons

Harpoons are traditionally used for whaling, but other species could also have been caught using these artifacts. The scarcity of archaeological evidence for whaling in the *Fretum Gaditanum*, or indeed elsewhere in the Mediterranean and the Atlantic, has recently been pointed out (Bernal, 2009b, 274-276). We know of a bronze example in *Emporiae* (figure 24A), dating from the sixth century BC (Castanyer, 2006, 22), and another one in the Dramont G shipwreck (Beltrame, 2002, 66), to which we could add the arrow points of the Macalón type, traditionally linked only with military weapons but which also appear in halieutic contexts, such as the Punic factory "P-19" (figure 24D), also dating to the fourth century BC (Arévalo, Bernal & Torremocha, 2004, 140). We also know of some small bone examples, from the preserves factories of *Iulia Traducta* (figure 24B), in the Bay of Gibraltar and *Castrum Perti*, in the Ligurian coast (figure 24C).

Thus their use is attested in the western Mediterranean in Antiquity at least from the sixth-seventh centuries BC. However, although the use of harpoons could be expected to bear evidence for whaling, the bone examples from *Emporiae*, "P-19", *Traducta* and *Castrum Perti* due to their small size rather point towards the capture of medium sized species of fish.

Although it is uncertain, there seems to have been a tradition of "heroic" and "prestige" fishing practices with harpoons during the Roman period: some evidence points in that direction, such as the stamp for cakes or bread from *Tamuda*, on the Mediterranean coast of *Tingitana*, which represents a scene in which a fisherman rides a *hippocampus* over a sea populated by sharks, dolphins and a whale (Bernal, 2009b, 271). This sort of tradition is indeed common in other regions and periods, for example, Egypt, where the Pharaoh is frequently depicted brandishing a harpoon in a defiant attitude (El Shahawy, 2005, 236). It is also important to try to define the chronology of the different raw materials used for harpoons, since there seems to be some discontinuity between the Upper Palaeolithic (Cleyet-Merle, 1990, 96-98) and the Roman period, and materials and shapes seem to evolve (and overlap?) throughout Antiquity.

Octopus fishing with clay pots

The traditional fishing making use of pots attached to a line to capture Cephalopods, especially octopus, has long been practiced from the Maghreb to the Atlantic (figure 25A). These pots, with an approximate capacity of 5 litres (30cm of maximum length and 10cm wide at the rim) have a characteristically sinuous profile and a flat base, where the orifice through which the line is inserted – also useful to evacuate the water when the pot is hauled in – opens.

A recent collection of 23 single-handled pots, found underwater off the Gulf of Naples between Ischia and Amalfi, has been linked with this technique (Rendini, 1997, 75). Typologically, they divide into two groups: I) ovoid in shape, 13-17cm in length and 6.7-10cm wide at the rim, with an overarching handle and a centred orifice at the base (figure 25B and D); II) these have an open rim, one handle and two perforations, one at the base and the other below the rim (figure 25C); their chronology could fall in the late Roman period (sixth-seventh century AD), according to typological parallels, but the authors remain cautious due to the lack of context for the find (Rendini, 1997, 77).

Attempts to find pieces with these features in the plain pottery repertoire from the Strait of Gibraltar have so far been unsuccessful; vases of Roman date with perforations in the base have indeed been found, but relate to farming – re-potting – and those found in fishing contexts all contain residues of fish fatty acids, suggesting their use as filters for *garum* and other fish preserves. Their small size (10-15cm high) and their truncated cone shape also argue against their use as octopus traps (Bernal & Sáez, 2006).

Given the lack of convincing evidence, and since the Tyrrhenian materials have no reliable context and lack other features for a secure attribution, it could be argued that this technique could have been carried out in the Roman period with tackle made of perishable materials which have not survived. It is possible that some of

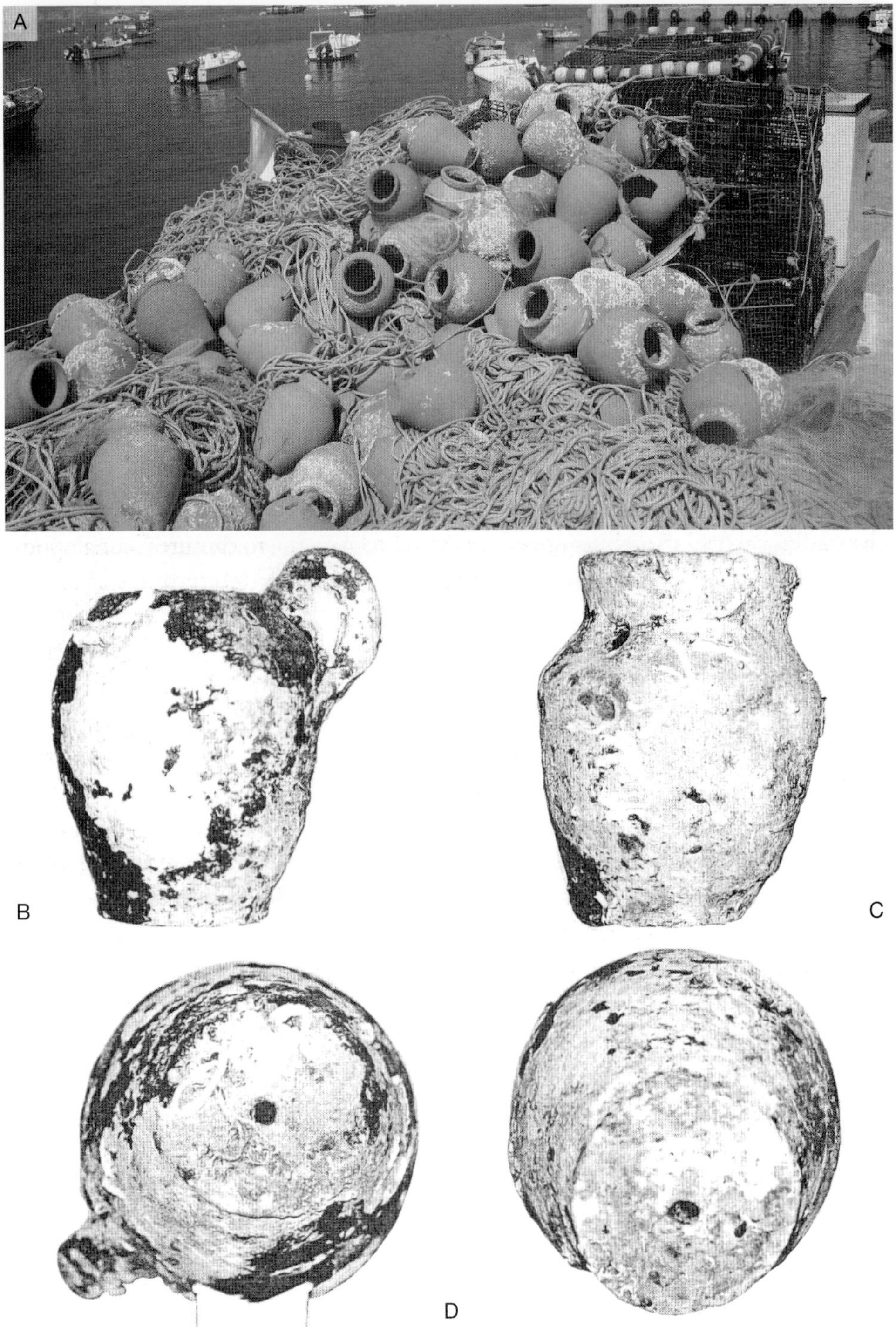

Figure 25. Modern vases linked by a line for octopus fishing in Sagres, Portugal, 2006 (A), and perforated vases from underwater sites on the Tyrrhenian coast (B, C and D) (Rendini, 1997, 75, no. 1 and 76, no. 3-5).

the net fishing practices shown in the iconography, such as the known Tunisian mosaics – for example El Alia, in Bardo Museum – actually depict this practice. Recently it has been proposed that mosaics showing fishermen emptying amphora-shaped fish-traps tied on a line in fact depict clay jars, not wickerwork creels.

Hand nets

The ancient iconography proves the use of circular or semi-circular, hand nets, to capture small shrimps or other species, as recently shown with a number of examples (López Monteagudo, 2006, 243). Red-figure Greek vases are particularly prolific in depicting these devices, such as a scene on a *kylix* from the fifth century BC showing a character carrying a long pole ending in a net (figure 26A) and another on a vase from the same century in the Kunsthistorisches Museum, Vienna (figure 26B); the frequent representation of these scenes suggests that the practice was very common. The Orpheus mosaic, at *Leptis Magna* (second century) showing a fisherman using a hand net, is yet another example (figure 26C). A silver plate in Milan (López Monteagudo, 2006, 254, pl. XIV, 1), dating to the third century AD, shows the ovoid shape of the instrument (figure 26D). The wide distribution of these devices throughout the Mediterranean is, however, only documented through the iconography, as the perishable materials of which they were made have not survived.

Wood seems to be the preferred material, explaining the ovoid shape shown on the plate in Milan and the extreme length of the pole in the other examples mentioned. We know that metal prevailed in the modern period; some nets could be submerged attached to ropes, to lure the catch within with bait (Sáñez Reguart, 1791-1795). Such metal rings are to date absent from the archaeological record, with the exception of a lead example, 50cm in diameter, found in Sardinia and used to fish coral, and a possible parallel found on the French coast (Galasso, 1997, 122, figure 4).

Fishing rods

Despite the richness of textual (Oppian, *Halieutika* 3.74-75) and iconographic evidence (especially Tunisian mosaics and Vesuvian wall-paintings) for the Roman period, no archaeological remains of fishing rods have been found, even in shipwrecks, due to the vegetal material used for their manufacture (Beltrame, 2002, 66). Other indirect evidence, such as canes used as hook cases in *Herculaneum* (Stefani, 1990, 10), seems to confirm the use of this material in fishing practices, for example for making rods.

The case of *Hispania* is only illustrated through the indirect evidence offered by the mint of *Carteia* on the Bay of Gibraltar. Some of the bronze *semisses* of the so called

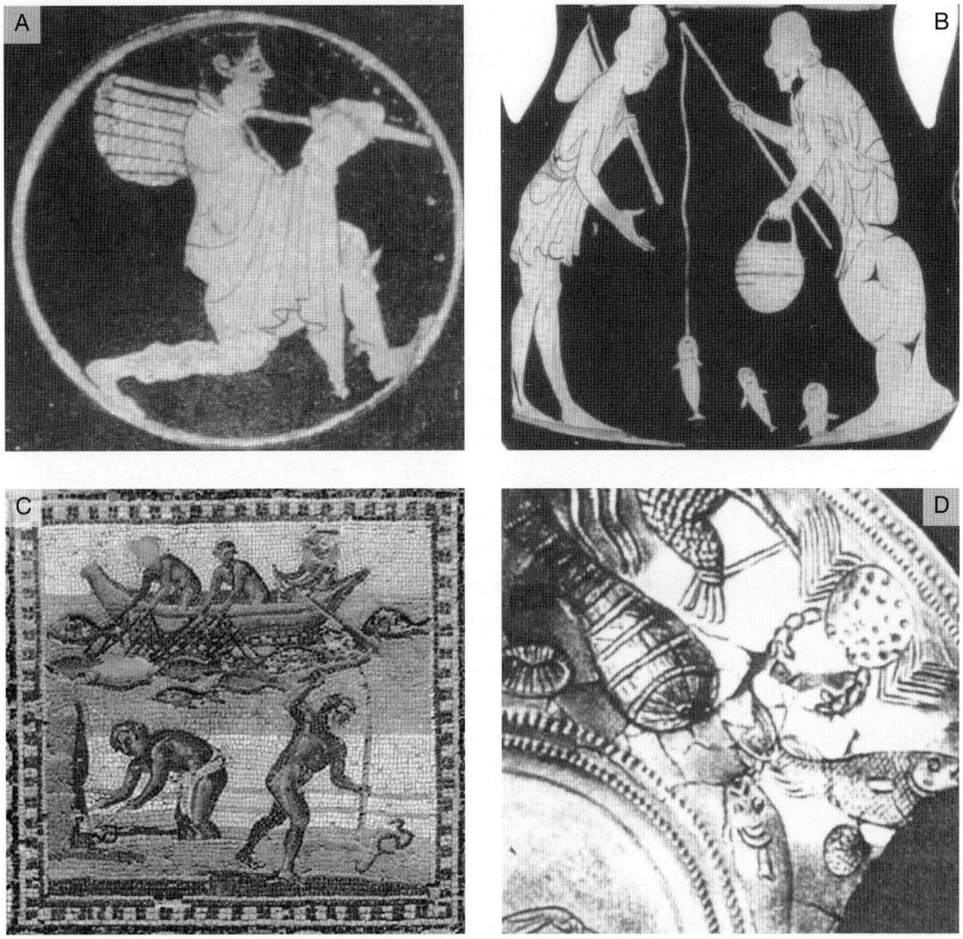

Figure 26. Red figure scenes from Gela Museum (A) and Vienna's Kunsthistorisches Museum (B); Orpheus Mosaic, Leptis Magna (C); and partial view of decoration on a silver plate in Milan (D), after López Monteagudo (2006, pl. VIII, 4, VIII, 3, XIII, 4 and XIV, 1).

28th issue, during the Augustan or the Tiberian period, show a fisherman holding a rod in the reverse (figure 27) (Arévalo, Bernal & Torremocha, 2004, 106; Chaves, 1979). Obviously, the choice of a such city emblem, reflecting the local fishermen's daily life, has its roots in the city's long tradition of fishing and fish processing.

Coral fishing

The evidence for Spanish coral fishing during Antiquity is scarce, and has not been subject to any comprehensive study to date. The evidence is scattered between the Neapolis of *Emporiae* – the excavations carried out in 1936 found many coral branches used as raw material for jewels and other decorative items (Castanyer,

Figure 27. Semis from the mint of Carteia (40 BC-15 AD), in the National Archaeological Museum at Madrid (Arévalo, Bernal & Torremocha 2004, 107).

2006, 15) – and the Strait of Gibraltar, for example in fishing contexts of imperial date in *Septem Fratres* and *Iulia Traducta* (Bernal, 2007, 101); some other unpublished examples found in the Bay of Cádiz, for example, in the Calle Sagasta, Cádiz, are being studied within the framework of the SAGENA project.

That coral fishing took place elsewhere in the Mediterranean is clear from finds in Pompeii confirming Pliny the Elder's observations about coral fishing in the Bay of Naples (Stefani, 2001, 13), and the excavation in Marseille (Place Jules-Verne) of the sixth century BC fishing boat containing coral (Gianfrotta, 1999, 11). Coral was widely used in antiquity for jewellery (Morel, Rondi-Costanzo & Ugolini, 2000), and the evidence suggests its exploitation by coastal communities. The frequent presence of coral in Spanish coastal sites of the Roman period suggests its economic exploitation, although it is difficult to ascertain the possibly accidental nature of the captures with large size nets. In fact, some of the most fructiferous coral areas were on the Atlantic shore of the Strait of Gibraltar, running into the Mediterranean up to Tabarca, in Tunisia, where coral fishing is fully attested for the medieval and modern periods (Russo, 2005, 17).

Regarding tackle used for coral fishing in Antiquity, the evidence is scant. The south-western Sardinian underwater site of Secca di Cala Piombo (Golfo di Palmas) offered an extraordinary find. It is a square-shaped artefact carved from calcareous stone, 50cm wide and 14cm thick, weighing 50kg and sporting 5 perforations, 4 of them on the angles and a single, bigger one, around the middle, ovoid in shape and with an angular bottom edge (figure 28A). Along with this device were found a large iron ring attached to a bolt, the remains of a chain, and a Baetican amphora fragment of early Imperial date (Dr. 8), confirming a Roman date (Galasso, 1997, 122, figure 3). Several attempts of reconstruction have been made, based on ethnographic parallels, and the most likely results in a device which, suspended from above, would carry four mesh baskets, hanging from lead rings (figure 28B), one example of which was found a few meters away.

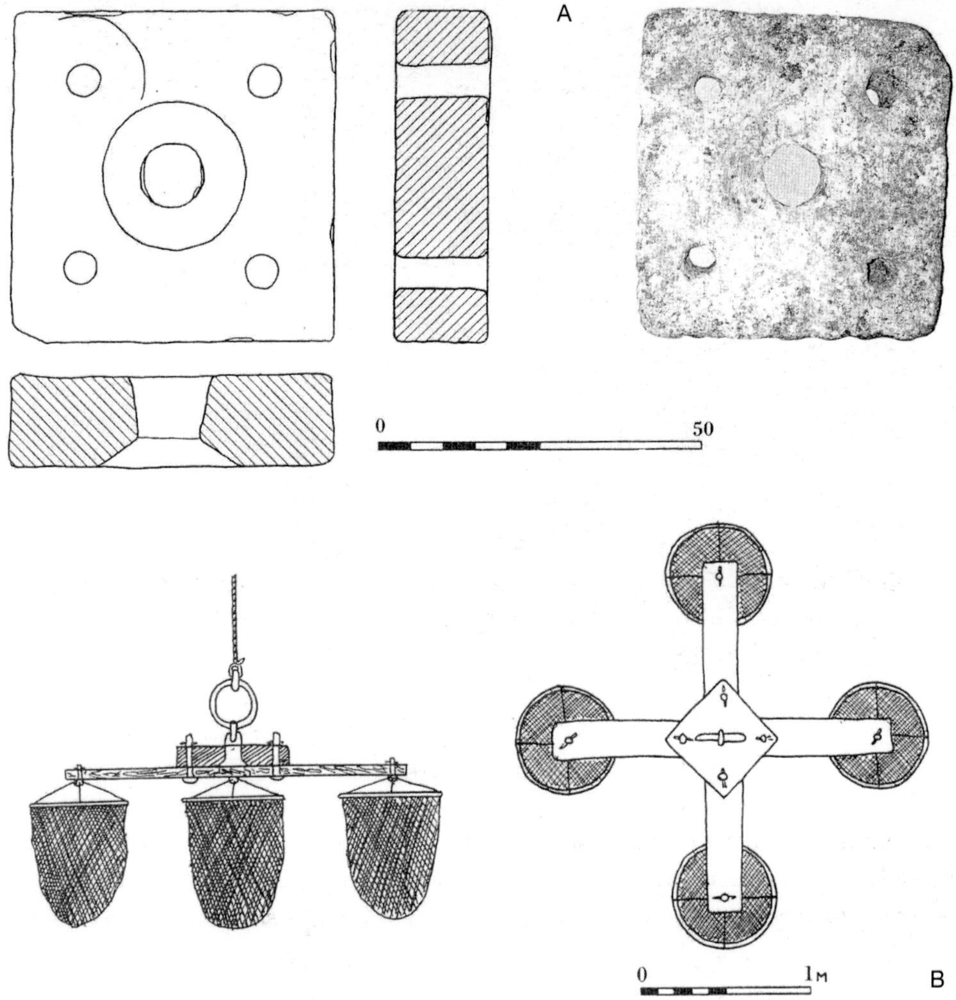

Figure 28. Stone device found in the Cape of Palmas, Sardinia (A) with proposed reconstruction (B) and use for coral fishing (Galasso, 1997, 122 and 124, figures 1 and 6-7).

Although the excavators prefer not to relate this fishing device to the shipwreck loaded with Baetican amphorae of the Claudian period (Dr. 8 and Dr. 20) found nearby, it is possible that the vessel, which had set sail from the Strait of Gibraltar, carried a variety of fishing tackle, as the underwater record shows to be common practice, this specific device perhaps being typical of *Baetica.* At any rate, it is exceptional in being the first piece of evidence for coral fishing clearly datable to the Early Imperial period (first century AD).

There are many other examples, made of stone, as from Île Maïre (near Marseille), Syracuse, Lavezzi, Corsica, Sardinia; maybe an oval shaped example from Israel; or a lead object, also perforated, as found at Île Riou, de Cassidaigne and Sassari, all of them described by Gianfrotta (1999, 16). In *Baetica* a circular stone with 5 holes

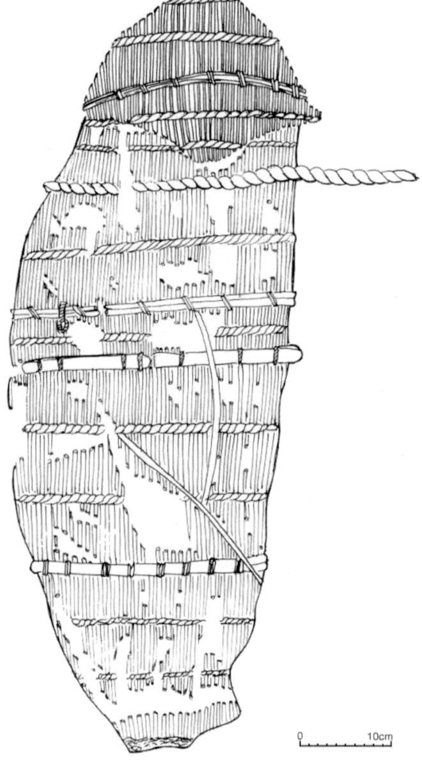

Figure 29. Roman creel from San Rossore-Pisa (Grandinetti, 2000, 115, figure 8).

was found in the Bay of Cádiz, Caño of Sancti Petri area, still unpublished (oral reference by C. Alonso Villalobos, archaeologist from the Centre of Underwater Archaeology at Cádiz).

Creels or baskets

The written sources confirm the use of creels, passive tackle, during the Roman period; for example, Oppian, in the second century AD: "Others again have their minds set rather upon weels, which bring joy to their masters while they sleep at ease, and great gain attends on little toil" (*Halieutika* 3.85-87, translated by A.W. Mair). Representations in mosaics, also demonstrate the use of creels/baskets, linked to a line and a sinker, for example, in Sousse (García Vargas, 2004, 224, photo 1), and also coinage, not discussed here in detail.

The archaeological evidence is restricted to underwater sites, such as the remains of a creel in the stern of the Comacchio shipwreck, associated with molluscs (Beltrame, 2002, 68-69). The recent finds in San Rossore, in Pisa, have recorded a creel made of longitudinal ropes linked with more slender lines, the structure reinforced in the weakest points with small and springy wooden rods and rolled rope (Grandi-

Figure 30. Creels in Sagres (Portugal); inside each of them, a rock used as a sinker (August, 2006).

netti, 2000, 113), confirming the use of the technique in the central Mediterranean. Its use in *Hispania* cannot be confirmed as yet due to the lack of specific evidence, but we consider it very likely. Any attempt at constructing typological definitions is also deterred by the fragmentary state of the evidence, although it should be noted that the Pisan specimen is rather large (around 100 × 30cm) and of pseudo-cylindrical shape with tapered ends (figure 29).

Nowadays, many creels are sunk with stone weights, individually attached (figure 30). This raises the possibility that stone weights, pebbles (PIII1) or masonry (PIII2) are evidence not only for nets, but also for creels.

Night fishing with torches

Night fishing is known from several sources, such as Oppian:

> At earliest dusk of night with lighted torch the fishers steer their hollow boat, bringing to the resting fishes a darkling doom. Then do the fishes exulting in the oily flame of pine rush about the boat and, to their sorrow seeing the fire at even, meet the stern blow of the trident (*Halieutika* 4.641-646).

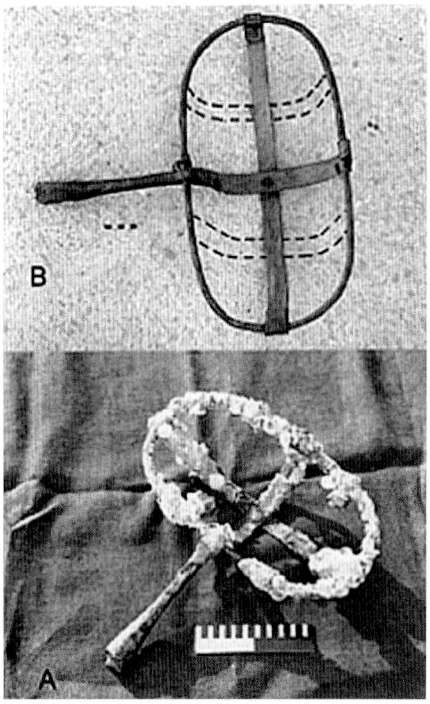

Figure 31. Metal lantern from a seventh century AD vessel in Dor, Israel (Galili & Rosen, 2008, 71, figure 7).

The first archaeological evidence for this practice, attracting fish to a vessel with lights on the prow or the stern, has been found recently in an underwater site off Israel, documenting this technique, at least in the east and during late Antiquity. The Byzantine Dor shipwreck was found 25km to the south of Haifa, and dated with precision to AD 665, according to the numismatic evidence (Galili & Rosen, 2008, 67), that is, shortly after the Arab conquest. A remarkable collection of fishing tackle was recovered, namely 159 lead weights – used for casting-nets, among others – a trident, a sounding line and an artefact interpreted as the "fire-basket of a lantern" (Galili & Rosen, 2008, 70-71). Elliptical in shape, and of medium size (46cm long, 24cm wide and 10cm deep, weighing 1.7kg), its central space is covered by a mesh made out of thin metal rods with a metal handle, which attached to a wooden piece linking the device and the ship (figure 31). The boat must have been of a medium size, according to the metal bolts found, for no remains of the hull survived, and sailing for fishing (or at least carrying a smaller boat for that purpose) (Galili & Rosen, 2008, 73-74). Galili and Rosen offer iconographic evidence for this technique from at least the eleventh century, showing that it was practiced exclusively with nets and tridents, as confirmed by the Dor shipwreck, in which no hooks were found, and indicating that lighting instruments may have been misinterpreted in other underwater sites (Galili & Rosen, 2008, 75).

No evidence has been found to date in *Hispania*, and it is therefore very important to keep a watch for possible evidence for lighting devices in the underwater record.

Traps made of wood; stone yards, so-called "corrales"

The use of passive traps is a fishing technique of ancestral origins; building structures where the fish gets trapped in rivers, lakes or the sea is easy and very effective. In Andalusia we have the familiar "corrales", stone structures of complex, but roughly semi-circular, layout, built in the intertidal area and where fish, Cephalopods and molluscs are trapped at low tide. The remote origin of these techniques has generated much commentary, from the descriptions of Columella in his *De Re Rustica* (8.17) onwards; their origin is securely attested from the sixteenth and seventeenth centuries (Arias, 2005, 89-97), although, cautiously, a Roman origin has been suggested for those in the coast around Barbate (Moreno & Abad, 1971). However likely this seems, especially considering Aelian's commentaries in the third century AD (*N.A.* 13.28), we must remain cautious until secure archaeological evidence suggesting a Roman date for these stone structures is found.

One wooden structure, however, has been chronologically defined beyond all doubt. It is an "armadilha" or trap discovered in 1983 in the ancient lagoon of Ovil, on the Portuguese coast, south of Espinho (Alves, Dias, De Almeida, Ferreira & Taborda, 1988-89). Three structures were identified, perpendicular to the coast and constructed by setting wooden stakes – rectangular in section and 6-8cm thick – in the soil, supporting a mesh of vine shoots (figure 32A; Structures I and II); the first of them, 4 × 6m, two curvy and symmetrical rows of oak (*Quercus robur L.*) stakes, bent outwards, was accompanied by a very clear stratigraphy – clay, clay/peat and sands. One of the stakes was tested for chronology (10-240 AD cal.), with results that were consistent with the stratigraphy, giving a Roman date for the structure (Alves, Dias, De Almeida, Ferreira & Taborda, 1988-89, 192-196). The trap had been active in the lagoon from the Roman period until it was overgrown by vegetation in the ninth and tenth centuries, followed by desiccation and silting. The stratigraphic studies confirmed that the structure had been lined with branches of rush of varying thickness (figure 32B). The other two structures (II and III) were not as well preserved; the first was a double, almost parallel, line of stakes (the radiocarbon dates offered a similar chronology) not so deeply stuck into the sediment. It is agreed that these structures would take advantage of the tidal sequence, a practice copiously attested in the ethnography. Therefore, they must have been common in Portugal, although the only archaeological parallels are to be found in the British Isles in the Middle Ages (Alves, Dias, De Almeida, Ferreira & Taborda, 1988-89, 221-222).

The importance of this find rests in the fact that it bears evidence for the use of this passive fishing technique from the first and second centuries AD. Traps must have been very common in the lagoon areas of the Iberian peninsula, but the perishable material they were built with has rarely left archaeological traces. We cannot be sure whether these structures precede the stone yards in the Bay of Cádiz and

Figure 32. Roman wooden fishing structure found on the Portuguese Atlantic coast, along with the plan of the so called Structures I and II (A), and suggested reconstruction (B), following Alves et alii *(1988-89, 212, figure 20; 216, figure 25).*

other points on the Atlantic coast up to Brittany (substituting wood for stone; maybe even in Antiquity?), or were contemporary. This point can only be clarified through future studies.

Tridents

The use of tridents is one of the best attested fishing practices in Antiquity, as argued recently by a number of scholars (Bekker-Nielsen, 2005, 89-90). The trident is a multi-functional tool whose halieutic origins are nevertheless obvious. Normally, they are made of iron, although there are also bronze examples, with votive as well as fishing purposes, as in the Turkish Uluburun wreck. They are known at least since the end of the second millennium BC, as shown again by Uluburun (Pulak, 1998, 211; 214). In the Roman period, these artefacts are normally found in shipwrecks, for example in Cavallo 1 (Beltrame, 2002, 66).

To date, no typology has been established for Greek or Roman tridents, a task that must be undertaken in the future. Some differences have already been outlined, however, the most obvious being the number of prongs; three in most Roman and pre-Roman specimens (figure 33A and B). Some examples with five prongs are also known, for example in the Dor wreck, in Israel (Galili & Rosen, 2008, 70, figure 6), in a context clearly related with fishing (figure 32C). The five tridents or "fiocine" found in the Fucine Lake are exceptional; used for fishing in the lake, they sport 3, 5, 7, 9 and 10 prongs, but their chronology cannot be ascertained due to the lack of a defined context (Saladino, 2001, 72). Indeed, tridents with more than five prongs and "complex" shapes seem typical of modern and contemporary examples, as shown by known parallels in ethnographic collections (Ferro, 2007, 93-103, no. 8-39).

Some tridents, showing a particularly slender outline and long prongs, should also be highlighted, for example, in the previously mentioned Uluburun wreck (figure 32A), against the wider and less developed types, for example, the one found in Venice lagoon, of uncertain chronology (D'Agostino & Fozzati, 1997, 297). Indeed, the imprecise chronology of all known examples is one of the biggest problems regarding the study of tridents. The most common type must have been simply driven into the wooden haft, being cast in a single piece with a pointed base, for example, a Greek specimen of the sixth century AD (figure 32B), whereas the most elaborate would end in a plate, bent to form a hollow fitting (figure 32C).

Probably tridents were not common in Spain, since only one example is known, from *Baelo Claudia*, with three prongs and a triangular shape. An interesting future task will be to analyze their distribution, and compare them with other gladiatorial and votive specimens, particularly regarding their chronology. Tridents are well documented in Late Antiquity and, iconographically and textually, also in the Imperial period.

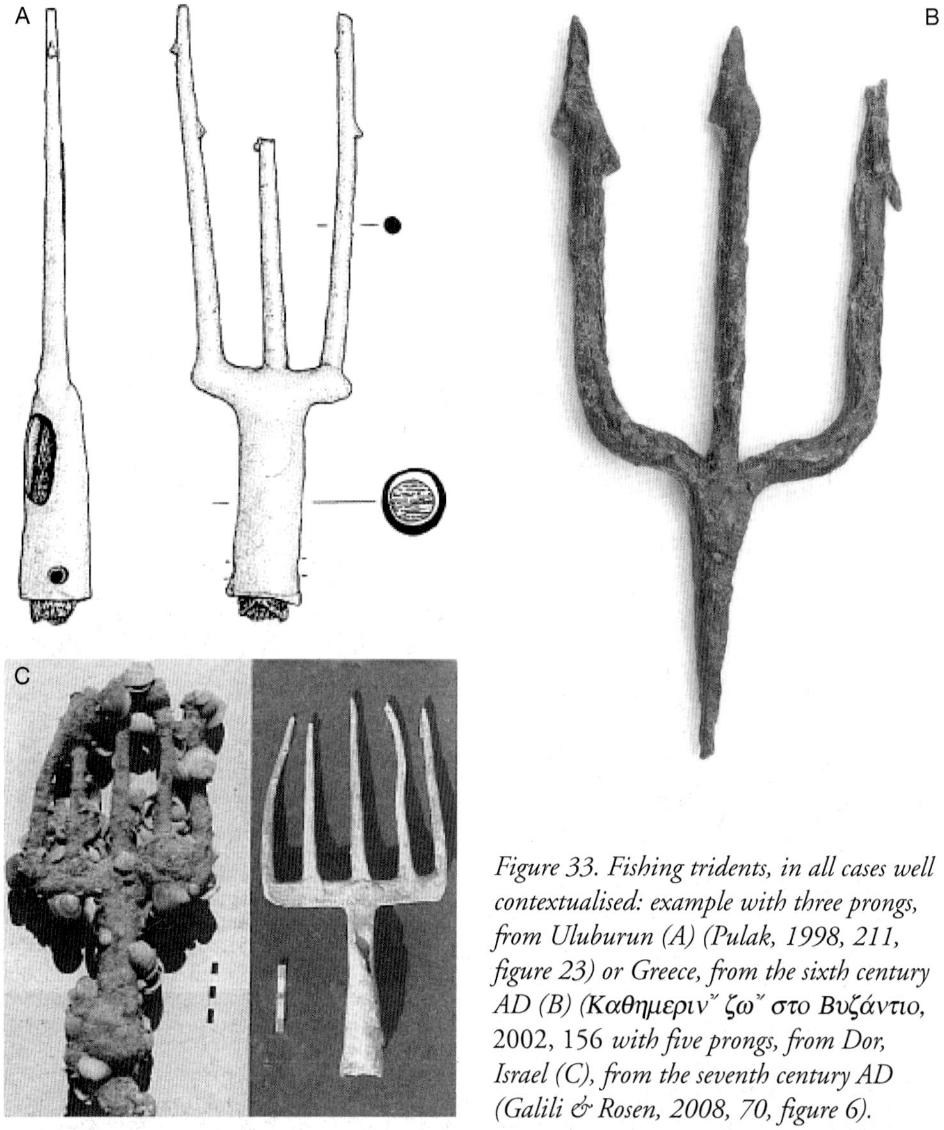

Figure 33. Fishing tridents, in all cases well contextualised: example with three prongs, from Uluburun (A) (Pulak, 1998, 211, figure 23) or Greece, from the sixth century AD (B) (Καθημεριν" ζω" στο Βυζάντιο, 2002, 156 *with five prongs, from Dor, Israel (C), from the seventh century AD (Galili & Rosen, 2008, 70, figure 6).*

A *corpus* in *crescendo:* future tasks

As mentioned at the beginning of this paper, it will only attempt a first, general classification of fishing tackle based on the archaeological evidence available, with the middle-term aim of comparing the results with the technological inferences made on the basis of the textual and iconographic evidence (Bekker-Nielsen, 2005).

Throughout the previous pages, the complexity of the topic has been documented against the oversimplifications and excessive generalisations which have dominated the published research to date, especially in extrapolating conclusions drawn from the Tunisian mosaics and the Vesuvian wall-paintings to the whole of the Empire, and assuming that only those fishing techniques depicted in these were practiced in real life. The answer, as some evidence already suggests (such as the absence of clay weights in Israel or the use of clay weights with lateral grooves only on the Atlantic coast) to the need of regional classification is to be found in the archaeological evidence, through the elaboration of complete catalogues for the Mediterranean and the Atlantic coasts. The archaeological record already indicates the existence of techniques other than those found in the pictorial record, for example, coral fishing with complex devices and trolling with livebait.

We are aware that the results are incomplete and also of the shortcomings of the typology presented. With the exception of the types of stone and lead sinkers in Galili, Rosen and Sharvit, all the types proposed are new, based on the available evidence at the time of writing.

The tremendously specialised nature of fishing in the Roman period, many aspects of which still elude us, must be underlined. Multiple hooks for squid fishing, off-centre leads to avoid line muddles, chained hooks, ringed nets or night fishing, all bear witness to the richness and development of fishing techniques in Rome, while offering a trail for future research to follow. At least 22 types of weights (with 50 sub-types) have been identified, a marked improvement on previous classifications. In addition, each type has been tentatively associated with net fishing (CI to CIV; PIII; PLI, PLII1-2 and PLIX), rod or line fishing (CV; CVI; PLII3-4; PLIII to PLVIII; PLX to PLXIII) or both (PI; PII).

Finally, it should be stressed that other types of fishermen's implements have intentionally been omitted from this study. These include tools for repairing nets (needles or shuttles); reels (as the above mentioned example of *Saguntum* and some others in Beltrame, 2002, 66-67); to haul the big catches aboard vessels, such as long hooked poles locally known as "bicheros" (Arévalo, Bernal & Torremocha, 2004, 140-141); to store and transport tackle, such as baskets (Beltrame, 2002, 66-67; Stefani, 1990, 14), or hook cases, for example, in Port-Vendres II, Cap Camarat B, Grado (Beltrame, 2002, 66) or *Herculaneum*; to transport the fish, with bone devices to be introduced under the fish gills; or to signal from one boat to another, for example shells, normally *Buccinum* (generally of the type *Charonia lampas*), well documented in Pompeii or the Cala Culip IV wreck. These are tasks for the future.

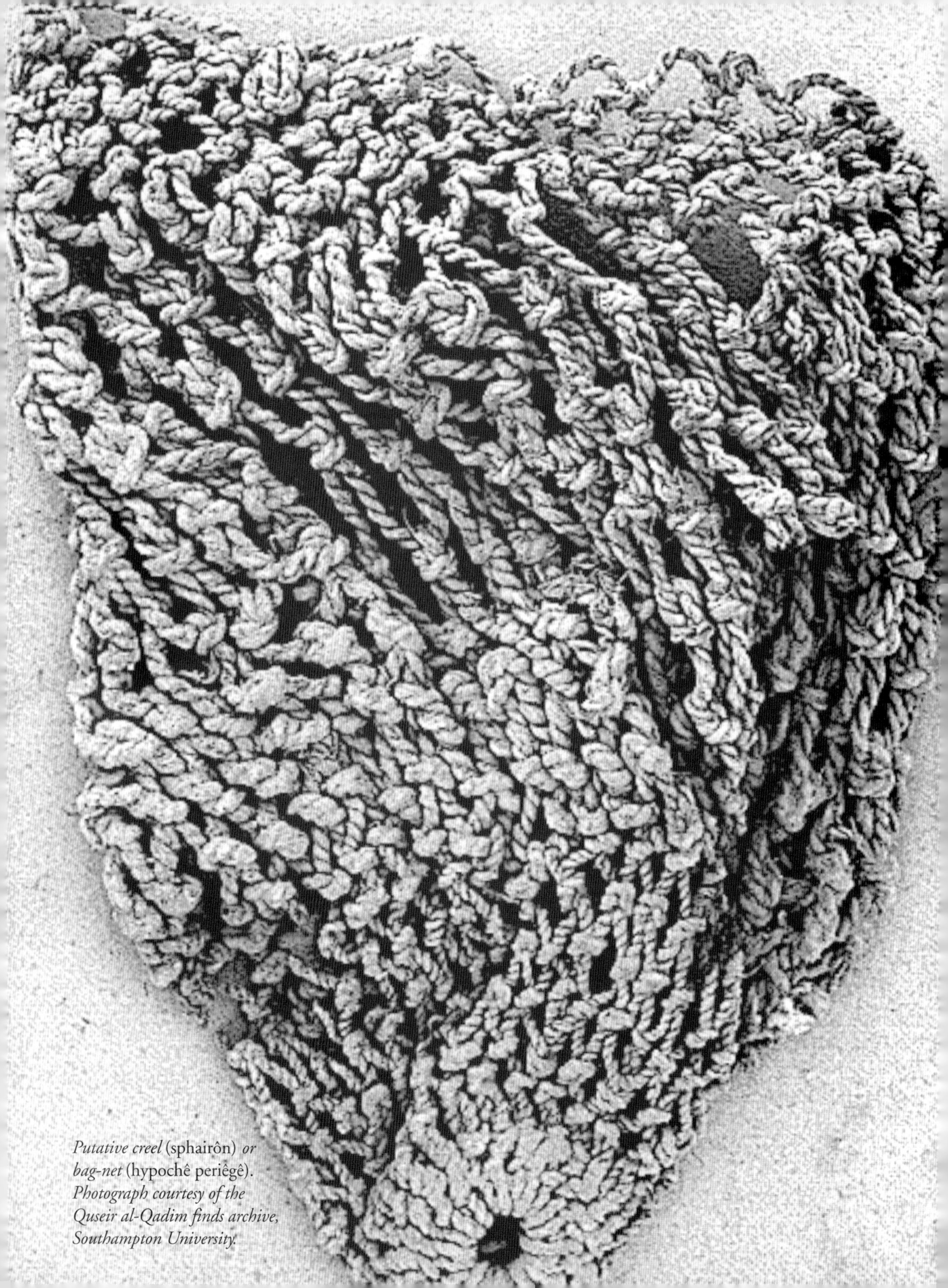

Putative creel (sphairôn) *or bag-net* (hypochê periêgê). *Photograph courtesy of the Quseir al-Qadim finds archive, Southampton University.*

4. Fishing Equipment from Myos Hormos and Fishing Techniques on the Red Sea in the Roman period

Ross Thomas

Introduction

Myos Hormos (modern Quseir al-Qadim, Egypt) was well known to Greek and Roman geographers for its role in the Indian Ocean trade from the late first century BC to third century AD (Peacock & Blue, 2006). Myos Hormos was on a stretch of western Red Sea coast inhabited by the *Ichthyophagi Arabaegypti* (Ptolemy, *Geogr.* 4.5), renowned for their fishing economy, where "Poseidon had assumed the function of Demeter" (see below). A combination of excellent preservation and extensive archaeological research in the region has enabled a thorough study of fishing on the Red Sea during this period. The artefacts preserved include weights, floats, nets and hooks that supply the main body of evidence for this paper, though historical, ethnographic and faunal data help illustrate how the fishing equipment was used (Wendrich & Van Neer, 1994). Faunal (Van Neer *et alii*, 2004) and epigraphic (Cuvigny, 2003b) data have also greatly increased our understanding of fishing methods and marketing in the Red Sea region during the Roman period. Fish (depending upon species) was marketed fresh, salted, dried, pickled, as sauce (fermentation) or smoked. In conclusion the archaeological evidence suggests there was a significant demand for fish products in the region that fuelled a diverse fishing industry at Myos Hormos (Thomas, 2007).

This paper will discuss the fishing equipment from Myos Hormos, and other contemporaneous finds from the Red Sea region. The combination of fishing artefacts complemented by epigraphic, historical, faunal and ethnographic data has provided us with a detailed insight into fishing gear, techniques, and the wider fishing industry in the Red Sea region during the first three centuries AD. This paper will also will discuss production, marketing and very briefly the social importance of this to the Red Sea port communities.

This paper is only possible because of a huge increase in archaeological investigation in the Eastern Desert and Red Sea coast over the last two decades (Zitterkopf

& Sidebotham, 1989; Sidebotham, 1991; Maxfield & Peacock, 1996; Peacock & Maxfield, 1997; Sidebotham, 2002; Cuvigny, 2003a; Peacock & Blue, 2006; Sidebotham & Wendrich, 2007). What was thought of as an empty desert marked with Roman and Hellenistic roads and forts, is now understood to be a busy landscape incorporating numerous types of sites, from ephemeral sites such as rock art, inscriptions, burials, camps and middens to ports, quarries, mines and forts. The preservation of the sites and involvement of research specialists in the field has also greatly increased our understanding of the material. Fishing equipment often sits awkwardly across various finds classes. Often this results in maritime artefacts being overlooked. However at Quseir, I was able to converse with and access information from the various finds specialists who worked there.[1]

Following the annexation of Egypt in 30 BC, Rome acquired control of the Eastern Desert, renowned for its gold and emerald mines, quarries and Red Sea ports. The most important Red Sea ports were Berenike and Myos Hormos, although I shall also mention comparative material from Abu Sha'ar just north of Myos Hormos in this paper. From the Ptolemaic period to the third century AD, Myos Hormos was well known to Greek and Roman geographers for its role in the Indian Ocean trade. This region was known for its mines, quarries and trade with India. Archaeologists are beginning to realise that fishing industries were important for providing sustenance for the growing population working in these industries (Van Neer *et alii*, 2004; Hamilton-Dyer, 2007). Myos Hormos supposedly had a Ptolemaic foundation date, though after eight years of excavation by the Chicago and Southampon teams, only Roman structures have been found there (Whitcomb & Johnson, 1982; Peacock & Blue, 2006), suggesting that maybe the settlement was relocated in the Augustan period. The site was reoccupied in the Mamluk period, which will not be discussed in this paper.

1 I would like to thank the Southampton and Egyptian team members who worked at Quseir al-Qadim for their input and hard work. I would particularly like to thank Lucy Blue and David Peacock for involving me in the project and their ongoing support in my researches in this region and period. Sheila Hamilton-Dyer (faunal remains) and Roberta Tomber (ceramics) have also been very generous with sharing their ideas and data that have been influential in this publication. I also greatly appreciate the help of the finds specialists Marijke van Der Veen (wood), Fiona Handley-Earl (cordage and textiles), Penny Copeland (metal), Wilfried van Rengen (epigraphy) and Jill Phillips. Conversations with the Berenike Team have also been very informative and I would particularly like to thank Steve Sidebotham, Willeke Wendrich, Andre Veldemeijer, Martin Hense and Caroline Vermeeren for answering my queries and sending me off-prints. Finally I would like to thank Athena Trakadas, Julian Whitewright and Lucy Blue for their feedback on this paper, given at various stages of its development.

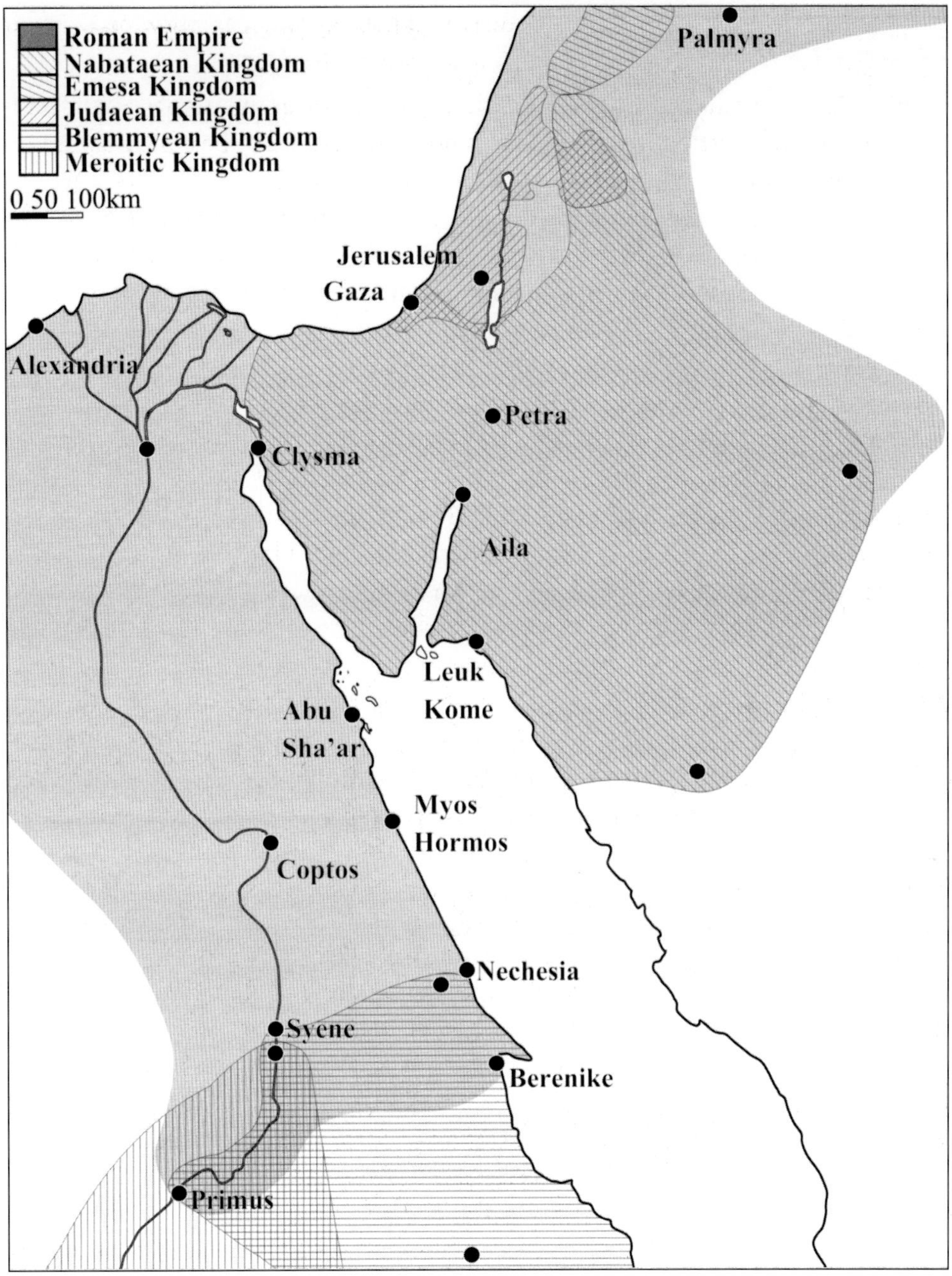

Figure 1. Map of sites and approximate extent of polities in the region (modified from Ball, 2000).

Following the annexation of Egypt, the political situation was relatively stable after initial conflicts with the Meroitic (or Kushite) kingdom to the south (figure 1; Anderson *et alii*, 1979), although there appear to have been increasing conflicts in the second century AD with nomadic groups in the region that the Romans la-

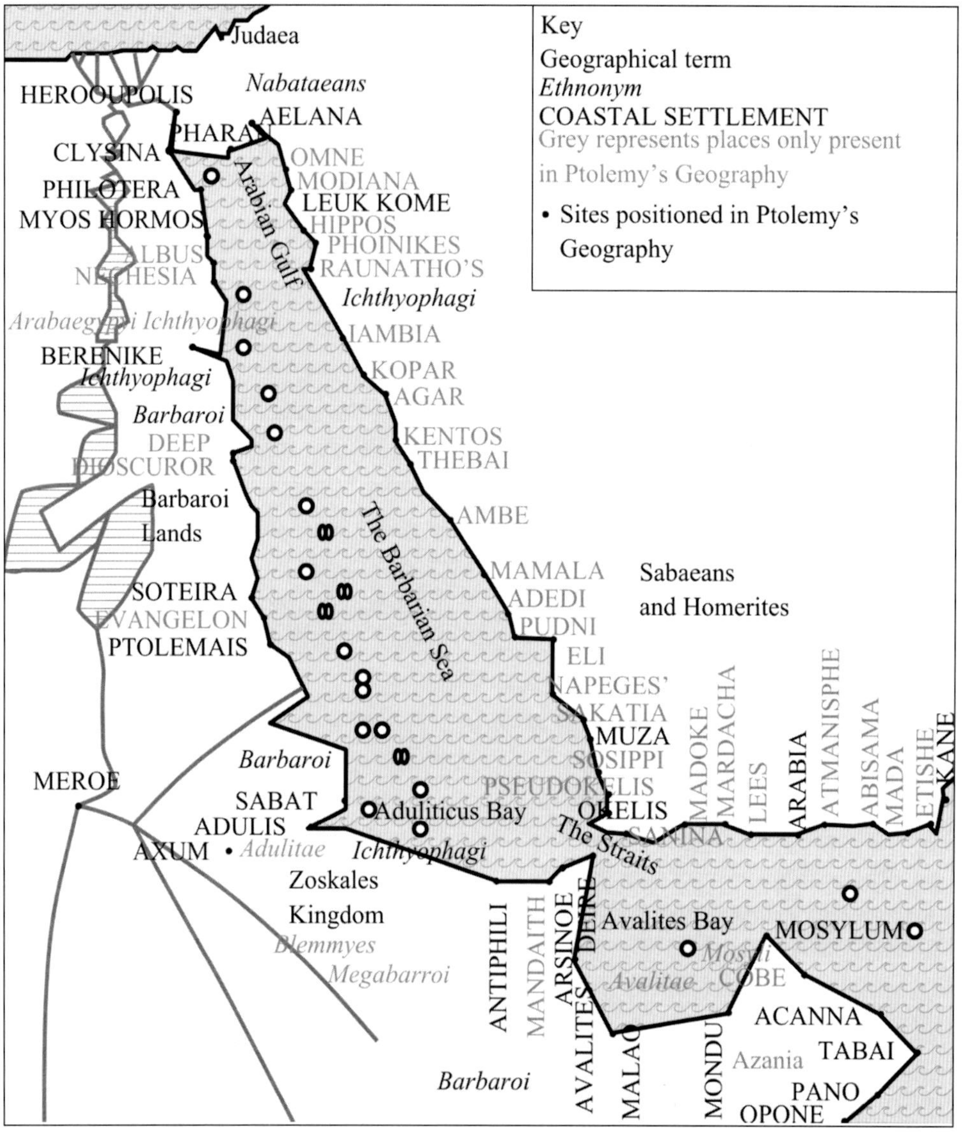

Figure 2. Cartographic representation of the Red Sea by Ptolemy with additional details from the Periplus Maris Erythraei (based on coordinates from Ptolemy).

bel *barbaroi* (De Romanis, 2003) and were also known as *Trogodytes, Megabaroi* or *Blemmyes* (Cuvigny, 2003b; Barnard, 2005). In the third century AD, a kingdom of the nomadic Blemmyes acquired large sections of the Eastern Desert from the Romans, including its emerald and gold mines (Olympiodorus in Photios, 1.37). Also in that century Myos Hormos was abandoned (Peacock & Blue, 2006).

Cartographic representations of the Red Sea (figure 2), or Erythraean Sea, as it was known to the anonymous author of a traders' guide to the Red Sea, the so-called *Periplus Maris Erythraei* (first century AD; Casson, 1989) and to the Alexandrian

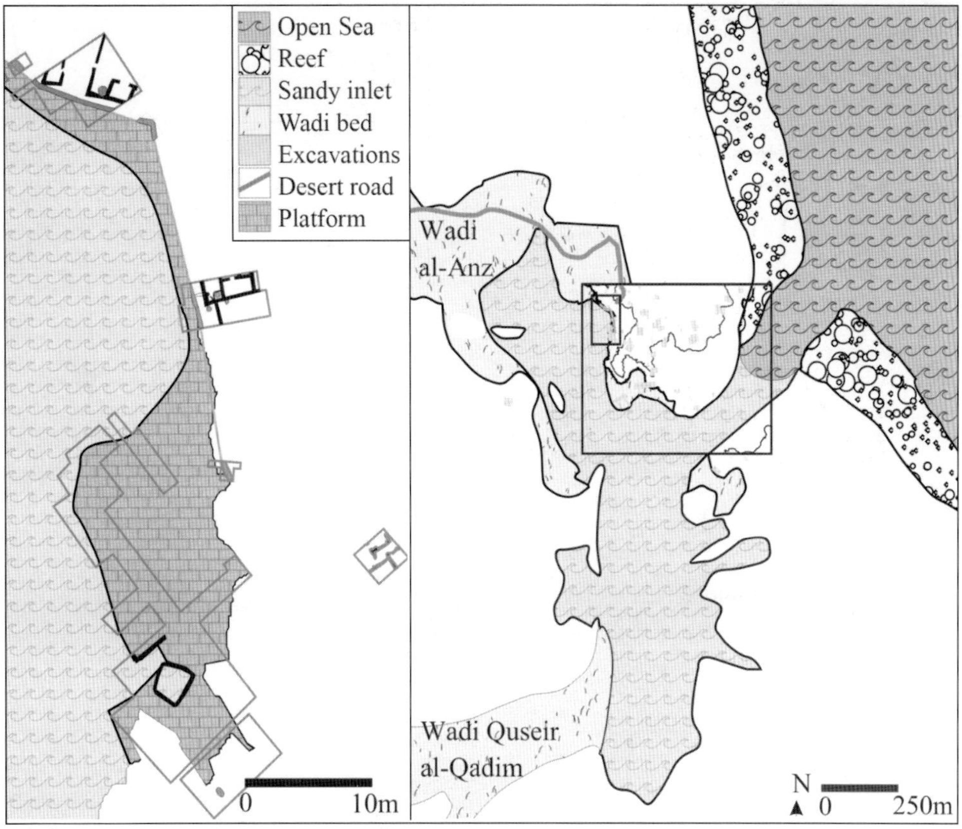

Figure 3. Location of Myos Hormos in Wadi al-Anz and Wadi Quseir al-Qadim, with harbour excavations inset (modified from Blue, 2006a; Peacock & Blue 2006).

geographer Claudius Ptolemy in the second century AD, illustrate the extent to which the Red Sea was known to the Greeks and Romans during the early Empire. The northern section was known as the "Arabian Gulf", the middle section as the "Barbarian sea", then *Aduliticus* bay, the straits, *Avalites* bay followed by the rest of the Erythraean Sea – the modern Indian Ocean. The inhabitants of the coast were known to Ptolemy as *Arabaegypti Ichthyophagi* and to other geographers simply as *Ichthyophagi*, or fish-eaters.

Classical accounts of the *Ichthyophagi* frequently emphasise the fantastic. One such source, Agatharchides writing in the Ptolemaic period, states that for them "Poseidon has assumed the function of Demeter" (Agatharchides, 5.34, translated by Burstein, 1989), for they were renowned for and defined by their fishing economy. Myos Hormos was situated in this area and its demography must have been quite mixed, including traders, shipwrights and sailors from Hellenised Egyptian backgrounds, Roman soldiers, Imperial freedmen as well as indigenous *Ichthyophagi* (Thomas, 2007) and desert dwelling nomads or *Trogodytes* as recent ostraka evidence suggests they were known to traders (Tomber, 2005, 41; Barnard, 2005).

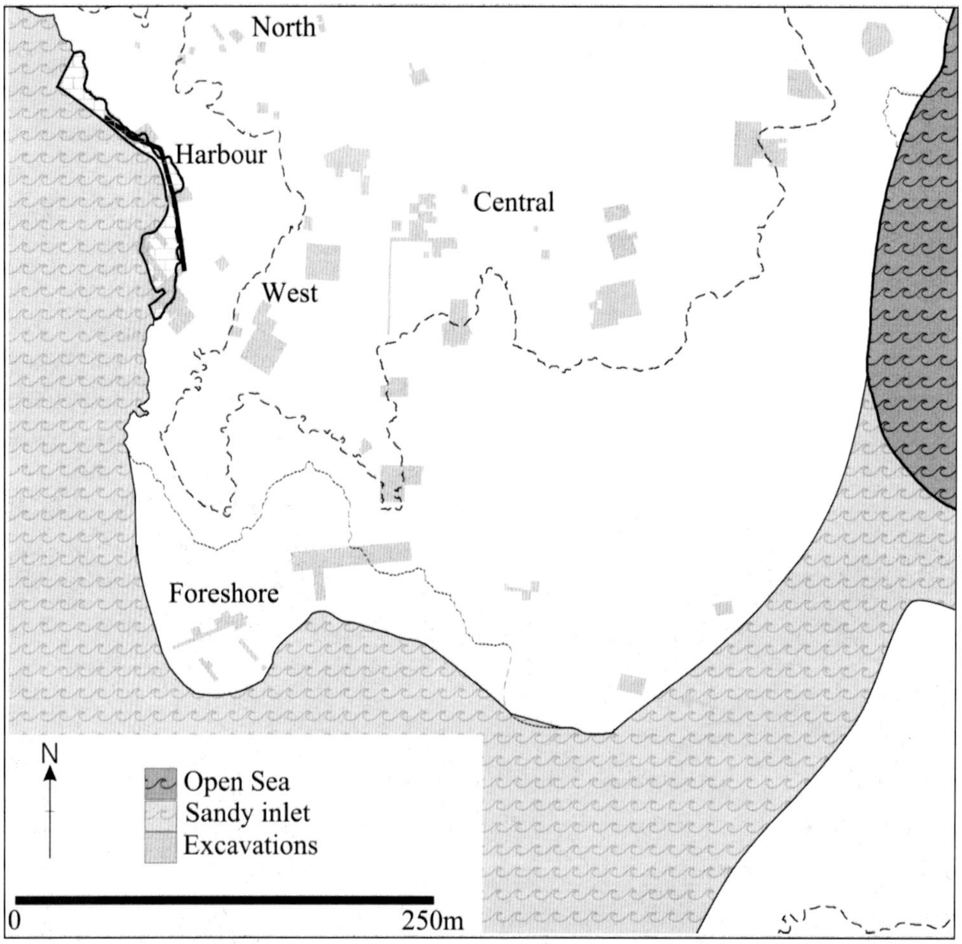

Figure 4. Myos Hormos settlement, showing extent of inlet and the areas excavated (modified from Peacock & Blue 2006).

Myos Hormos was built around a sheltered lagoon, which has since silted up as a result of long shore drift and wadi deposits from Wadi al-Anz and Wadi Quseir al-Qadim (Peacock, 1993; Peacock & Blue, 2006). The marine environment as it would have been in the first to third centuries AD has only been fully explored in the area defined by the black square, by Lucy Blue (figure 3; Blue, 2006a). The marine environment would have then consisted of three different marine environments, a sandy or silty lagoon area with shallow sloping beaches, a shallow reef with a vertical face adjacent to deep pelagic waters that rapidly drop to a depth in excess of 70m.

At the north-west of the site, the harbour was built in the Augustan period over a mangrove swamp (figure 4; Thomas, 2006). Amphorae and other broken pottery were used to create a platform with a trampled surface facilitating the loading and unloading of ships up until the second century AD (Blue & Peacock, 2006). Behind this surface was a stone sea defence joined with hydraulic cement that created a revetment on which further structures were built (Blue, 2006b). In and around this area, a mixture of maritime artefacts representing various fishing and ship maintenance activities – such as woodworking, sealing, hull sheathing, anti-fouling (Thomas, forthcoming b), rig and sail maintenance (Whitewright, 2007) were also found. This area was quite damp, and only in the waterlogged and desiccated deposits were organic materials found. Metal was also quite corroded from contact with salt and water.

Just under 120 artefacts from the Roman levels of Myos Hormos can be securely identified as fishing equipment (Thomas, forthcoming a). They are distributed unevenly across the site, with concentrations near the harbour and in certain domestic areas of the town, where they account for 2 to 5% of all artefacts found. Warehouses storing ship artefacts and wine amphorae, but no fishing equipment at all, were found in the central settlement. This suggests that these two maritime activities: fishing and trading, were undertaken by different people living in different areas of the site. On a ridge overlooking the harbour is a domestic area. On this ridge, excellent preservation allowed many organic artefacts to be preserved, including fishing equipment (Thomas & Masser, 2006). Adjacent dumps also provided numerous examples of fishing activity in the form of artefacts, faunal and epigraphic material, such as ostraka and papyri (Van Rengen & Thomas, 2006).

Fourteen floats were found, made from cork or wood (figure 5). Both types appear to have drilled holes. It is uncertain from the forms what type of fishing, net or line, they represent. The 11 weights found were made from stone, lead or ceramic. Three examples were made from rounded beach stones (L0115, figure 5). Examples made from coral (FR0069, figure 5) were found in mixed Roman/Islamic contexts. Methods of attachment included drilled holes, pecked grooves to facilitate tying or in this case, a neatly tied net within which the stone was contained. Unlike some ceramic and lead examples, a particular method of fishing cannot be confirmed, though line fishing is likely. However, due to their larger size, the coral examples may be for weighting something larger, such as a trap or net.

Many of the fishing artefacts were made from reused materials, and their excellent preservation serves to remind us how little would be allowed to go to waste, an element of site formation processes of which we need to be aware. Elements of ship maintenance such as the lead sheeting used to cover the hull (Thomas, forthcoming b) could be reused, with scraps folding into a round line sinker (five examples) or a rolled net weight (M0302, figure 6). In one trench overlooking the harbour we found a stack of discarded hull sheathing fragments, associated with fishing equipment, suggesting that this lead sheeting was being stored for this purpose (Thomas & Masser, 2006). Similarly, all the cork must have been imported, and the most likely source are the numerous wine amphora stoppers (Thomas, forthcoming c) found on site such as the ones (W0448 and W0477) pictured in

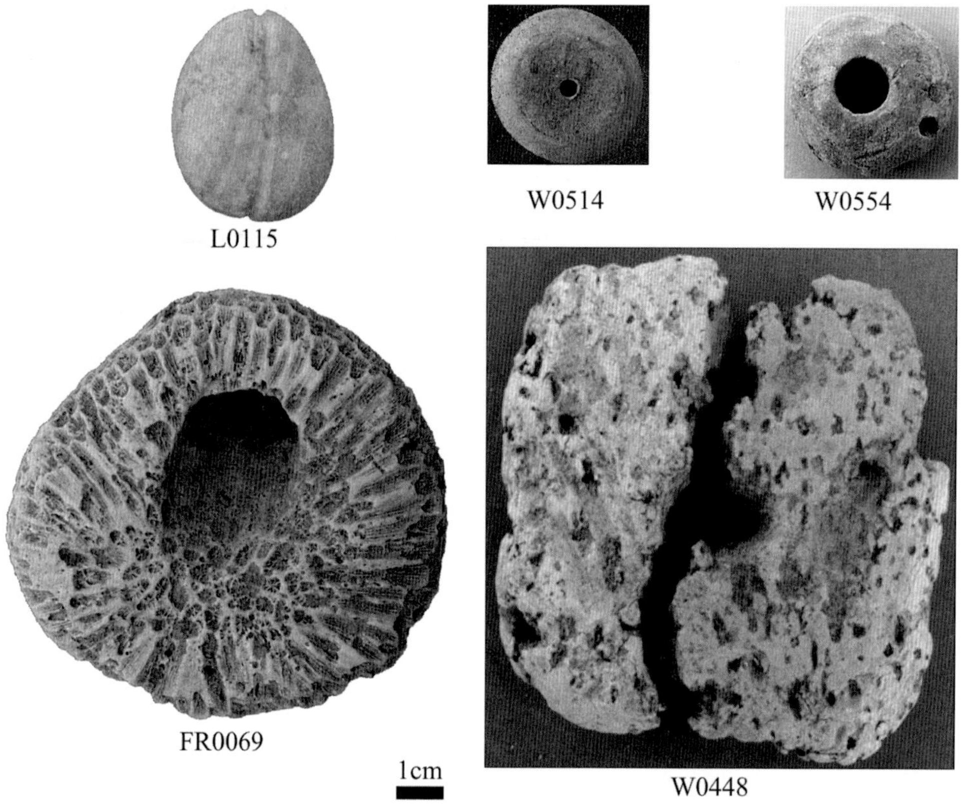

Figure 5. Weights and floats. Photographs courtesy of the Quseir al-Qadim finds archive, Southampton University.

figures 5 and 6. Many cork stoppers were already pierced with a handy hole, the original function of which was to allow the fermentation process to continue unhindered (Thomas, forthcoming c).

Usually archaeologists rely on indirect evidence to indicate the presence of fishing nets, such as the purpose-made tubular profile ceramic weights and a rolled lead weight illustrated above (C0036, C0046, M0302, figure 6). These are the only examples of net weights that can be confidently identified, and it is possible that indistinguishable stones or coral fragments may have been used as well, such as those found at Abu Sha'ar (Wendrich & Van Neer, 1994). The rolled lead weight is similar to those found on wrecks in the Eastern Mediterranean, such as the seventh-century Dor and Yassi Ada wrecks (Kuniholm, 1982; Galili *et alii*, 2002); their design clearly intended to limit fouling of the net. Net fragments are more common, with 27 found at Myos Hormos (Thomas, forthcoming a). Unfortunately, no examples were found attached to weights or floats, though the largest examples are up to 74cm long. Some were found in a secondary context, being reused in the making of sandals (CB0304).

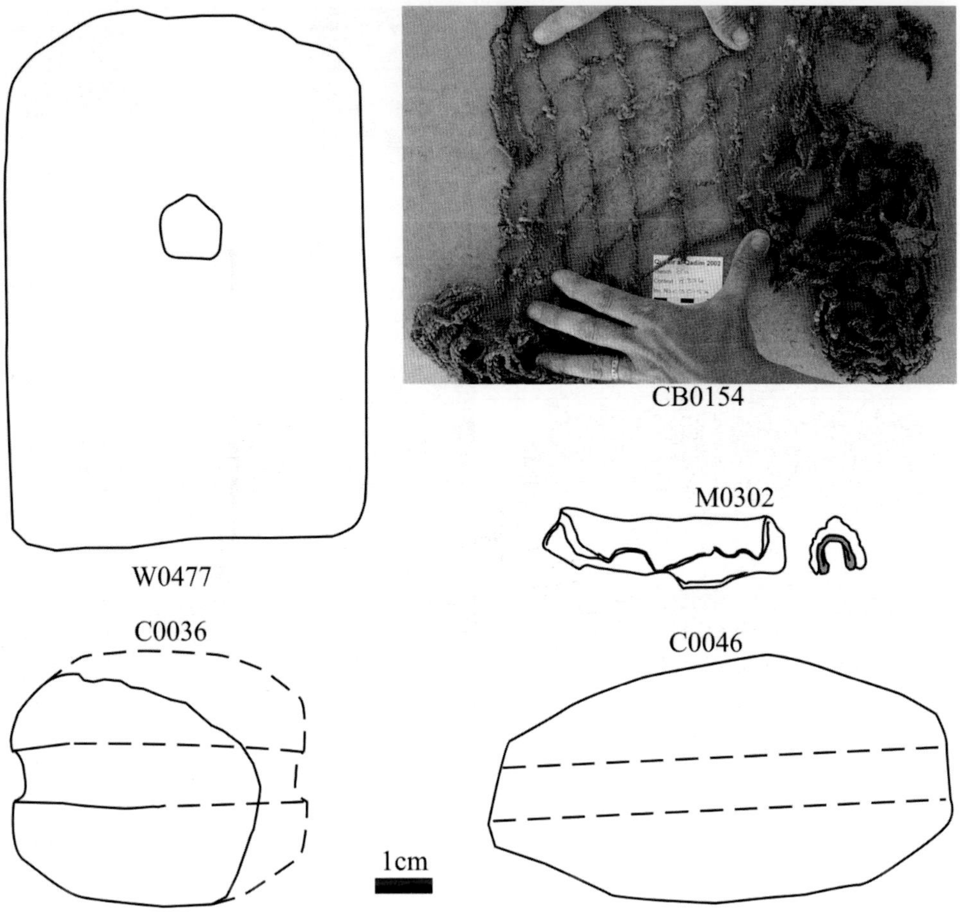

Figure 6. Fishing net equipment used at Myos Hormos (drawings by author, photograph courtesy of the Quseir al-Qadim finds archive, Southampton University).

Nets can be used for both passive and active forms of fishing. Passive net fishing is when nets are left unmanned to trap fishes whilst active net fishing techniques are when a net is dragged or thrown to hunt fish. Both are represented at Myos Hormos. Nets at Myos Hormos were commonly made from bast fiber, probably flax, just as they were at Abu Sha'ar (Wendrich & Van Neer, 1994) to the north and Berenike to the south (Veldemeijer, 2004; 2005a; 2005b; 2006). The nets came in two sizes: Three fine-meshed nets were found with string diameter of *c.* 1mm and mesh spacing averaging 12mm. These are likely to have been used as casting nets, either from the shore or boat, for catching small species of fish such as sardines. More common are coarser nets, represented by 20 examples that have string diameters of *c.* 3.8mm and mesh spacing averaging 35mm. These would have been used to target larger fish species and may have been used for a number of techniques, such as drag nets, as described by classical authors and depicted in mosaics (table 1).

Greek term	English term	Function	Material
Kalymma	Cover net, or veil net	Fine net for small fish	*linoi* (flax)
Amphiblêstron	Casting-net	Circular, cast from above	*linoi* (flax)
Griphos	Draw-net	Category of nets	*linoi* (flax)
(Griphos) gangamon	Drag-net	Small net drawn through water	*linoi* (flax)
(Griphos) hypochê periêgês	Round bag-net	Medium net? Or a scoop net on a pole	*linoi* (flax)
(Griphos) sagênê	Seine	Large net drawn to shore	*linoi* (flax)
Skolios panagros	Crooked trawl	Large drag net	*linoi* (flax)
Peza	Ground-net	Passive trap	*linoi* (flax)
Sphairôn	Ball-net, or creel	Passive trap	*linoi* (flax)
Kyrtos	Basket trap	Passive trap	Wickerwork

Table 1. Greco-Roman fishing techniques (Bekker-Nielsen, 2005; English terms from Mair's translation of Oppian).

Passive net traps were also used on the Red Sea. The best represented of these at Myos Hormos are four bag nets tentatively identified as "fishing pots" (CB0143, figure 7). They are bag nets made from grass, palm or bast fiber that can be set up as a trap, trailed behind boats or even used to keep fish alive (without losing them) whilst aboard boats (Bekker-Nielsen, 2005). Veldemeijer has persuasively argued that similar finds found at Berenike are carrier nets that were not used for fishing. This is because the materials (grass and palm fiber) and knots (reef and half knots) are not suitable, as they would absorb water, become heavy, would be damaged more easily and more difficult to repair (Veldemeijer & Van Rode, 2004). However, the materials used in construction may have been dependent upon availability and at Myos Hormos vary from find to find for this reason. Also, as a passive trap, weight and strength would not have been such an issue as with active net traps. Finally, these finds from Myos Hormos were associated with deposits that had extensive evidence of fishing activity (Thomas, 2007; forthcoming a).

Basket traps are a type of tidal or baited trap that are to this day made from woven palm fiber, attached by cordage to a float to mark their position that is usually at 4-12m depth (Beech, 2004). Whilst there are fragments of basketry, constructed from wood, grass and cordage, preserved across the site of Myos Hormos, none have been positively identified with known basket trap construction techniques. Frag-

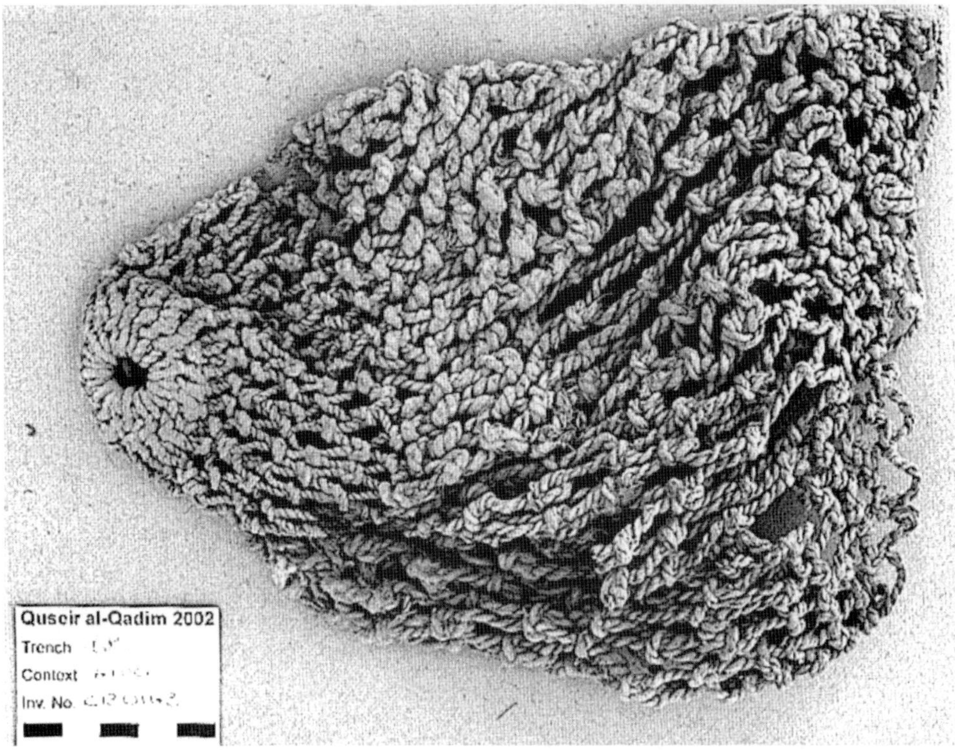

Figure 7. CB0143, putative creel (sphairôn) *or bag-net* (hypochê periêgê) *(photograph courtesy of the Quseir al-Qadim finds archive, Southampton University).*

ments of basketry found at Abu Sha'ar have, however, been positively identified as belonging to two types of basket trap built from rushes – the conical "four systems" trap and the spiral construction "twinned" trap – and it is likely that this fishing method was used at Myos Hormos as well (Wendrich & Van Neer, 1994). More permanent or movable unmanned tidal traps are also likely to have been used, but have left no trace in the archaeological record at Quseir al-Qadim. The indigenous people of the Red Sea, the *Ichthyophagi*, are described as having built stone traps across channels and inter-tidal pools by the Greek writer Agatharchides (5.32-36; GGM 1, 130-133), who describes the traps as being near huge shell middens. Large shell middens are found in the Berenike region (Tony Rouphael, pers.comm.).

> The homes of these aforementioned peoples are located along rocky shores, which have deep depressions, irregular ravines, narrow channels and curving inlets. Where these exist suitable to their need, they place rough boulders in the depressions so as to form several narrow passageways. Then when the tide is borne in to the land from the sea, which happens twice each day ..., the sea covers the whole rocky shore, It also brings with its surge from the straits many fish which remain near shore browsing in the sheltered recesses for food. But when

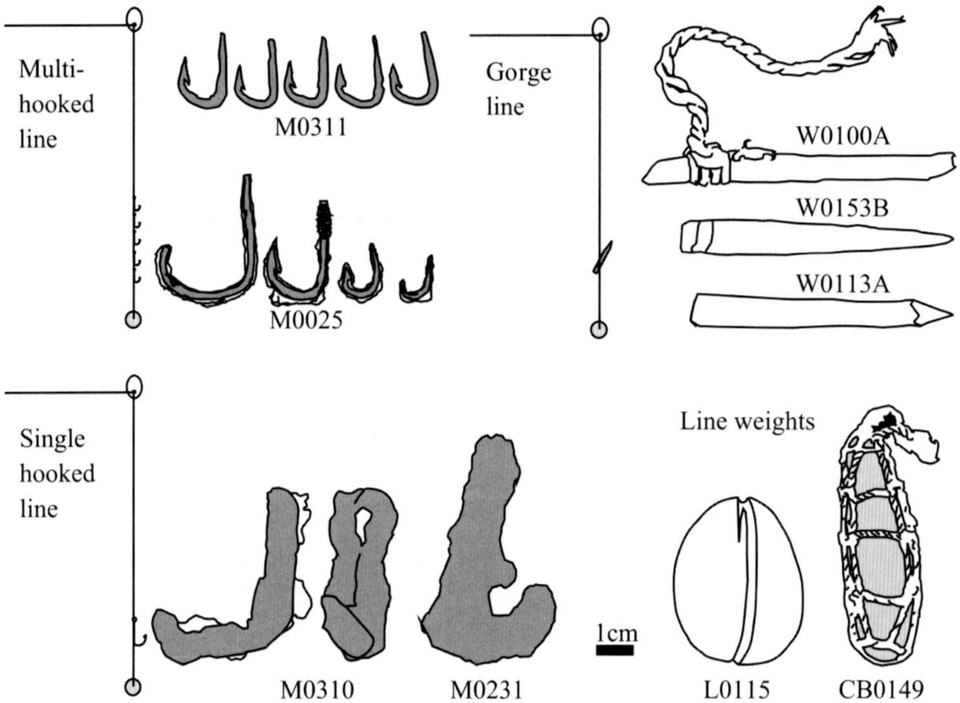

Figure 8. Fishing line techniques used at Myos Hormos. Drawings by author.

> the ebb tide occurs again, the water flows back through the stones and passageways to the depths that attract it, but the fish that remain behind in the hollows are easy prey and food for the Ichthyophagi. The other fish … are easily subdued, but when dogfish, large seals, sea-scorpions, eels and all creatures of this sort fall into the trap, the enterprise becomes dangerous (Agatharchides, 5.32-33, Burstein's translation).

Diodorus elaborates on the description of Agatharchides, informing us that weapons made of stone and horn were used to subdue these animals (Diodorus, 3.15.5-7). Such traps survive today near the modern Ababda village called Qulun, south of Quseir al-Qadim (Tony Rouphael, pers.comm.). There the technique is remembered, though no longer used by the Ababda nomads (Tony Rouphael, pers.comm.). Finally, pit traps may have been used, dug into the foreshore to trap fish in ponds at low tide, or to keep caught fish and shell fish alive, also discussed by Agatharchides (Agatharchides, 5.34). A number of man-made pits of unknown function were found on the foreshore and a stone-lined pond was built into the platform of the Myos Hormos harbour, paved with amphora and other pottery fragments (figure 3, see also Peacock & Blue, 2006).

In addition to the weights and floats that may have been used in line fishing, there were also many examples of fishing hooks found at Myos Hormos. Small barbed copper alloy hooks, 1cm by 2cm long without eyes, predominate with 53 examples (fi-

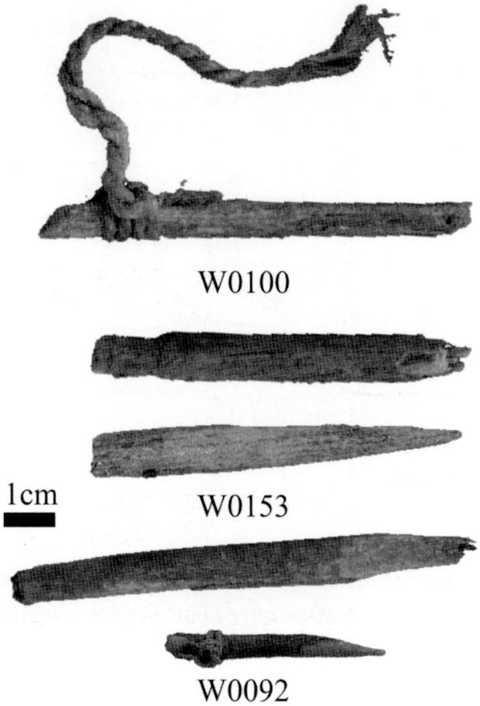

Figure 9. Gorges, photographs courtesy of the Quseir al-Qadim finds archive, Southampton University.

gure 8). They even appear to have been made locally, in a metal-working area of the site on the southern foreshore. These hooks were often found in groups (M0311, M0025, figure 8), sometimes corroded together, suggesting they may have been used for multiple-hooked lines. Four examples of larger, 3cm by > 5cm long iron hooks were also found, though only in isolation and in a poor state of preservation. These are likely to have been used for single-hooked lines, baited for larger fish species. Both small and large forms are found both barbed and non-barbed, presumably a difference that depended on the preference of the user, though barbed forms predominate. Barbed forms can hold a fish for longer, but would be slower to unhook, thus reducing the speed of fishing when directly over a particularly productive shoal (Beech, 2004). Larger iron hooks and gorges would have been baited for larger fish and shark species.

Eight gorges were found in secure Roman contexts, though many more were found in mixed Roman/Islamic deposits. A gorge is a straight piece of shell, bone or wood attached to a line, baited and laid parallel with the line. Fish or sharks taking the bait are caught by making the line go taut, causing the gorge to stick in the throat or belly of the fish (Beech, 2004, 68). This is not a technology found in the Mediterranean (as far as the author is aware) though gorges have been found on archaeological sites in the Persian Gulf, where they are made from wood or bone (Charpentier *et alii*, 2004). All the Myos Hormos examples were made from a twig of locally available wood such as tamarisk (Hiebert, 1991, 155), or may have been whittled down from a larger piece of wood until they were approximately 4 to 8cm long. They were

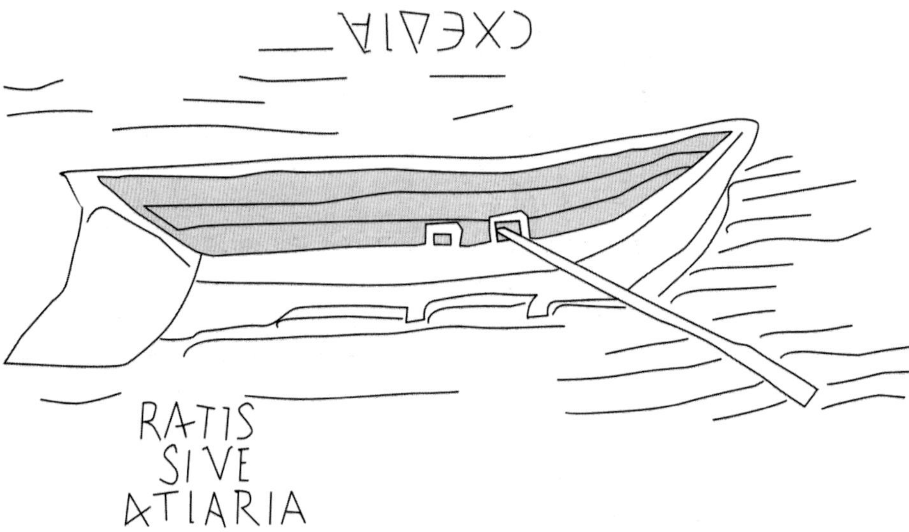

Figure 10. The third-century AD Althiburus mosaic, depicting the fishing boat (schedia). Line drawing by author, after photograph of the Althiburus mosaic in the Bardo Museum, Tunis.

sharpened at one end and notched at the other to attach the line (figure 9). Gorges are likely to have been used to catch larger fish and sharks, and so represent the use of a specific technique to fill a specific demand (Hamilton-Dyer, pers.comm.).

Absent from the archaeological record is evidence of indigenous fishing boats of the Red Sea region. There is plenty of evidence for the maintenance of large ships, but none for fishing boats. Fortunately ostraka from both Myos Hormos and from a fort, called *Maximianon* on the Myos Hormos-Koptos road do mention small boats called *schediai* used in fishing activities. One ostrakon is a permit for an individual calling himself *Pakubis Ichthyophagos* to move a number of *schediai* to *Philoteras*, a settlement to the north of Myos Hormos (Van Rengen, 2002: *O.Myos*, 512), and another is discussing an order of fish for one *Ioulius Maximus* from *Maximianon* that was delayed because the *schedia* (see below) had not yet returned (Bülow-Jacobsen *et alii*, 1994: *O.Max.* 175).

The literal meaning of *schedia* (and its Latin equivalent, *ratis* or *rat[i]aria*) is raft, or flat-bottomed boat. It is a term that is frequently used in the description by ancient geographers of indigenous Red Sea boat types (Casson, 1989; *Periplus Maris Erythraei* 7; Strabo 16.4.18). The classical accounts and iconography represent all the fishing techniques represented at Myos Hormos, except gorges, which appear to be a specifically Red Sea and Persian Gulf phenomenon. These sources also illustrate the importance of small boats in the ancient fishing economy. A *schedia* is labeled on the Althiburus mosaic of the third century AD (figure 10) and depicted as a small, flat-bottomed rowed boat. Although it may not resemble what was actually used on the Red Sea, it is likely to have had much the same function as a small fishing boat.

Published early Roman fishing boats from Fiumicino (Boetto, 2006a) and Galilee (Wachsmann, 1995) may be contemporaneous with Myos Hormos, though do not necessarily resemble what was used on the Red Sea during the early centuries AD.

The preservation is such at Myos Hormos that the most fragile fish bones, scales, cartilage and even flesh have been preserved, including a complete parrotfish that has been sliced in half (Hamilton-Dyer, 2003). The faunal studies by Sheila Hamilton-Dyer (2003) suggests that fish were a major part of the diet in Myos Hormos as well as being exported to quarries such as *Mons Claudianus* (Hamilton-Dyer, 2001) and *Mons Porphyrites* (Hamilton-Dyer, 2007) in the desert.

The species fished were numerous, and from various different environments; we find species from pelagic open water, benthonic sandy bottoms and the reef environment. Due to the species' different diets and habitats, different fishing techniques were required and used to catch them. The most common were parrotfish, though grouper, emperor, triggerfish, shark, sea bream, jack and trevallie, surgeon and unicorn, snapper, wrasse, goatfish and mullet were all also popular. Gar pike and needlefish, squirrel fish, various *perciformes*, moray, various *gerres* species, boxfish, porcupine fish, pufferfish and sardine were also found (Hamilton-Dyer, 2003), though possibly not specifically targeted. There were three different marine environments adjacent to the port, placing the inhabitants of Myos Hormos perfectly to target easily the different fish species available on table 2.

Ethnographic studies of traditional and modern fishing techniques in the Red Sea region and the Persian Gulf have also provided useful information as to how fishing equipment relates to faunal remains in the archaeological record (*cf.* Morales in this volume). From this it is clear that a mixture of line, net and trap techniques must have been used in a number of different environments (lagoon, reef and open-water) to catch the large range of species consumed at Myos Hormos. Line techniques would have been used to catch both reef and open-water species such as shark, but would also have been effective for catching barracuda, jacks and trevallies. Net and trap techniques must have been used to catch parrotfish, as well as mullets, sardines and a range of reef species. Commonly consumed reef fish such as grouper, emperor, trigger, sea bream, snapper and wrasse could have been caught with a variety of trap, net and line techniques, which may explain why they were so commonly caught.

Combining the archaeological, faunal and ethnographic data confirms the use of many fishing methods at Myos Hormos. These include the various net fishing techniques outlined in Bekker-Nielsen's study of Greek fishing terminology described by Oppian (Bekker-Nielsen, 2005), with the addition of Red Sea specific technology, such as gorges. It even confirms the material being used as flax bast fiber (Veldemeijer, 2004). At Myos Hormos, it is unlikely that seines could be maneuvered in the lagoon, though fine-meshed veil nets, casting nets, drag nets, ball-nets and traps are all likelihood represented.

Common name	Tidal trap	Basket trap	Trawl net	Casting net	Hooked line	Multiple line	Trolling line
Parrotfish		B	W	W			
Grouper		B			B,W	B	
Emperor		B			B,H,W		
Trigger				W	W		
Shark					W	B	
Sea bream	B	B	W		B,H		
Jacks and Trevallies	B	B	W		B,W		B
Surgeon and Unicorn			W	W			
Snapper	B	B	W		H		
Wrasse			W		W		
Goatfish				W			
Mullet	B			W,B			
Barracuda	B		W	W			B
Garpike and Needlefish	B			W			
Squirrel			W	W			
Gerres	B						
Sardine				B			
Rabbit	B	B		W			
Grunt		B	W		B,W		
Mackerel					B		B
Tuna							B

Table 2. Ethnographic data on species caught with different fishing techniques in the Red Sea and Persian Gulf (source listed in table as W: Wendrich & Van Neer, 1994; H: Hamilton-Dyer, 2001; B: Beech, 2004). Fish species listed in approximate order of occurrence at Myos Hormos (Hamilton-Dyer, 2003).

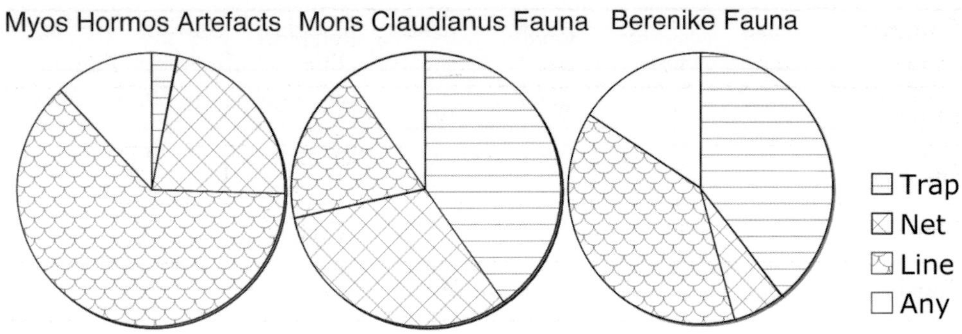

Figure 11. Fishing techniques represented by artefact and faunal evidence. (Van Neer & Lentacker, 1996; Van Neer & Ervynck, 1998; Van Neer & Ervynck, 1999; Hamilton-Dyer, 2001; Thomas, forthcoming a). Note the discrepancy between the faunal evidence for fishing techniques from the fish supplied to Mons Claudianus and the artefacts from Myos Hormos. Also regional differences with Berenike can be observed.

Evidence for how prevalent different fishing techniques were comes from artefactual, faunal and epigraphic evidence. There were major differences between what the fishing equipment found at Myos Hormos and what the faunal evidence from *Mons Claudianus* (supplied by Myos Hormos) suggest (figure 11). Fish species commonly caught in traps were common in the faunal record, whilst fishing hooks are more prevalent artefacts. Fishing traps and nets would have been more common than the finds suggest, they are just less likely to be preserved due to their organic nature, reuse of materials and the fact that fishing hooks, weights and floats are less easily lost than a large net or trap. Clearly trap and net techniques were more commonly used than even the excellent preservation conditions of Quseir suggest. These are biases influenced by human agency and site formation processes. Figure 11 also highlights regional differences between the early Roman period fishing undertaken in Berenike and that in the Myos Hormos region.

Demand for fresh fish (*opsaria*) included the ordering of specific species that were clearly popular, such as parrotfish (*skaros*) and rock mullet (*triglia*), and an as yet unidentified species (*glaukiskapin*), as well as small fish (*ichtydia*) and shellfish (*balanos*) have been recorded in the epigraphic record (Bülow-Jacobsen, 2003; Leguilloux, 2003). However the warm environment of the Eastern Desert was such that some form of preservation was usually required to prevent putrefaction, and some of these forms of preservation became quite popular for their peculiar taste (Curtis, 1991; Trakadas, 2005; Van Neer *et alii,* 2004; Wilson, 2006; Van Neer *et alii,* 2007; Van Neer & Parker, 2008). The methods of preservation varied, and not all are precisely clear as to how they were carried out (table 3). Not all these methods are archaeologically visible, indeed human agency can be difficult to identify as the environment naturally preserves through the process of desiccation. The Greek terms most commonly used for preserved fish products include *temachia*

English	Greek	Reference
Shellfish	*Balanos*	Bülow-Jacobsen, 2003
Unknown fish	*Glaukiskapin*	Bülow-Jacobsen, 2003
Parrotfish	*Skaros*	O.Max. 793, O.Krok. 1, O.Krok. 63
Salted/dried parrotfish	*Temachia karou*	O.Max. 793
Mullet	*Kestreis*	O.Krok. 1, O.Krok. 63
Rock mullet	*Triglê*	O.Max. 707, O.Max. 1300, O.Max. 869
Fish (fresh)	*Opsarin*	O.Krok.1, O.Krok. 63, O.Max. 1138, O.Max. 1463, O.Claud. 241
Tarichos pickled in brine	*Tarichou*	O.Claud. 1264
Temachion salted or dried	*Temachia*	O.Max. 876, O.Claud. 233
Hallex fish sauce	*Hallex*	O.Max. 279, O.Max. 1512
Garum fish sauce	*Garos*	Tomber, 2006
Filleted and "cooked for preserving" (smoked?)		O.Max. 639
Little fish	*Ichtydia*	O.Claud. 242

Table 3. Fish species and fish products exported from Myos Hormos to settlements in the Eastern Desert. Ostrakon number given where available from source sites Maximianon, Krokodilo (Bülow-Jacobsen, 2003; Leguilloux, 2003) and Mons Claudianus (Hamilton-Dyer, 2001; Tomber, 2006).

(sliced and either dried/and or salted), *tarichos* (pickled in brine), *allex* or *garos*, Latin *garum* (fish sauce produced through fermentation of fish and fish guts) and possibly, even one case of smoked fish.

The effort and expense involved in developing and practicing a variety of fishing techniques implies a significant demand for a variety of different fish species. Evidence for the production and marketing of fish from Myos Hormos mostly comes from other sites, where faunal remains, ostraka, papyri and the dipinti on fish transport amphorae have been studied (Bülow-Jacobsen, 2003; Cuvigny, 2006). This complementary evidence of faunal and epigraphic data helps us realise the extent and distance of trade in both fresh and preserved fish from Myos Hormos. Much of this was concentrated along the Myos Hormos road leading to Koptos and

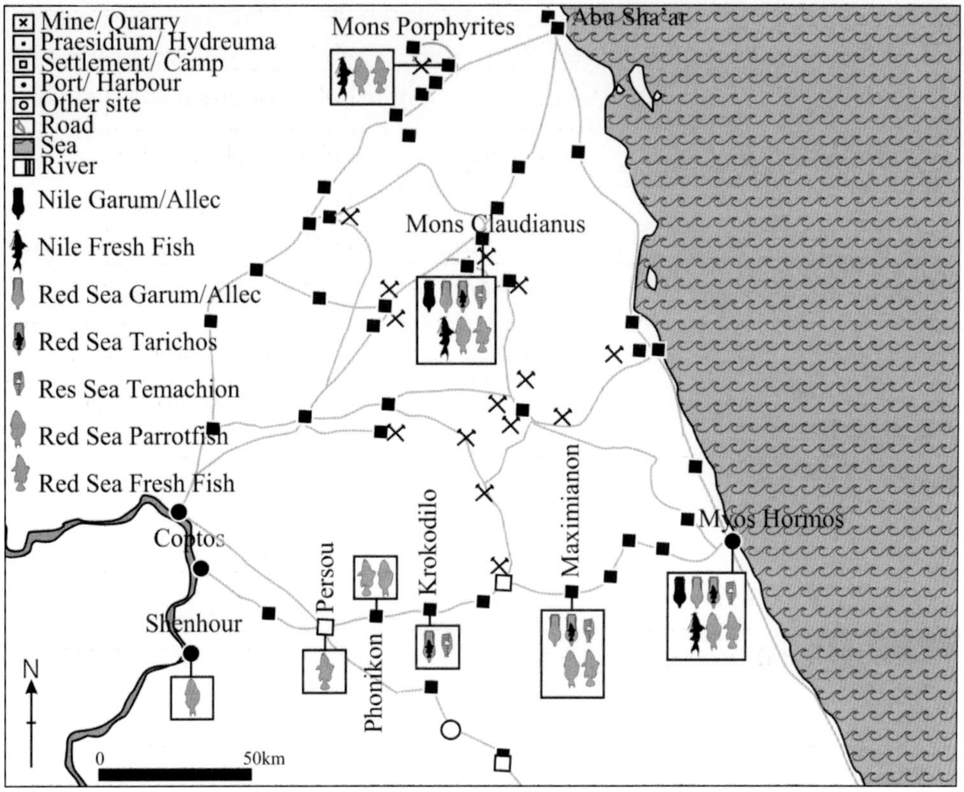

Figure 12. Exportation of fish and fish products (data after Hamilton-Dyer, 2001; Bülow-Jacobsen, 2003; Leguilloux, 2003; Van Neer et alii*, 2004; Tomber, 2006; Van Neer* et alii*, 2007; Van Neer & Parker, 2008).*

the Nile (figure 12, southern road) and to the quarries, mines and forts, in order to feed the greatly increasing population of the Eastern Desert. It was this trade that stimulated the use of the fishing techniques described above.

The various preservation techniques used would have allowed fish caught at Myos Hormos to be exported to the Nile. This is confirmed by parrotfish remains found at the late Roman site of Shenhour (Van Neer *et alii*, 2004), as well as references to drying or salting in the ostraka record (Table 3). Whilst it is likely that *Mons Porphyrites* and *Mons Claudianus* were supplied by more local sources in the Abu Sha'ar region, as yet no site from the first three centuries AD has been identified there and so it is likely that Myos Hormos, as the largest port in the region, would have supplied them. From broader studies of the region, we now know that the port sites of Berenike and Aila were also exporting Red Sea fish and fish products into the Eastern Desert and the Levant (Van Neer *et alii*, 2004). The role of fishing in the Red Sea ports cannot be overestimated. Despite the consumption of Nilotic and Eastern Desert fauna in settlements in the Eastern Desert, including ports such as Myos Hormos, marine species from the Red Sea – fish and shellfish (and to some degree turtle and sea mammals also) were important to most people's diet at these sites (Hamilton-Dyer, 2001; 2007; Thomas, 2007). Whilst it is likely that there were individuals who occasionally fished to supplement their diet, there were certainly specialist fishermen present at Myos Hormos. This is confirmed by *Pakubis Ichthyophagos'* permit request and *Ioulius Maximus'* complaint to *Gaius Apoliuanus* about the fishermen in their *schediai.* At *Mons Claudianus*, fish were supplied to the fort by people the Greek speakers there labeled *Arabs* (Cuvigny, 2003b), whilst *Pakubis* identifies himself as *Ichthyophagos*, a fish eater (Van Rengen, 2002; Thomas, 2007). The epigraphic record confirms, then, that the specialist fishermen were, in these examples, the indigenous inhabitants of the Red Sea coast known to the classical geographers as the *Arabaegypti Ichthyophagi* (Ptolemy, 4.5.101).

Conclusions

The maritime artefacts show us the range of fishing techniques available to, and used by the fishermen of Myos Hormos. These included angling with multiple hooked lines, single baited lines, and gorge lines, to catch a range of fish of different sizes. Evidence for the hooks, the weights and the floats were found, as well as in some cases, the line itself. Nets were used, made from flax, and came in two different sizes, fine mesh for small species and coarser mesh. Net weights made from lead and ceramic and possibly floats were also used on nets. It is likely that both casting and drag net techniques were used.

Fishing traps were almost certainly used to catch the popular parrotfish. This is likely to have included traps made of basketry, that may be represented amongst the many undiagnostic small fragments of basketry found at Myos Hormos, as found at Abu Sha'ar (Wendrich & Van Neer, 1994). Also permanent traps, such as that found at Qulun may have been used, but are no longer archaeologically visible. Finally, bag nets are likely to have been used as both passive traps and possibly for keeping fish alive whist on boats, though this is a contentious issues (Veldemeijer, 2004).

The epigraphic evidence from two ostraka confirms that the fishing industry was reliant on small boats called *schediai* and that net, line and some of the traps would have been used with these boats to greater effect than if shore-based strategies alone were used. In conclusion, a variety of advanced technologies and techniques were used in combination to provide a variety of fish to feed the large demand for fish products in the wider region. This contradicts the "primitivist" accounts given by Greek and Roman geographers (Agatharchides, 5.32-40; Strabo, 16.4.5-20; Pliny, *N.H.* 6.176), accepted uncritically by some academics (see critique in Bekker-Nielsen, 2002b; Bekker-Nielsen, 2005). Some species such as parrotfish were particularly sought, traded inland as far as the Nile. It should also be noted that fishing activities were regulated and controlled through permits and taxation as represented by the *Pakubis Ichthyophagos* permit.

Transportation required preservation and desiccation through the salting and drying of fish slices. This is epigraphically attested, as is bottling in brine. Production of sauces that have been reduced through controlled rotting were also clearly produced and sold as *allec*. One ostrakon from an Eastern Desert fort also states that the fish should be preserved through "cooking", which has been interpreted as smoking (Bülow-Jacobsen, 2003). Production of such fish preserves and their marketing is widespread in the northern Red Sea region during this period, and can be recognised in Aila and Berenike also. Because of this it should be recognised for its economic and social importance to the indigenous Red Sea inhabitants, for integrating them within a wider Roman sphere of economic and cultural connection and exchange. Despite this, other research the author has undertaken has recognised distinctly different expressions of identity within the various port communities of the Red Sea, where intra-site analysis shows what and how people were eating was markedly different in different areas, and that this bears a direct relation to what maritime activities were (or were not) undertaken by individuals, illustrating the social as well as the economic importance of fishing on the Red Sea (Thomas, 2007).

Neptune mosaic from Itálica (Seville) (photo Junta de Andalucía).

5. Nets and Fishing Gear in Roman Mosaics from Spain

GUADALUPE LÓPEZ MONTEAGUDO[1]

Some of the most frequent representations in Roman mosaic art are related to water and fishing activities: live marine fauna as a motif of banquet scenes (*xenia*); different styles and techniques of fishing: from boats, land, or by diving, and to a lesser extent involving oyster farms; the marketing and consumption of sea food. Thus, Roman mosaics, together with other art forms such as painting, ceramics, glassware and metalwork, can be seen as important documents, along with literary sources, that shed light on the different aspects of fishing in Roman times.

This type of pavement was generally used in *piscinae* and fountains, covering not only the floor itself but also the walls; the effect was vivid, above all in the case of stepped areas, where the movement of the water further enhanced the realism of the aquatic scenes.

In the early third century AD, Aelian (12.43) – complementing Oppian (3.73-78, 80-87) who around 180 AD had written a treatise on fishing, *Halieutika* – described the four fishing techniques used in his time in the following order: nets, harpoons, fish traps and hooks. All these, as well as lobster pots, were already documented in Egyptian art dating back to the second half of the third millennium, in Greek ceramics, and were particularly abundantly represented in Roman mosaic art, especially in *Africa Proconsularis* and in *Tripolitania* (López Monteagudo, 2006, 219-267).

The most comprehensive representations of the different modes of fishing are found in the two mosaics in the museum of Sousse (ancient *Hadrumetum*) on the Tunisian coast (Sousse Museum, inv.no. 57.204, 57.095; Foucher, 1960, pl. XLVI, XXI): the mosaic from the *hypogeum* of *Hermes* (late second century) and that from

1 This work has been undertaken under research projects HUM2004-01056 and HUM2007-61878.

the "House of Virgil" (*c.* 200-210). The former depicts (see p. 192), in one and the same scene, four pairs of fishermen in four boats floating on a sea crowded with a variety of marine fauna: fish, crustaceans and molluscs. Three of the boats are anchored and the fishermen are using nets, rods, and a series of fish traps resembling woven baskets to catch octopodes. In the fourth boat, whose anchor has been raised, one man is rowing and the other is about to throw a round casting-net. In the mosaic of the House of Virgil (also known as the "Arsenal House") which is very similar to the previous one, fish traps (creels) and harpoons are also shown.

The different modes of fishing mentioned in the sources also appear on the Tunisian mosaic from the triapsidal hall of the "House of the Chariot of Venus" at *Thuburbo Maius* (end of the third or beginning of the fourth century) and now preserved at the Bardo Museum. Here, several boats are depicted on a sea replete with icthyofauna as well as several fishermen with nets, round-nets and rods (CMT II, 3. Tunis 1987, 83-88, fig. 8, pl. XXXIV-VII and LXI). A variant with *erotes* fishing is found in a mosaic at Piazza Armerina (fourth century) in which there are several fishing scenes with rods and line, nets, harpoons, lobster pots and fish traps in front of a sea-side *villa* (Carandini *et alii*, 1982, 249-258, figures 149-152, 154-155, pl. XXXVII, 79).

According to Aelian (12.43), nets were made of plant fibres, in general esparto or white and black linen fibres. Oppian (3.80-84) distinguishes two types of net. One of them is the casting net (*amphiblêstron*), a conical net that is cast from a boat and closed by means of a drawstring once it had reached the bottom. Among others, this type of net is depicted in the Sousse and Piazza Armerina mosaics mentioned above. In the Algerian mosaic of the Marine Venus of Kamissa, conserved at the Guelma Museum, third to fourth century AD (Lassus, 1965, 175-192, figure 4; Ferdi, 1998, 140-141), there is explicit documentation for this type of net: it has just been closed by pulling on the drawstring and brought out of the water by the young fisherman resting on a rock, while another three blond, naked fishermen are fishing in a sea replete with all kinds of marine creatures: with a rod and line, one of them has just caught a large fish; another has harpooned a squid with a trident, and is holding a lobster pot in his left hand, while the third, standing on the Goddess' shell, is casting a round net; the last one, also on a rock, is pulling on a net full of fish that he has closed with the drawstring.

The other type of net described by Oppian is the drag-net (*griphos*), a rectangular fabric with weights made of pieces of ceramic or stone and disc-shaped floats made of wood or cork. It is also documented in the Orpheus mosaic from the Roman factory of *Leptis Magna*, probably from the end of the second century. In one of the depictions accompanying the central mythological scene, there are three fishermen in a boat, one of them rowing and the other two pulling on the drag-net, full of fish; in the foreground, another two are fishing, one with a rod and line, with which he has captured an eel, while the other has caught some kind of crustacean with a lobster pot (Aurigemma, 1960, 52-54, pl. 107-111). The same type of net is depicted on an Algerian mosaic from the "House of Isguntus" in the port city of Hippone (ancient *Hippo Regius*) dated to the first half of the third century. The mosaic shows a fishing scene in a sea replete with marine life, with a city visible on the coastline; from a sailing boat that is towing a smaller vessel, of the *rostrado* type, two fishermen are drawing on a drag-net in the presence of another man dressed in a short tunic – perhaps he is the owner or a person in charge of controlling the fishing activities (Marec, 1958, 109-112).

Both types of net were used in a strictly littoral setting and are widely documented in mosaics from all periods, as is the mode of fishing involving the setting of nets between two boats. In the *Odyssey* (22.384-387), Homer mentions fish brought to the shore by fishermen using "nets of many meshes". Harpoons or fishing spears (*Od.* 10.124-125) and underwater fishing (*Il.* 16.746-748) are also mentioned by the Greek poet.

Harpoons and tridents, reported by Aelian (12.43) as fishing techniques carried out from boats "by sturdy arms", from a shallow reef or from the shore, were used to spear fish, octopi and squid, even sea urchins. In some mosaics this type of fishing gear is combined with others (the mosaics at Sousse, Piazza Armerina and Kamissa, mentioned above) and also with other activities related to fishing. Some Tunisian examples are conserved at the Bardo Museum; among these, the pavement from El Alia is outstanding. This has been dated to the beginning of the second century and depicts fishing scenes on a bay, with buildings on the shore. Fishing is taking place using a trident from a boat, while from the shore, four fishermen are pulling on the rope of the net, aided by two oxen, while another person is transporting the catch in two baskets suspended from a pole across his shoulders (Yacoub, 1993, 144, figure 112). In a mosaic from the "House of Ulysses" in Dougga (ancient *Thugga*) dating from the mid-third century and conserved at the Bardo Museum, two mythological episodes – Dionysios and the pirates of the Tyrrhenian, Odysseus and the Sirens – are interspersed with scenes of real-life fishing.

In the Dionysios scene, two fishermen in a sailing boat are struggling to pull the net out of the water, while a third harpoons an octopus; to the other side, two *erotes* cast their fishing-traps into the water. In the lower part, we see a fisherman sitting on a shallow reef or rock fishing with a hook and lobster-pot; at the centre, a comrade in a boat is pulling up the anchor to begin fishing with a hook and line, while a third one to the right is mending his net. In the panel of Odysseus and the Sirens, a fisherman is standing in a boat, holding a large lobster; the lower part shows a naked fisherman kneeling on a rock and harpooning a squid with his trident. Another naked fisherman is launching his boat out to begin fishing, as one deduces from the fishing rod in the boat, next to the catch basket, while a third one is about to cast a round net (Poinssot, 1965, 219-230, figures 16-20).

An extremely interesting example is the Tunisian mosaic from the "Sidonius Baths" at Sidi Abdallah (near Bizerta) of the late fourth or early fifth century, which depicts several fishing scenes from a shore, lined with the huts of the fishermen. There are four fishermen in a boat pulling on a large net full of fish, and a man on the top of a rock or cliff is harpooning a squid (Yacoub, 1993, 143, figure 111). A trident and harpoon are also being used by the fisherman on a pavement in the Greek island of Kos (now at the Castle of Rhodes) in his attempt to spear a large fish; at his feet is a basket for the catch (Kankeleit, 1999, 69-79, pl. XIV). Sea urchin fishing with a harpoon is documented in a mosaic at *Utica* from the beginning of the third century, preserved in the British Museum (Hinks, 1933, 119, no. 44, figure 136).

On other occasions, it is *erotes* who are involved in the different fishing operations, such as the above-mentioned mosaic of the *erotes* of Piazza Armerina, in which one of them is trying to spear a fish with a trident.

The fish-traps (*kyrtoi*) cited by Oppian (3.85-87) were made from dried marine rushes and seaweed and was cast from boats, equipped with cork floats and weights, as documented on the pavements in Sousse (from the catacomb of Hermes and from the House of Virgil/House of the Arsenal; in the latter case, fish traps are used in combination with harpoons), and in a mosaic on a pool wall from the same city (now in the museum in Sousse), first half of the third century, featuring two *erotes* (Inv. Sousse 57.159; Foucher, 1960, pl. XXXV b-c). This fishing gear has its antecedent on a Greek red-figure kylix by Ambrosios, *c.* 480 BC, now in the Museum of Fine Arts in Boston, which also depicts a fisherman using a hook, with some fishes in a fish-trap in the water beside him, and is documented as far back as the second half of the third millennium BC in a painting in the tomb of Ti at Saqqara (López Monteagudo, 2006, pl. VII,1).

Regarding fishing with hooks, according to Oppian (3.73-78) and Aelian (*N.A.* 12.43), the rods were made of plant material, while the line was horse mane, boar bristles, esparto grass, wool or flax. Oppian tells us that several hooks could be attached to a flax line at the same time (*polyankistron*). This type of multiple-bait fishing gear is documented from prehistoric times and also represented in ancient Egyptian art in the painting from the tomb of Idut at Saqqara, *c.* 2300 BC. In some fishing scenes in Roman mosaics, the rod has been replaced by a simple stick, as may be seen in the mosaic of Orpheus in *Leptis Magna*, or in the mosaic of *Invidiosus* at Ostia, first half of the third century, conserved *in situ* (Becatti, 1961, 218-19, pl. CLXIII). Both depict the technique described by Homer in the *Iliad* (24.80-82) as the most elementary form of fishing gear, a fine line equipped with a weight and a hook at the end, made of ox horn, as repeated in the *Odyssey* (12.25), or without a rod and perhaps with a small stick at most. This is what is show on the above-mentioned red-figure kylix by Ambrosios. This rudimentary mode of fishing is still practised today, generally by children, who roll a line around a piece of cardboard tube or similar object.

According to Aelian (12.43), fishing with a hook is the most perfect and most appropriate mode for free men, and indeed we know that it was one of the recreational activities of the Roman elite. Pliny the Younger (*Ep.* 9.7.4) fished for sport at his *villa* on lake Como, using a rod and line directly out of his window, *paene etiam de lectulo,* "almost from his bed"; this must have been a common practice, as we learn from the ironic words of Martial (*Ep.* 10.30) concerning other people. The use of this type of fishing gear, depicted in mosaics, was the most frequent technique and was practiced both from boats offshore and from the shore, and it is seen not only in realistic scenes of fishing but also in those in which the agents are *erotes, putti* or pygmies.

In Hispanic-Roman mosaic work, fishing scenes are not very frequent and here we must refer in particular to scenes of fishing performed with the modes described in ancient literary sources, since the marine fauna is well represented in many mosaics from all periods and geographic zones of *Hispania,* both bichrome (Povôa de Cos in the Lisbon Museum, Córdoba, Gilena and Cuesta del Rosario in Seville) and polychrome (Córdoba, preserved at the Archeological Museum, the *Villa* de Santa Rosa in Córdoba, *Tarraco,* La Vega Baja de Toledo), sometimes combined with fantastic species (the bichrome mosaic at Cortijo del Alcalde in Córdoba), or accompanying gods and people related to water: *Oceanus* (Carranque, in which real fish share a place with mythological animals, and Écija), *Thetis* (Bruñel), Neptune (Itálica), Europe (Écija), nereids (Écija and Mérida), marine centaurs (Conímbriga), tritons, etc., motifs that recur throughout the rest of the Empire (San Nicolás Pedraz, 2004-2005, 301-33; López Monteagudo, 2008, 2547-2568). All these scenes document the ancient familiarity with the different species as well as the importance of fishing and fish consumption in the economy of Roman *Hispania* – apart from the artistic value of such mosaics as decorations for places featuring water since, as in the rest of the Empire, this type of pavement was generally used in baths (*thermae*) and in peristyles, *impluvia,* fountains, *piscinae* and ponds, covering not only the floor but also the walls. This is seen in the examples at Milreu, Pissões and Balazote, dating to the third-fourth century. It may be assumed that these water-related areas must have been very impressive, especially when the pools contain steps, as in the House of the Waterfall in *Utica,* second to third centuries AD (CMT I/1, 19-56, pl. VIII-IX, XVII, XXV, LXIII). The effect produced by the water running down the different levels generated the impression that the water was full of live fish and marine animals in constant movement, producing greater realism.

Whether from the realistic or the mythological point of view, the representations related to fishing activities in Roman mosaic art of *Hispania* cannot be compared either in number or in variety with those found in North Africa, in particular Tunisia. As in the rest of the Empire, fishing scenes in Hispano-Roman mosaics also feature fantasy figures, such as *putti, erotes* and pygmies.

Currently, depictions of fishing modes are documented in only six mosaics from the second to fourth century in Spain, and not specifically in coastal zones, as might be expected: all of them are at sites inside the Peninsula, the pavements at Itálica being those closest to the coast. Nearly all of them, however, are related to places with water: baths, ponds, fountains, *piscinae*. Among the types of fishing gear, there is a predominance of the use of rods and lines from the shore and boats, but also nets; there are few scenes, however, depicting the use of tridents or harpoons or in which lobster pots or fish traps are being used. Also, as in other zones of the Empire, in *Hispania* the different types of fishing gear may be combined in a single mosaic. Fishing with rod and line is depicted in the mosaics at Itálica, Seville, Balazote, La Vega Baja de Toledo and Noheda. Nets appear in those at La Vega Baja de Toledo, Balazote and Noheda. The trident and harpoon are represented at Itálica and Noheda. The lobster pot or the fish trap at La Vega Baja de Toledo. In these scenes, the aquatic medium is indicated by straight or wavy lines or black tessellae, stepped lines, broken or toothed lines, zigzags, E- or F-shaped lines, triangles formed by parallel lines, swirls, the so-called water flies, etc.; all of them figures generated by water coming and going, although there are also neutral-coloured or slightly bluish-tinged floors, and the lines of water in which the figures in the pavement of the boats at Toledo swim.

One of the most complete Hispanic mosaics with regard to marine activities is that in the Roman villa of La Vega Baja de Toledo, third-fourth centuries, which paved the bottom of an octagonal *impluvium* or *piscina*, most likely of an *atrium* or peristyle (CMRE V, 1982, 36-40, no. 25, pl. 16-19 and 4; López Monteagudo, 1993, 1241-57). On this deteriorated pavement, there are different scenes of ports, fishing and transport in littoral waters (figure 1). In the part of the mosaic that has been preserved, we see a semicircular construction on pillars, flanked by two towers, which could be interpreted as the quay, since at one end lies an oared ship; the pier and lighthouse; a naval trophy; and three obelisk-like structures or commemorative columns; three quadrangular edifices with angled roofs and steps down to the sea, very probably sanctuaries; a circular, crenellated edifice, possibly intended to represent an amphitheatre; a cylindrical construction in rough-hewn stone with a conical roof, probably a fisherman's hut of the kind represented in the mosaics of Neptune at Itálica and at the House of the Amphitheatre in Mérida. On the shore or on reefs in the sea there are five fishermen. Four are angling with rod and line, two sitting on the pier and lighthouse, another two standing up on a reef in front of the naval trophy. Meanwhile, a fifth fisherman is about to cast an object, a very large spherical lobster pot or perhaps a circular scoop net, from the small lighthouse or column at the entrance to the port.

Regarding the sea, full of fish of different sizes, there are two figures swimming between the horizontal lines that represent the water, an *ostriarium* or *cetaria*, and eight fishing or marine transport boats, four of them with different types of sails, one with two people, and another towing a smaller vessel. A boat without sails is propelled by four oarsmen sitting down, directed by a man standing up in the poop. A six-oared boat is tied to the quay and from another, of which only a piece of prow is left, we see the fishing net (figure 4). It may be assumed that the owner

Figure 1. Mosaic from the Roman villa "La Vega Baja de Toledo". Toledo, Archaeological Museum (photo G. López Monteagudo).

of this villa was in some way engaged in trade along the *Tagus*, perhaps he was a *navicularius*, since the anchor is represented in another mosaic of his mansion.

Among the boats, two are directly related to fishing, since in the others it is not possible to identify elements characteristic of fishing, as distinct from transport, vessels. Likewise, other Hispanic mosaics show boats, but these cannot be specifically identified as fishing or transport vessels. This is the case with the bichrome mosaic at Mataró (near Barcelona), dating from the end of the second or the beginning of the third century, that paved a large semicircular room surrounded by a bench; only a few fragments with a figured scene have been preserved, showing a vessel whose prow has the shape of a bird's head (Barral i Altet, 1978, no. 118, pl. LXXIII). In a mosaic fragment from Milreu preserved at the National Archaeological Museum in Lisbon there is a rowboat, probably related to fishing or trade in salted goods (e.g., *garum*) in the area since the archaeological site in question has provided many mosaics, most of which are conserved *in situ*, decorated with fish (Hauschild, 1994, 285-291, pl. CLXXXVII-CXC; Dos Santos Duran Kremer, 1999, 509-519; Teichner, 1997, 106-162) (figure 2). Another vessel was represented in a mosaic from Fuente Álamo (Puente Genil, Córdoba) that is now lost and known only from a drawing preserved at the Royal Academy of History (figure 3). The details of its dis-

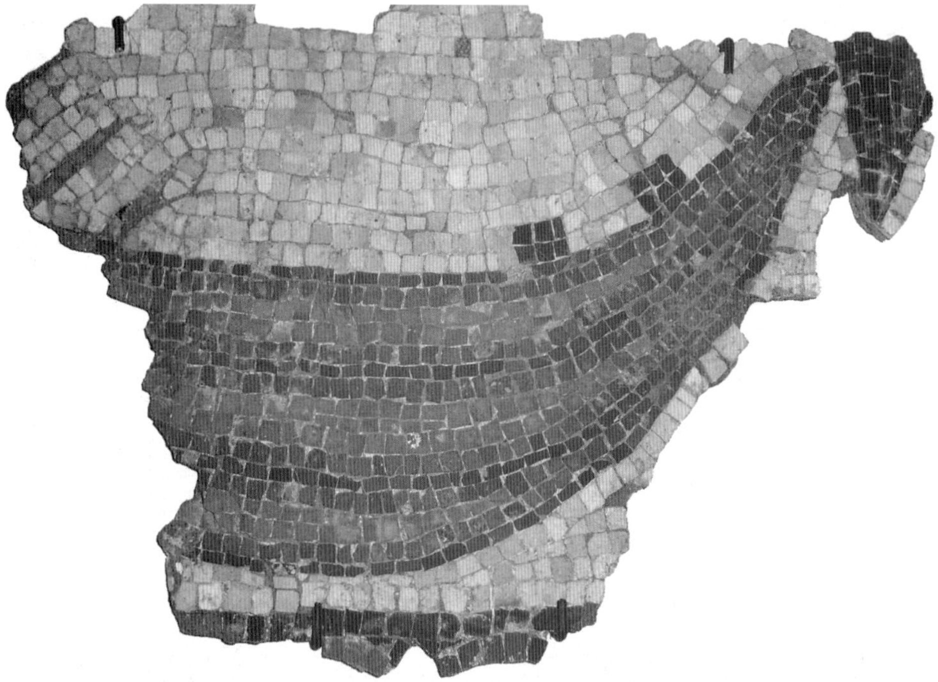

Figure 2. Mosaic fragment from Milreu. Lisbon, National Archaeological Museum (photo G. López Monteagudo).

Figure 3. Lost mosaic from Fuente Álamo (Puente Genil, Córdoba). Drawing preserved at the Royal Academy of History.

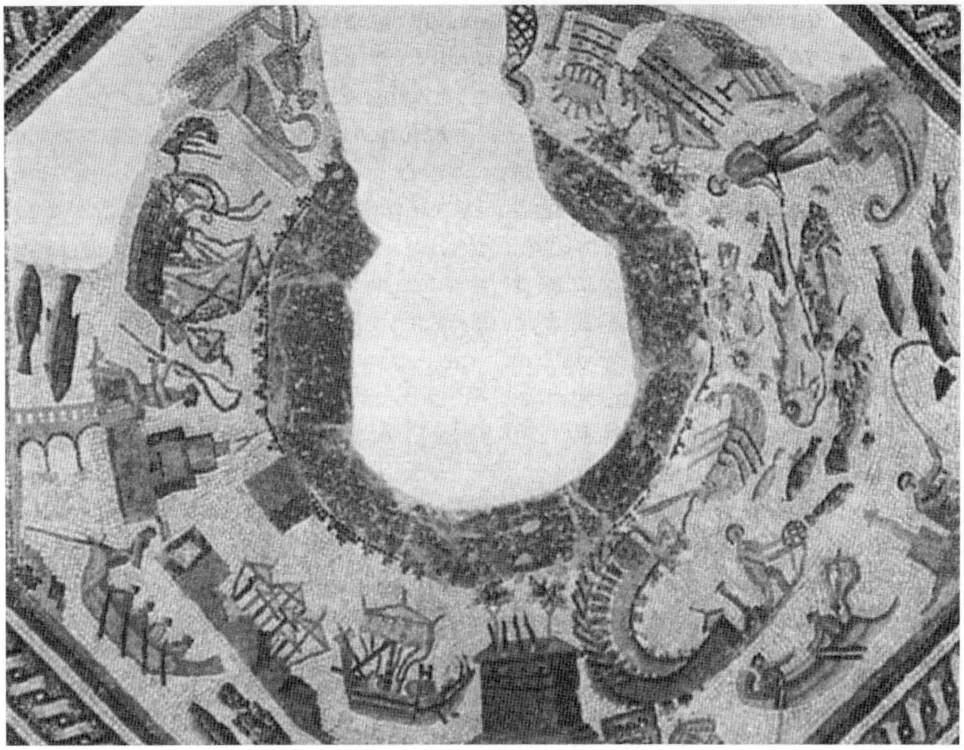

Figure 4. Mosaic from the Roman villa "La Vega Baja de Toledo". Toledo, Archaeological Museum. Probably fishing scene with an enclosure-net (photo G. López Monteagudo).

covery, which occurred in the second half of the nineteenth century, are unknown, although in view of the theme it can in all probability be related to the bath complex of the Imperial period, antedating the construction of the fourth-century *villa*. The depiction is one of mythological marine figures (a triton, a sea-horse, perhaps ridden by a nereid) and the sail of a vessel (López Palomo, 2007, 145-156, pl. I). It is possible that a fragment of mosaic with the representation of a fish, preserved at the Museum of Puente Genil, formed part of this pavement.

Outside *Hispania*, the theme is found on a Libyan mosaic in *Tripolitania* (third century) showing several fishing episodes in a sea populated by real and fantastic fauna, in which several fishermen can be seen. Two of them are aboard a vessel whose sails are swollen by the wind, one of them handling the steering-oar and the sail while the other is catching a large fish with a hook. Another two fishermen are in a launch rowed by one while the other is about to use his trident. Nearby, a further fisherman, standing on a reef with a rod, line and basket, has just caught a fish; the same mosaic depicts a basket with the catch (Aurigemma, 1960, 44, pl. 69-70).

In the most deteriorated part of the Toledo pavement, there are the remains of a fragmented scene of great interest, since it is possible to identify a fishing scene involving two vessels using a seine (figure 4). Scenes of the same type, showing seine fishing, can be found in other mosaics of the *pars occidentalis* of the Empire, such as that from the House of *Scorpianus* in Carthage (second quarter of the second century), now in the Museum of Carthage, in which different types of fishing gear are represented (net, hook and trident, or harpoon). In the background, lacustrine vegetation is shown, with two water fowl; at the sides, we see one man with a fishing rod and another spearing an octopus with a harpoon; in the centre, a seine fishing scene involving two vessels. The net, which is full of fish, has been set by two fishermen in two boats, each propelled by two oarsmen (Gauckler, 1910, 819). In the similar third-century Venus mosaic from Sousse the figures are *erotes* (Inv. 57.025; Foucher, 1961, pl. III) as in the previously mentioned fourth-century mosaic from Piazza Armerina, in which the different types of fishing gear are also combined (Carandini *et alii,* 1982, 249-58, figures 149-52, 154-155, pl. XXXVII, 79). Another scene showing seine fishing, this time performed by *putti*, can be found in the mosaic of the Triumph of the Sea Venus at Djemila (fourth-fifth century), Of special interest in this mosaic is the scene depicting a seine of the tuna-fishing type: a large net has been cast from two vessels, each holding three naked fishermen, apparently *putti.* The net is almost full of fish, which can be seen through the mesh; on each side, two people are pulling on the rope passing along the upper edge of the net that serves to close it (Blanchard-Lemée, 1975, 65-69, pl. XI).

In the mosaic of Vega Baja de Toledo, mentioned earlier, there are four fishermen using rods; two are sitting down fishing, one of them on the quay, next to the lighthouse, is just about to cast, the other is sitting down next to a light-house or a column; a third is on a reef next to a boat, hauling a large fish out of the water with his line. Yet another is standing on a reef crowned by a naval trophy. In Roman mosaics of all periods, the figure of the sitting fisherman is repeated over and over, following the model that had already appeared in a Greek red-figure *pelike* by the Painter of Pan (fifth century BC) found at Cerveteri and now in the Kunsthistorisches Museum in Vienna.

In North African mosaic art, to cite but a few examples, we see such fishermen sitting on the quay and often wearing a straw hat in the Tunisian mosaics conserved at the Bardo Museum, from the House of Fishing at Althiburos, end of the third or beginning of the fourth century (Yacoub, 1995, 236-38, figure 120 a-b); from the House of Neptune at *Thuburbo Maius*, of the same period, showing a fisherman wearing some kind of headgear, using a rod and line to fish from a rock (CMT II/1, Tunis 1980, 158-161, figure 9, pl. LXIV-V and LXVII); in a Libyan mosaic, (late second or early third century) from the Villa of the Nile in *Leptis Magna* there is a scene of intense fishing activity. Among and between the reefs, we see two fishermen with rods and lines. One is sitting down baiting his hook, taking the bait from a small basket resting on a rock to his left. Next to him, another fisherman, dressed similarly and with his head protected by a broad felt hat (*petasos*), has just caught a large fish with the rod he is holding in his left hand and is

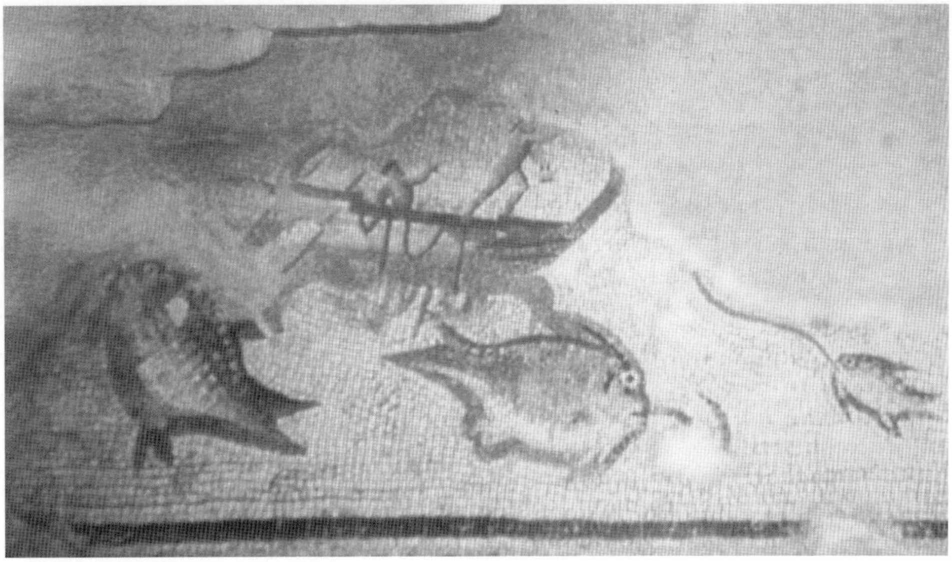

Figure 5. Mosaic from the Roman port of Hispalis, Seville. In situ (photo G. López Monteagudo).

now trying to transfer it to a scoop net that he is holding in his right. In the background, a third fisherman, in the water, is pulling on a net, the floats on the upper edge clearly visible. Another three naked fishermen are struggling to pull a net full of fish on to the rocky shore (Aurigemma, 1960, 49, pl. 94-97).

Similar scenes, showing fishermen sitting or standing, fishing with rod and line from the shore, are repeated in two Algerian mosaics of the Triumph of the Sea Venus (Ferdi, 1998, 100-1, 108). One of the third or fourth century, was found at Khenchela and is now at the Museum of Constantine: in it, we see the typical anglers, one of them old and the other young, wearing *petasoi*, standing on reefs in a sea teeming with fish, molluscs and a lobster. The old man seems to be resting, the young man is watching him while casting with the road and holding in his left hand a basket filled with their catch. The other mosaic paved the *frigidarium* in the roman Baths of Sétif and is dated to the fifth century. This time the two fishermen have cast lures, one of which has been taken by a bream-like fish *(dorada)*, while the other one is apparently being shaken by the fisherman who has the basket in his left hand. The same iconic type of a sitting fisherman with rod in hand appears in a mosaic in the House of the Desultor at *Volubilis*, conserved *in situ*, where the legs of the fisherman can be seen through the water (Limane *et alii*, 1998, 38).

Fishing with rods was also done from boats, where the fishermen are portrayed either sitting or standing, sometimes with a hat for protection. This type of rod fishing from a vessel is documented in two Hispanic mosaics from Sevilla *(Hispalis)* and Balazote (Albacete, figure 6).

Of the marine mosaic discovered in the zone of the Roman port of *Hispalis*, dated to the second or third century, only a fragment remains. This represents a scene

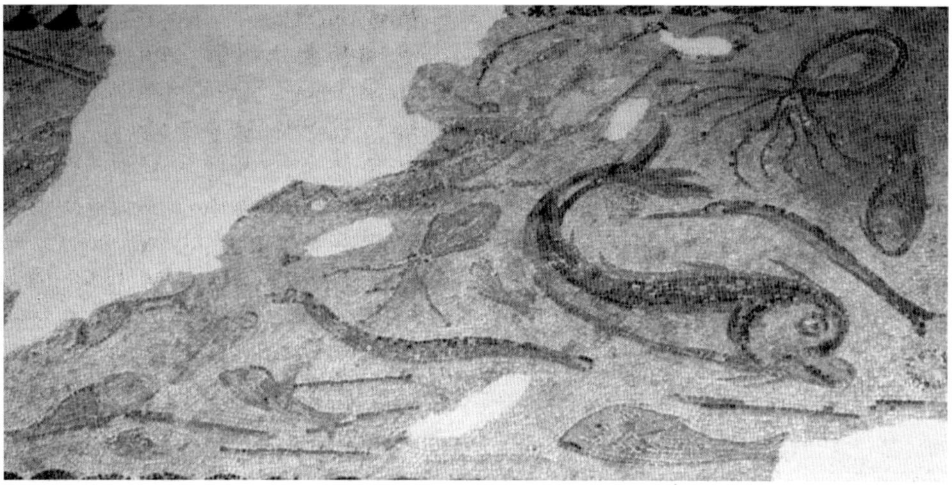

Figure 6. Mosaic from the Roman villa "Camino Viejo de las Sepulturas", Balazote (Albacete). Albacete, Archaeological Museum (photo G. López Monteagudo).

with two fishermen in a boat, one of them sitting and who seems to have cast the nets, and the other, with a brimmed hat (*petasos*), probably of straw, is standing up and holding a basket in his left hand. He has caught a large fish (López Monteagudo, 2006, 238-239, pl. VII, 2) (figure 5). The pavement formed part of the *thermae* discovered when the Los Seises Hotel was being built in the neighbourhood of the Cathedral of Seville, at the site where the Roman port must have been, judging by the recent find of important architectural structures indicating the existence of a large building of the early Imperial period, along with remains of wooden building elements, pavements, decorated stucco, and plaques of marble, as well as later constructions from the Severan period, when the area was given over to industrial activity. These finds fit well with a public setting of commerce and goods storage in the port area of *Hispalis* and point to the vitality of the place between the end of the first century BC and the beginning of the third century AD (Ordóñez Agullá, 2003, 59-79). The large number of inscriptions referring to *scapharii, navicularii, diffusores oleum, corpus oleariorum, negotiantes olearii*, etc. in the zone harbouring the Cathedral of Seville led A. Blanco to consider that this place probably hosted a "Forum of Corporations" or harbour *forum*, similar to that at Ostia (Blanco, 1972, 3-22; Blanco, 1979, 133-35).

In the pavement from Room XXXIII in the roman Baths of the "Camino Viejo de las Sepulturas *villa*" at Balazote (Albacete), third to fourth century AD, there is a large, although deteriorated, fishing scene. Fishing is taking place from two boats in a sea containing a large number of marine species, among which we can identify a moray, a dolphin, an eel, an octopus, a squid and a sea urchin (figure 6). Also preserved is a part of a vessel, located to the centre-right of the composition, from which someone is angling (this figure has not been preserved, but judging by the

Figure 7. Mosaic from the Roman villa "La Vega Baja de Toledo". Toledo, Archaeological Museum. Fisherman casting a large lobster pot (photo G. López Monteagudo).

position of the line, the person must have been sitting in the boat) and perhaps part of another vessel that would have occupied the centre of the lower part. The water is represented by short lines, some continuous and others broken up with black *tessellae* (CMRE VIII, 1989, no. 31, pl. 12, 23-25).

Similar scenes, in which fishermen are angling with rod and hook, either standing or sitting, with or without *petasoi*, are seen in several North African mosaics, sometimes combining different fishing techniques in the same manner as the mosaic from Seville. Among these are the Tunisian mosaic of Orpheus and Arion from La Chebba conserved at the Bardo Museum (Tunis), first half of the third century (Dunbabin, 1978, 254) and the above-mentioned pavement in the triapsidal room of the House of the Chariot of Venus, also at the Bardo, from *Thuburbo Maius* (above, p. 162). The same kind of fishing, but performed by *erotes*, appears in the Late Imperial Italian mosaics at Piazza Armerina (above) and *Aquileia* (below). Many boat fishermen wear a hat, almost certainly made of straw, such as the man in a *vegeia* or *placida* boat in the only fragment, now preserved at the Sousse museum, of an apsidal mosaic discovered in the excavations of a house in *Hadrumetum* in 1952, decorated with a fishing scene from the end of the Severan period showing four fishermen in boats (Inv. no. 57.261; Foucher, 1961, pl. LXII a).

On other occasions, it is the fishermen on the shore who are wearing hats for protection. Among others, we find them on one of the panels depicting a fishing

Figure 8. Mosaic from the Roman villa "La Vega Baja de Toledo". Toledo, Archaeological Museum. Swimmers and ostriarium (photo G. López Monteagudo).

scene accompanying the above-mentioned Tunisian mosaic of Dionysius and Ulysses (found at Dougga and now in the Bardo Museum) where it is clearly possible to see the straw used to make the hat. The fisherman is using a rod, which he holds in his right hand, while the left-hand one holds a lobster pot, and at his side is the basket for his catch. In a mosaic from Khenchela, third-fourth century, preserved in the Museum of Constantine, the anglers, one standing and the other sitting on some rocks, are wearing a curious pointed hat, in all probability made of straw, that resembles the shells of certain molluscs (Ferdi, 1998, 143-44).

The Toledo pavement shows another way of fishing: using a fish trap (*nassa*) or lobster pot. We see a fisherman about to cast a large lobster pot into the water from a small lighthouse or column situated on a stepped base at the entrance to the port (figure 7). These also appear in some North African mosaics – following same model as a red-figure Greek kylix in the Museum of Gela – for instance, those appearing on the panel at Dougga (above), on the mosaic portraying the *erotes* at Sousse and that represented on a mosaic fragment in the Bardo Museum, in which a fisherman has just caught a very large lobster (Donati & Pasini, 1997, 61).

In another of the maritime scenes of the Toledo mosaic, two people are swimming in the sea, the water being indicated by bluish lines (figure 8). In our opinion, these characters, depicted with the iconography elsewhere used to represent the River

Figure 9. Nilotic border of the Neptune mosaic from Itálica (Seville). Pygmy fishing with a rod and line. In situ (photo A. Blanco-J.M. Luzón).

Orontes (LIMC I, s.v. "Antiocheia"; LIMC III, s.v. "Axios"; LIMC IV, s.v. "Fluvii"), should be interpreted as sponge- and oyster-divers, a practice documented in literary sources. Their use must have been common, as seen from a famous passage in the *Iliad* (18.414), in which the smith-god Hephaistos is cleaning soot off himself with a sponge before welcoming the sea-goddess Tethys. During such immersions, oysters were gathered, as we also learn from the *Iliad* (16.745-748). Both sponges and oysters are represented in the Hispano-Roman mosaics among the varieties populating the marine depths – especially oysters, which appear in nearly all fish mosaics from the third and fourth century, above all from Galicia, the Portuguese littoral and around Milreu on the Algarve coast. However, oysters are also found on pavements in the interior of the Peninsula, for example at Noheda (Cuenca). Oyster consumption is documented, too, in Hispanic *xenia* mosaics from the first century in Marbella (Málaga) (CMRE III, no. 55, figure 23, pl. 62); Campo de Villavidel (León), fourth century (CMRE X, no. 5, pl. 4, pl. 25) and Vega de Ciego (Asturias), fifth century (CMRE X, no. 32, pl. 20). Outside *Hispania*, oysters are shown being consumed, together with the *garum* contained in amphorae, in the mosaic showing three port scenes and the head of *Oceanus*, paving the apsidal *triclinium* of the third-century Roman *villa* at Bad Kreuznach in the *limes* zone (López Monteagudo, 2006, 262-63, pl. XVIII, 5; López Monteagudo, 2008, 2547-2568).

Within the context of fishing, figures swimming in water are seen in Roman mosaic art from an early date (first to fifth centuries AD). In the *caldarium* of the baths at the House of Menander in Pompeii, there is a very interesting almost bichrome pavement, dated to the end of the first century BC, representing a swimmer and another person fishing with a trident (Maiuri, 1933, 146, figures 68-70). Swimmers also appear alongside dolphins, together with both realistic and fantastic fauna, in other bichrome mosaics on the Italic peninsula, such as that from Este, (first half of the first century AD), several pavements of marine *thiasos* at Ostia, the mosaic of the Baths of the Cisiari in Ostia (120 AD) and the Triumph of Neptune mosaic at Ostia (mid-third century) (Becatti, 1961, 26-27, 173-174, pl. CLXI-CLXII, CLXIV-CLXV). North African mosaic art also shows swimmers in the polychrome Nilotic mosaic from El Alia (above, p. 163), in this case a male and female swimmer; in the bichrome pavement of the Museum at Constantina (Algeria), second or third century (Ferdi, 1998, 175); in the above-mentioned Tunisian polychrome mosaics of Dionysios and the pirates of the Tyrrhenian Sea from Dougga, and finally from Sidi Abdallah with the portrayal of a coastal scene showing some buildings and a shore. Apart from the fishermen it is possible to see people swimming, one of them being swallowed by a large fish and others jumping into the sea from the cliffs. The image of the person being swallowed by a fish in the mosaic of Sidi Abdallah was used in Christian mosaic art to represent the biblical episode of Jonah being swallowed by a whale, twice represented on the pavement of the Basilica of Aquileia (Marcuzzi, 1993, 20-22).

To the upper left of the swimmers of Vega Baja de Toledo (figure 8), a quadrangular object can be seen. This is identified as an *ostriarium*, a floating oyster farm (López Monteagudo, 1993, 1251-1257, figures 3-4) of the type of those represented and identified by their name on the Puteolanean glass vessels produced at the end of the third and beginning of the fourth century and found at Populonia, Rome, Odemira, Ampurias and Prague, where the inscription *ostriaria* appearing over the installations leaves no doubt as to their use as oyster farms (García y Bellido, 1954, 212-226, figures 1-5). Real *cetareae* or *vivaria*, some of them devoted to the production of the famous *garum*, are documented in coastal towns of *Hispania*, such as Vilamoura, in the Portuguese Algarve, and also at *Baelo Claudia*, in the province of Cádiz (Arévalo & Bernal, 2007). Oyster nurseries and fish-ponds date back to the late second century BC, and are referred to by Pliny the Elder, Cicero and other authors. It seems to have been *Sergius Orata*, the owner of a *villa* at *Baiae* – whose *cognomen*, according to Varro, was taken from his favourite fish, the goldbream (*Sparus aurata*) – who began building *vivaria* around 108 BC, using the coastal lagoons of the Phlegrean Fields near Naples, in particular the Lucrine lake. Several politicians, among them *Licinius Murena* (whose cognomen was also related to a fish species) are known to have been involved in fish farming; Cicero maliciously referred to them as *piscinarii* (*Ad Att.* 1.19.6; 1.20.3).

However, the Romans of the *pars occidentalis* did not only know how to breed oysters, their favourite mollusc, and other fish species but also how to preserve the fish. In general this refers to exotic species brought from other part of the Empire, which were kept in fishponds to ensure their availability regardless of the season of the year good for fishing, since it is known that during winter navigation and fishing activities were suspended. It was therefore crucial to keep the fish alive so that they would always be available fresh, although they were also kept in brine as another type of preserve. According to Pliny the Elder (*N.H.* 9.171), the first person to organise a *vivarium* for morays was Gaius Lucilius Hirrus, a gentleman from Campania. His success was so great that he was able to provide six thousand of these for the public banquets organised by Caesar to celebrate his triumph in 45 BC. And although the golden age of the *vivaria* was between the end of the first century BC and the end of the first century AD – which is when fish figured among the most important elements of banquets, what Tacitus called the *luxus mensae* (*Ann.* 3.55) – fishponds continued to be used throughout the Empire until the start of the fifth century AD. They were built on rocky coastal zones; that is, protected but at the same time open to the sea currents so that the water would be replenished, as is done today. They were open pools, divided into sectors to separate the different species or different developmental stages of the fauna, and they were equipped with niches, sand, rocks and algae. In *Hispania*, there are marine examples of this type at Campello (Alicante). The first record of the existence of *vivaria* dates back to the third century BC (Varro 3.10; Columella 8.16) and Plautus was the first to use the technical name of *piscinae* to designate the *vivaria piscium* in lagoons, ponds and coastal lakes.

Live fish served not only to bring life to the ponds and pools of gardens; fish were also kept in tanks close to the *triclinia* or in large vessels, such as those found at Pompeii and its surroundings, decorated with fish motifs. On the Iberian peninsula, a tank or pool covered with *signinum* is conserved at the third-century House of the Jets in Conímbriga (Correia *et alii*, 1942, 115; Bairrão Oleiro, 1992, 19-20); this surrounds the head of the *oecus-triclinium* on its eastern side and has been interpreted by excavators as a type of fishpond, with the peculiar feature of amphorae built into its walls, used to keep the fish in while the pool was being cleaned (Correia *et alii*, 1942, 115; Bairrão Oleiro, 1992, 19-20). A similar arrangement is documented at a Gallo-Roman *villa* at Vaison-la-Romaine (Roland, 1958, 406-412). At the third-century Roman *villa* in Pisões near Beja, the floor of the basin in the *triclinium* – probably another fishpond and preserved *in situ* – is covered by a polychrome mosaic featuring fish, including a moray eel, a sea-bream, and a squid (Vargas Costa, 1985, 95-135, figure 9B and 14).

In these *piscinae* and *vivaria* – both fresh and saltwater – the live fish were "caught" directly by the owner of the house and his guests recreationally so that they could test their skills with the hook and line. This is because fishing, like hunting,

Figure 10. Neptune mosaic from Itálica (Seville). In situ (photo Junta de Andalucía).

was the remit of the important landowners who would enjoy catching fish either with a hook and line or with a lobster pot, as is done today with the fish and crustaceans in nurseries.

In Spain as in the rest of the Empire, some representations of fishing have a mythological content and those portrayed are not real persons, but *erotes*, *putti* and pygmies, fishing with nets or hooks. Among the older mosaics, two Nilotic bichrome examples from the second century AD are outstanding; they were found in Itálica (Santiponce) and both are connected with hydraulic installations (Blanco & Luzón, 1974, pl. IV, XII, XVIIB, XVIIIA, XXA, XXIA; García y Bellido, 1979, 94-102, 132, figure 34, pl. XI). In the border of the Neptune mosaic, which paves the main hall of a domestic suite of baths and features a typical Alexandrine landscape, we see a pygmy (figure 9) fishing for eel or lamprey with a rod and line next to a typical fisherman's hut, similar to that portrayed in the Vega Baja de Toledo mosaic; he is wearing a hat and is holding a basket on his left arm; another two pygmies, mounted on two large aquatic birds, are harpooning a crocodile (figure 10). Copied from or inspired by the figures of the previous one, the Nilotic mosaic that paved a small fountain in the House of the Exedra, also preserved *in situ*, portrays a scene with pygmies, cranes and fish in typically nilotic surroundings, all around a large squid at the centre of the composition; a pygmy with a basket in his left hand is sitting fishing with a rod and

Figure 11. Nilotic mosaic from the House of the Exedra at Itálica. In situ (photo G. López Monteagudo).

hook; another one astride a crane is harpooning a dolphin (figure 11). Outside *Hispania*, the same theme of pygmies fishing is especially popular in mosaics from North Africa (Foucher, 1965, 137-146).

In the polychromatic pavement of the so-called House of Hippolytus in *Complutum* (Alcalá de Henares, Madrid), coming from what has been interpreted as the *frigidarium* of the *balneum* of the sumptuous residence of the family of the *Annii* and dated to the end of the third or beginning of the fourth century, there are three *putti* fishing with a large net from a *vegeia-* or *placida*-type vessel (figure 12). The vessel is at the centre of an aquatic scene (the water is indicated by straight, stepped, and zigzag lines) populated with numerous marine species, among which are molluscs, crustaceans, sea urchins, up to a total of 24; ten tunas, a swordfish, a grouper, three eels or morays, a shrimp or lobster, a sea urchin, two cuttlefish, two octopodes, while a dolphin, a lobster, and two tunas are trying to escape from the net (Rascón *et alii*, 1995-1997, 39-62; García Entero, 2004, 143-158, figures 4, 5 and 13; Rascón, 2007, 119-152).

Realistic scenes of net fishing being carried out by adults are very frequent in mosaic art, above all in North Africa and the Italic peninsula, and are always, as in *Complutum*, characterized by a sea crowded with marine species. Some examples have already been described, such as the mosaic from the triapsidal hall of the House of the Chariot of Venus at *Thuburbo Maius*, now in the Bardo Museum (above, p. 162). In this mosaic, there are several fishermen in a boat, one of them rowing and the others drawing a net full of fish from a sea teeming with ichthyic fauna, as at *Complutum*, although the closest parallel is probably the one at Sidi Abdallah, mentioned above, portraying a marine scene at whose centre there are four fishermen casting their net from a boat.

Figure 12. Mosaic from the House of Hippolytus, Complutum, Alcalá de Henares. In situ (photo G. López Monteagudo).

Outside *Hispania*, *erotes* and *putti* fishing with the different types of fishing gear described by ancient sources, generally in an aquatic environment filled with fish and other marine fauna, are very numerous, especially in North Africa and Italy; some of them have already been mentioned. In this connection, let us recall the mosaics of a pool in Sousse; the border of the Venus mosaic from Djemila; in Italy, the mosaics from Piazza Armerina and those of the Roman *villa* in Desenzano, also of the late Imperial period, with *erotes* fishing (Scagliarini Corlaita, 1992, 57, figures 28, 37-38). The same motifs persist in the art of the Paleo-Christian basilicas, such as that in Aquileia (Marcuzzi, 1993, 21, 23).

Recently, an extraordinary find has been made in the Roman *villa* at Noheda (Cuenca). This is a very large polychromatic pavement of 300m^2, which originally covered the floor of a triapsidal room, in which glass and gold *tessellae* were used. This has been dated to the end of the fourth century. The pavement, which is still being excavated, comprises three zones; the upper two are decorated with mythological scenes while the lower features fishing and pugilistic themes, an almost certain indication that this was the floor of a bath (Sarmiento, 2007, 56-62). In the upper panel, a Bacchic triumph is portrayed. The god is shown in the frontal position, drawn by two pairs of centaurs playing musical instruments; the central part

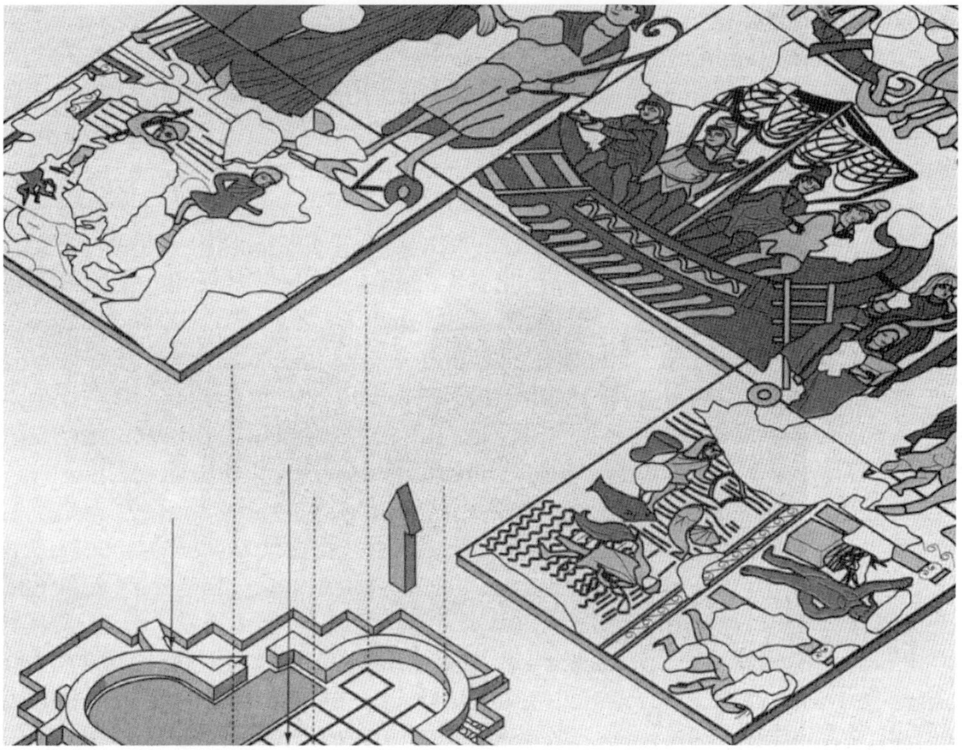

Figure 13. Fishing scenes of the Roman mosaic from Noheda (Cuenca). In situ (M. Sarmiento).

is decorated with two scenes, difficult to interpret, of a boat with the sails unfurled. The lower part comprises several panels, one depicting athletic games and the others containing different representations of fishing carried out by *putti* standing on a reef, each of them using a different type of fishing gear in a sea full of different marine species (figure 13). To one side, a *putto* is harpooning an octopus, while the other one must be casting a rod and line or a net. On the other side, a character seems to be using a round or conical casting-net, while the fourth, wearing a straw hat, in his right hand holds a large fish that he has caught with his rod, which he is holding in his left hand with the line and hook furled around it (figure 14). It is likely that in the central zone, which has still not been uncovered there will be an enclosure-net fishing scene.

The interplay between real and mythological themes and the combination of different types of fishing gear, seen in the Noheda mosaic, is common in North African mosaics, such as the above-mentioned Tunisian pavement of the House of Ulysses at Dougga, where the mythological episodes of Ulysses and the Sirens, and Dionysius and the pirates of the Tyrrhenian Sea, are intermingled with real-life fishing scenes. However, the closest parallels for the fishing scenes of the Noheda mosaic can be found in some Algerian pavements of the same period from Kamissa and Djemila, in which fishing scenes are likewise inserted into mythological themes, in particular those of the Triumph of the Sea Venus, suggesting a relationship between the Goddess and the economic activities of fishing.

The fishing scenes from Kamissa, preserved at the Museum of Guelma, form part of a mosaic with the theme of the Triumph of the Sea Venus, dated to the third-fourth centuries AD. These are of great interest because they show the use of different types of fishing gear by young, blond, naked fishermen in a sea brimming with all types of fish: one has just caught (with a rod) a large fish; another is harpooning a squid with a trident and holds a lobster pot in his left hand; the third one, standing up on the goddess' shell, is about to cast a round net. The last one, resting on a rock, is hauling in a net full of fish that he has closed by pulling on the drawstring (Lassus, 1965, 175-192, figure 4; Ferdi, 1998, 140-141).

Similar scenes are found in the border of the above-mentioned representation of the Triumph of the Sea Venus from Djemila. A fisherman is about to draw a net full of fish into his boat, which is close to capsizing; nearby, another fisherman, standing on a rock, is using a harpoon to spear a large fish, out of which streams of blood gush very realistically; in other scenes, *putti* are angling with rods and lines from a boat and from a reef.

Finally, Hispano-Roman mosaic art shows fishing with the different types of fishing gear used at the time and also, interestingly, the result of the consumption of fish. A fish skeleton is represented in the top right corner of the lower panel of the mosaic of Venus and Erotes in the House of the Amphitheatre at Mérida (*Emerita*), with a third-century scene of grape picking – the result of a restoration carried out in antiquity (CMRE I, 1978, no. 39, pl. 72-74)[2] (figure 15). In terms of imagery, the fish skeleton is comparable to those appearing in the so-called "unswept room" (*asarotos oikos*) mosaics, of which a good example is the *Heraclitus* mosaic conserved in the Vatican Museums (Andreae, 2003, 46-51, figure 49), a copy of the mosaic of Sosos in Pergamon mentioned by Pliny (*N.H.* 26.184). Fish skeletons also appear in a mosaic from *Aquileia* of the first century BC (Cuscito, 1989, figure 54), in the Tunisian *asarotos oíkos* pavements from Oudna, towards the end of the first century or the beginning of the second (Yacoub, 1995, 99-100, figure 38), and in the House of Months at El Djem from the first half third century, contemporary with the Mérida mosaic (Foucher, 1961, 291-97, pl. XI-XVIII).

Through the images provided by Roman mosaics it may be observed that in Roman art fishing did not have same the aristocratic connotations as hunting, enjoyed by kings and the privileged classes, rich urban *villa* owners and wealthy farmers – even though we know from the sources available that these also fished for sport and exer-

2 I am grateful to Ms Mary Paz Pérez, a restorer at the Institute of Archaeology of the Consortium of the City of Mérida, for these observations and the photo.

Figure 14. Mosaic from Noheda (Cuenca). Putto angling with a rod. In situ (photo M. Sarmiento).

Figure 15. Venus mosaic from the House of the Amphitheatre at Augusta Emerita, Mérida. Fish skeleton. In situ (photo M.P. Pérez).

cise. However, the fishing images that appear in Roman mosaics do not represent a recreational activity confined to members of the elite, who in hunting scenes are shown ostentatiously attired, mounted on richly bridled horses; instead, those using the different types of fishing gear are clearly no more than fishermen, as we see from their clothes. They were humble people, dressed in short tunics and wearing *petasoi* or conical hats, to protect themselves from the sun. Despite all this, the large number of fishing scenes and marine fauna represented in the mosaics, together with the amount and variety of fish depicted; the fact that they use mythological persons or episodes (Oceanus, Tethys, Venus, Europe, marine thiasos, etc) in order to introduce scenes of fishing from daily life, everyday images, realistic scenes, along with unreal figures such as *erotes*, *putti*, and pygmies fishing. The pictures of *xenia*, in which images of fish and oysters are shown together with animals and products of the land, the processing of marine products and their commercialization indicate the huge importance of fishing in Roman culture and in the diet of classical times, thereby highlighting one of the key sectors of the Roman economy.

Regarding *Hispania*, although mosaics devoted to fishing are not very widespread in Roman mosaic art, especially when compared with North Africa, the large number of mosaics related to fishing itself (even though on occasions the persons fishing are *erotes*, *putti*, or pygmies, and not real people) and also the many mosaics depicting fish, either as a main or secondary motif, in realistic scenes, as motifs of *xenia* or in mythological contexts, is a clue suggesting that we should not overlook the importance of fishing and the consumption of fish and sea food in Roman *Hispania* and also confirming that the Hispanic pavements formed part of the artistic *koinê* comprising Roman mosaic art throughout the Empire, both in the *pars occidentalis* and the *pars orientalis*.

Mosaic from the "Maison de la Cascade", Utica (detail).

6. Fishing in the Roman World

Tønnes Bekker-Nielsen

According to Aristotle's famous dictum, man is an "animal that lives in cities" (Pol. 1.2 [1253a]), and animals need to eat. Eating is a fundamental body function, and if it ceases, all other corporal functions will eventually cease as well. Humans form part of the same food chain as small fish, big fish, and marine mammals (figure 1). But catching and eating fish is not merely a question of nutrition, of finding something to eat. In a coastal environment, there are other, simpler, ways of obtaining animal protein, e.g., by collecting shellfish or the eggs of seabirds. Only in exceptional environments, such as subarctic hunter-gatherer communities or among shipwrecked sailors on a small island, will humans attempt to survive on an exclusively marine diet.

In most societies, fish form a supplement to other sources of nutrition such as grain, vegetables, meat etc., and the place of fish in the diet, and hence of fishing and fish processing in the economy, is culturally defined. Far from being a simple two-way interaction between the old man and the sea, fishing takes place within a wide-ranging and complex pattern of interaction involving social, economic, cultural, biological and environmental aspects among which organization and technology, the topics of this workshop, are only two. But even if we reduce the focus to these two aspects it still includes a number of issues that are common to both, for instance the problem of manpower. Thus we cannot discuss fishing gear without considering fishing organization, or vice versa.

Fortunately, we have a range of sources at our disposal, ranging across many genres and periods and giving a varied and colourful, but sometimes contradictory impression of the way Romans fished. Furthermore, the interpretation of these sources rests on a series of *a priori* assumptions about the nature of Roman society, economy, law and mentality. Thus a researcher needs to be familiar with the problems and pitfalls of many different sub-types of historical sources, and be ready at each step to question the apparent truths put forward by previous researchers in the field.

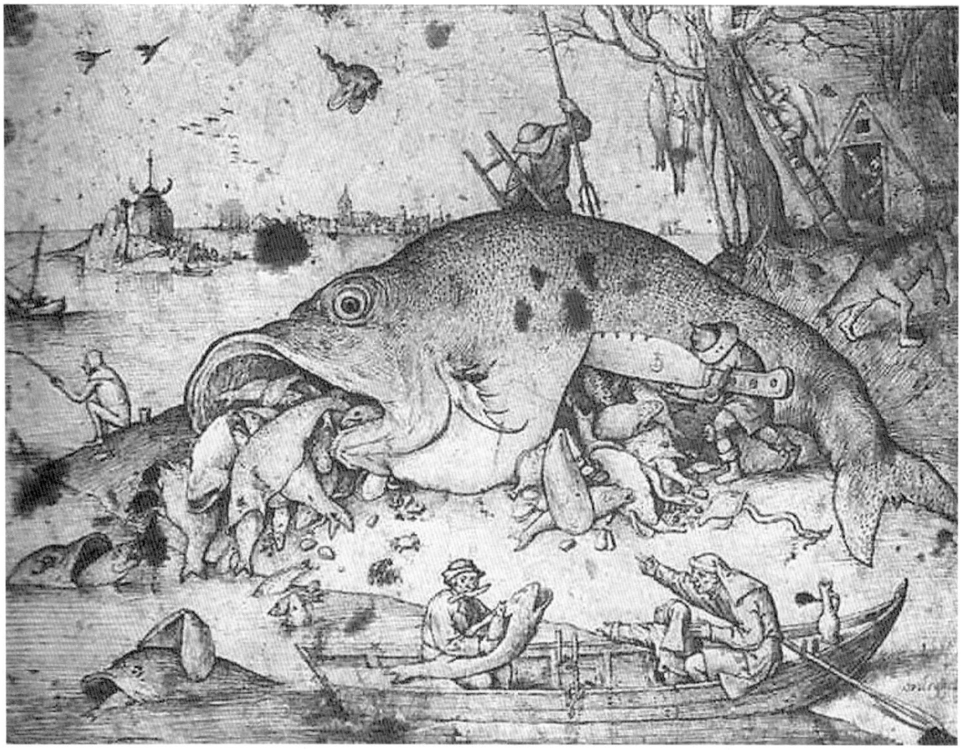

Figure 1. Food and eating were favourite themes of the Flemish artist Pieter Brueghel the Elder (d. 1569), often as an ironic commentary on society and the human condition, as in the ink drawing of Big Fish Eating Little Fish (1556) (Wikimedia).

Technology

Let us start with the question of fishing gear. Romans used a range of different types, adapted to different categories of fish (pelagic, benthic, migratory) in different environments (coastal, shallows, deep water, straits). On the well-known mosaic from Sousse (figure 2) we see four basic types in use: hook and line; seine; casting-net; traps. Traps, pots and creels are passive implements that require no constant manpower input, or as Oppian so poetically expresses it, "they work while their master sleeps". Other forms of fishing involve active implements, and some of these require the input of more than one person or supporting technology such as boats. We can subdivide the Roman fishing techniques that are known to us into nine main categories (see table 1).

The prime determinant is the nature of the waters being fished. In still, shallow water, the fisherman will have the choice of a wide range of passive or active implements; at greater depths in or more turbulent waters, his choices are restricted. Again, within each category, some implements are more suited to shallow water, others to deep water fishing.

manpower input	productive capacity: low → high			water
low	Baskets, creels and pots	Stationary nets	Traps in migration routes	shallow
↓	Spear, harpoon, trident	Casting-net from shore or boat	Beach seines, dragnets	↓
high	Hook and line	Seines worked from one boat	Seines worked from two boats	deep

Table 1. Roman fishing techniques.

As we move from the upper left to the lower right in the table, the required capital and manpower input increases, but so does the productive capacity. Although no catch data are available from antiquity, observations of modern fishermen using analogous techniques, e.g., the *almadraba* (García Vargas & Florido del Corral, this volume), indicate that production rises more rapidly than manpower, that is to say, that when combining the efforts of many fishermen in, for instance, fishing with a beach seine, the team will achieve a greater total haul than if each of them fished with his own casting-net. Indeed, if this were not the case, we would never see ancient fishermen joining up in teams to work seines requiring dozens of hands at once.

I stress this somewhat banal observation because it is sometimes claimed that ancient fishing was an activity of last resort for coast-dwellers to fall back on during periods of famine or poor harvests, and carried out only on a small scale. This might seem to find some support in the *Geography* of Strabo, who, in the opening chapter of the sixth book, relates that the inhabitants of *Elea* (Velia) on the Tyrrhenian coast "because of the extreme poverty of the soil mostly devote themselves to the pursuits of the sea, establishing fish-salteries (*taricheias*) and other businesses of that sort" (Strabo, *Geogr.* 6.1.1). However, it should be noted that Strabo's description of the Eleans' activities is not restricted to coastal fishing: he uses the verb *thalattourgein*, "to work the sea" or "to pursue occupations of the sea" and gives fish-salting as an example of business "of that sort", which could describe any activity from salt-extraction to piracy. Clearly, if they operated fish-salteries, the Eleans needed fish; thus they fished; but they also needed salt extracted from sea-water. On the Tyrrhenian coast, salines were operated over the summer, the evaporation taking place from late June or early July onwards (*cf.* Rutilius Namatianus, *De reditu suo,* 1.479-484). Operating a *taricheia* was integrated into an annual work cycle requiring advance planning and long-term investment in salines and salting-tanks; fish-salting was clearly *not* a short-term solution to food shortages and from Strabo's description, it would seem that the marine activities of the Eleians took place on a regular basis.

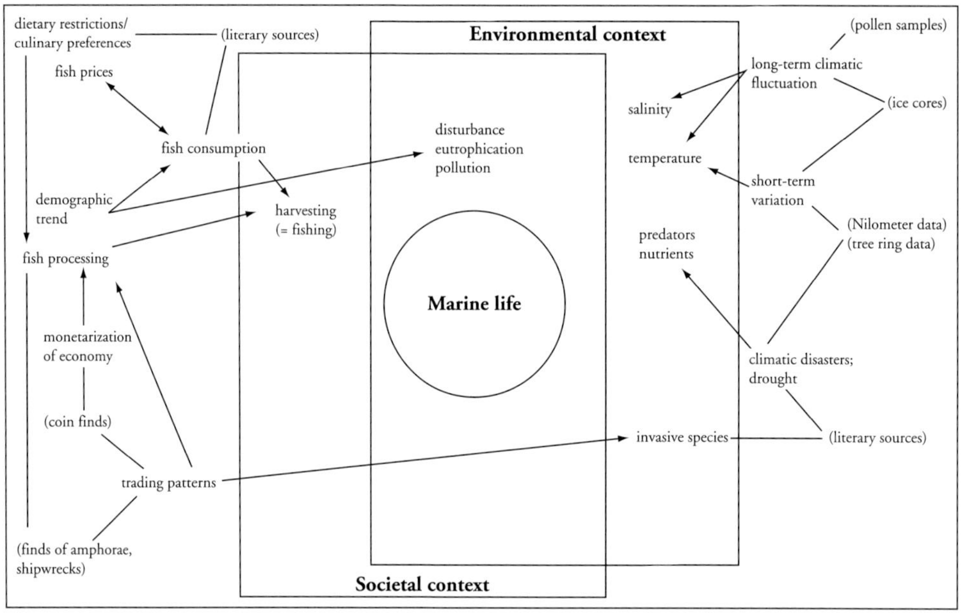

Table 2. Fishing in its wider societal and environmental context (Bekker-Nielsen, 2004).

The second point is that of scale. It is easy to be misled by our main category of pictorial evidence, the mosaics, which depict crews ranging in size from two to six persons; for instance, the mosaic from *Hadrumetum* (figure 2) shows two fishermen working a seine from their boat. This is, however, an obvious attempt by the artist to simplify his subject in a way that will permit it to be rendered effectively within the restricted area of the mosaic (and mosaic is not the easiest medium in which to depict fishing lines and net meshes!). Other images showing seine fishing in media that are not subject to the same limitations, e.g., reliefs in Egyptian tombs, show beach seines being hauled in by teams of four to nine persons at each end; in one case, no less than twenty-eight men are working together to draw a seine ashore (Sahrhage, 1998, 108). That this was also the case in the Roman period is attested by Aelian, who gives us this description of fishing with a beach seine off the coast of northern Asia Minor (*N.A.* 15.5):

> Each boat has six young men a side, stout rowers. The nets are stretched out; they are not light, held up by cork floats and weighted with lead; and the shoals of fish swim straight into these nets. (...) And the men fasten a very long rope to one of the posts of the watchtower, then row their boats in close formation, keeping near one another, since, as you can see, the net has to be distributed evenly between the boats. And the first boat lets go its section of net, then returns to shore; then the second, then the third; and the fourth lets its section out. But the rowers in the fifth boat wait, for they must not release the net yet. Then the others row in each direction and take in their section of the net, pausing from time to time.

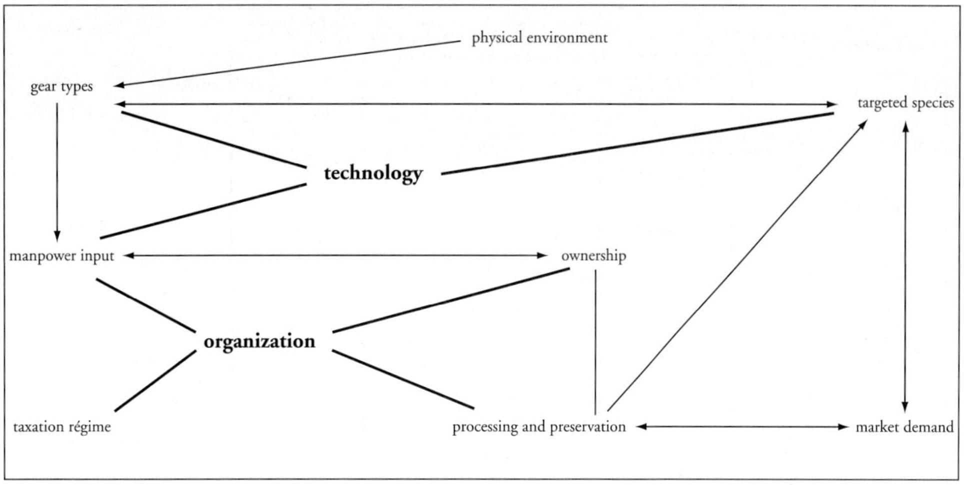

Table 3. Fishing organization and technology in context.

What Aelian describes here is clearly a major fishing operation, involving more than seventy persons and five large boats – not small fishermen's skiffs, but substantial, twelve-oared vessels. Aelian's picture is corroborated by a fishing guild record from Parion on the Hellespont, as we shall see below.

Of course, every fishing expedition did not involve seventy persons. There were individual fishermen, fishing either for food or for pleasure. Then as now, angling was the preferred fishing technique for recreational fishing, from the solitary fisherman casting from the shore to the "royal" fishing expeditions described by Oppian. Hook and line fishing can be an efficient commercial fishing technique, especially for larger fish. By using multiple hooks, the fisherman increases his chances of a good catch. This method was used in antiquity and described by Oppian; it is also attested by archaeological finds of large stocks of fishing hooks and sinkers.

For commercial fishing in general and especially for the smaller species, a casting-net (*amphiblêstron*) was more efficient. It is a simple technology that provides significant catches with a minimum of resource input. Its use is attested by literary sources and by finds of the characteristic pyramid-shaped or cylindrical net weights. The casting-net is weighted at the sides and sinks in the water, enveloping the fish before the net is closed with a drawstring and hauled into the boat or onto the shore. The various stages of its use – awaiting the cast, casting, and drawing in the net – are depicted on mosaics of the second to fourth centuries.

Observations of modern fishermen using the casting net from shore how that a skilful user targeting a shoal can take a large catch in a single throw of the net (figures 3-4). Photographs taken in Oman forty years ago show the casting-net being used from a boat (figures 5-6) in a manner closely corresponding to similar images from the Roman period (figure 2). One fisherman stands up when throwing the net, while another holds the boat on course.

Figure 2. Mosaic of the third century AD from Hadrumetum (Sousse) showing, clockwise from top: fishing with rod and line; seine; casting-net; creel.

The seine is a much larger net. It is not thrown from above, but let into the water where it hangs suspended between weights (at its lower edge) and floats (at the upper edge). In one mosaic (figure 7) we see the seine in a bird's-eye view, with the floats towards the centre of the picture. The upper edge of the seine is shown in a brown colour, with larger brown squares at intervals indicating the cork floats that give the net its buoyancy.

Laying a seine from a boat, then closing it by hauling at both ends simultaneously is technically simple but places a limit on the size of the seine, since a disproportionately large net will capsize the boat when the hauling process begins. Using two boats raises a new problem: unless the boats have been anchored beforehand, when hauling is commenced the boats will move towards the seine rather than vice versa. For this reason, a large seine is more easily worked from the beach. In that case, boats are used to lay the seine, which is pulled ashore by teams of fishers standing on the beach or in shallow water.

A fixed net – in Greek known as *peza* – minimizes manpower because the fish move towards the net, and not vice versa. It can be used for many species of fish but is especially effective as a trap for migrating species such as scombrids. For large species such as tuna, a round-the-clock watch is required: if the fish are left to their own devices af-

Figures 3-4. Net fishing from the shore using a casting-net, Oman, 2002 (photos Jørgen Christian Meyer, from Bekker-Nielsen, 2005).

ter being trapped in the net, they may damage it or escape – a problem also mentioned by Oppian (3.139-144; 775-782). The fixed net thus requires a large capital investment and constant manpower input, but offers the promise of very large catches.

Organization

This brings us to the second question, that of organization. How were capital and manpower marshalled for collective fishing projects? In situations where high capital investment and high risk were combined with high chances of gain – such as maritime trade or mining – the normal ancient response was to spread the risk by combining the resources of many persons into a guild (*collegium*) or a company of investors (*societas*). We are fortunate to possess the list of members of a sea fishing collective in Asia Minor and a fragment of an account ledger from a freshwater fishing collective in Egypt.

The city of Parion on the Hellespont was an important trading centre and enjoyed the status of a Roman colony. It was also located directly on a major migration route, an advantage that was exploited by a fishing collective whose members are known to us thanks to an inscription in honour of Priapos (*IGSK* 25.5). Like the fishing activity described by Aelian on the Black Sea, this was a combined terrestrial and marine operation, where fish-watchers on shore kept a lookout for signs of an approaching fish shoal, and when they saw one, alerted the others. We may assume that, as in Aelian's case, several boats were used to lay a single net, that is, a large seine. The number of boats involved is the same as in Aelian's description, namely five; and each has a boat-master and a net-master. The function of the *phellochalastos*, literally "the man in charge of loosening the cork (?)" is not quite clear, but again points to the use of one large seine held up by floats. The two pilots presumably supervise the navigational part of the operation.

The organization of the collective is quite complicated and not every aspect is clear from the inscription; we may, however, draw some general conclusions. The names reveal that many of the leading participants are connected by family ties or by manumission (García & Florido, this volume, 215-216), and some hold more than one office, which suggests that the proliferation of positions and functions may reflect not only the division of labour but also a desire to maintain a social hierarchy within the collective. The presence of a secretary indicates that membership is restricted and entails certain rights and duties, presumably including a pre-defined share of the joint catch. The net-masters clearly outrank the boat-masters; they appear at the head of the inscription, and one boat-master is the son of a net-master. The net-masters appear to be identical with the *telônarchontes*, the "tax managers" or possibly the "head contractors". The obvious implication seems to be that these are responsible for seeing that a duty or tax on fishing is paid – responsible, that is, both to the members of the collective and to the authorities collecting the tax. A parallel is the "chief leaseholder" (*manceps*) apparently acting on behalf of "the contractors of fishing" (*conductores piscatus*) mentioned in a second-century inscription from the Netherlands (*CIL* 13.8830 = *ILS* 1461). A second possibility, which does not exclude the first, is that the tax was levied not as a share of the catch but as a duty or *vectigal* on the net itself, in which case the net-masters would naturally be responsible for its payment.

The division between terrestrial and marine functions again implies that all members are entitled to a share of the joint catch. The collective has an official for almost every conceivable purpose but none responsible for gutting, cutting up or preserving the fish, so the catch was presumably divided and sold shortly after being landed.

We have a fragment of a fishing account preserved from the Roman period, in a papyrus record from Oxyrhynchus (*P.Oxy.* 3495). Although it concerns a freshwater fishing collective, it may also shed some light on sea fishing. The document details income and expenses over a period of three weeks, listing the proceeds of each throw of the net. In this case, the collective controls the entire process, from fishing right through processing into salt fish, *tarichos*. It is interesting to note that the fish set aside for processing are not those left over at the end of the day, but the product of a specific throw of the net, on some days the very first. We also observe that on a series of consecutive days with good catches, no fish at all go to the *taricheia* for processing; presumably the salting-vats were full.

Again, many details elude us but some general conclusions can be drawn. One net is used, and both the variation from catch to catch and the complicated bookkeeping operation imply that it is a large net requiring many hands to work. A part of the income, about 8 per cent, is set aside for the *theagos*; this has been interpreted as a tax, a *vectigal* for the lease of the fishing-rights, or as a rent for the use of boats. While freshwater fishing obviously cannot be directly compared with sea fishing, the points in common between Parion and Oxyrhynchus suggest that the two fishing collectives were organised along the same general lines.

As will be clear from the above, our knowledge of ancient fishing is quite extensive, both at the technological and the organizational level. There are, however, also a number of open questions that require further research if we are to understand ancient fishing in its societal context, and to understand the interaction between technology, organization and outside factors.

Taxation and regulation

The question of fishing organization at the private level – that is, within the fishing cooperatives – is directly linked to the question of organization at the societal level – that is, state control and taxation.

According to Roman legal theory, anyone was free to exploit the resources of the sea. A number of legal texts affirm that the sea and its resources *a priori* belong to no one and that he who catches them, owns them. As it is expressed in the *Institutes* of Justinian (2.1.1; 2.1.12, translated by J.B. Moyle):

> Thus, the following things are by natural law common to all: the air, running water, the sea, and consequently the sea-shore. No one therefore is forbidden access to the sea-shore, provided he abstains from injury to houses, monuments, and buildings generally; for these are not, like the sea itself, subject to the law of nations. On the other hand, all rivers and harbours are public, so that all persons have a right to fish therein. The sea-shore extends to the limit of the highest tide in time of storm or winter. Again, the public use of the banks of a river, as of the river itself, is part of the law of nations; consequently everyone is entitled to bring his vessel to the bank, and fasten cables to the trees growing there, and use it as a resting-place for the cargo, as freely as he may navigate the river itself. But the ownership of the bank is in the owner of the adjoining land, and consequently so too is the ownership of the trees which grow upon it. Again, the public use of the sea-shore, as of the sea itself, is part of the law of nations; consequently every one is free to build a cottage upon it for purposes of retreat, as well as to dry his nets and haul them up from the sea. (…) Wild animals, birds, and fish, that is to say all the creatures which the land, the sea, and the sky produce, as soon as they are caught by any one become at once the property of their captor by the law of nations; for natural reason admits the title of the first occupant to that which previously had no owner.

Similar views are echoed in the *Digest* of Justinian and in the *Institutiones* of the second-century jurist Gaius. When it came to the practical application of legal theory, however, the situation was less clear-cut. The *Digest* quotes an opinion attributed to the third-century jurist Ulpian in a case where the owner of two adjacent shore properties (*fundi*) had sold one of them, with the restriction that the new owner must not establish a tuna-fishing operation (*piscatio thynnaria*) on the property *contra eum*, i.e., offshore of the *fundus* retained by the seller (*Digest* 8.4.13; *cf.* Franciosi, 2002; Purpura, 2007). On the question whether such an agreement was binding, the opening phrase of the comment, as quoted in the *Digest*, is telling: *quamvis mari, quod natura omnibus patet, servitus imponi privata lege non potest,* "although the sea, which by nature is open to all, cannot be subject to restrictions imposed by a private agreement", yet because "good faith" demands it (*bona fides … exposcit*), the terms of the agreement are binding not only on he original parties to the contract, but on all subsequent owners of the property! The single word *quamvis* speaks volumes about Roman attitudes to the practical application of the law, and warns us to be careful about extrapolating from the letter or the theory of the law to the realities of everyday life. Another legal opinion (*Digest* 1.8.4.pr) quotes a rescript of the mid-second century AD – the rescript itself has not been preserved – in which the emperor specifies that although fishermen are indeed entitled to use the shore, they must keep their distance (*abstinere*) from private dwellings and public buildings (*aedificia et monumenta*).

In the case of other "fruits of the sea" such as flotsam, the situation is equally ambiguous. In most modern legal systems, goods lost in a shipwreck or thrown overboard in a storm become the property of their finder or the state. Not so under Roman law, where they remained the property of the original owner and appropriating them was punishable on a level with theft or robbery (*Inst.Just.* 2.1). Salt, arguably the most important resource that ancient societies extracted from the sea, was not a free resource; state control over the salt trade had been imposed in Rome as early as the third century BC (Livy, 29.37). Purple dye was another commodity from the sea that could not be freely used: the wearing of purple garments was circumscribed by the Roman state.

Given that exploitation of three categories of marine resources (wreckage, salt, purple) were to a greater or lesser degree restricted by the law, it is easier to accept the notion that a fourth resource (fish) may also have been controlled and taxed, legal theory notwithstanding. For instance, the theory of a sea "open to all" did not prevent the emperor Claudius from imposing a five-year *moratorium* on fishing *scarus Cretensis* in the Tyrrhenian Sea (Bullock, 2008; Pliny, *N.H.* 9.63; Macrobius, 3.16.10). Although the sea could be exploited by everyone, the stock of *scarus Cretensis* – which had recently been introduced from the Aegean – could not.

Within the Roman Empire, the most important direct tax on primary food production was the grain tithe, which was assessed when the harvest had been brought in, and paid in kind. For obvious reasons, a similar tax in kind on fresh fish would be impracticable; it would also be easy to evade. But there were various indirect taxes, assessed *ad valorem* and paid in cash, such as the *quadragesima Galliarum* on goods brought across the Alps. This was collected by the state, but cities could collect harbour duties (*portorium*) on incoming vessels, a category that might presumably include fishing vessels returning with a catch. Because the rates were low – the *quadragesima*, as its name indicates, was 2.5% – the incentive to evasion was not great.

Furthermore, regulation – even when it involved the payment of a lease or *vectigal* – might be the lesser evil. The owner of the two adjacent coastal estates did not want a competing *piscatio thynnaria* close to his own; presumably, a fishing guild would gladly pay a *vectigal* if it ensured them a monopoly on fishing within the body of water in question (*cf.* the case discussed in the *Digest* 43.14.1.7). The view that exploitation of marine resources might involve a tax or *vectigal* finds some support in the inscriptions from Parion and from the Netherlands mentioned earlier, and less clearly from a passing reference in Strabo's *Geography* to the tuna fisheries in the Thracian Bosporos "providing the Byzantines and the Roman people with a considerable income" (Strabo, 7.6.2). While it is clear that the Byzantines would profit from the fisheries, tax or no tax, in Strabo's text, the expression "the Roman people" (*toi dêmoi tôn Rhômaiôn*) can hardly be other than a metonym for the state. This is at present our best evidence for *Imperial*, as opposed to *local*, taxes on fishing. It has few parallels in Roman or early Byzantine sources and we cannot exclude the possibility that Strabo was misinformed on this point. Clearly, there is room for further work on the status of marine resources under Roman law and the ways in which they were taxed (*cf.* Fiorentini, 2003; Bekker-Nielsen, 2009, 292).

Culinary preferences and target species

Form follows function. To understand ancient fishing gear and fishing methods, we need to know their purpose: what species did the fishermen intend to catch? Scattered references in the ancient literary sources will inform us that one species was highly prized for its culinary or prestige value or that another was preferred for industrial purposes, but the species so identified are only a handful of the hundreds inhabiting the Mediterranean and the Black Sea.

More general overviews of fish species are provided by the *Natural History* of Pliny the Elder (first century AD), the *De alimentorum facultatibus* of Galen and the *Halieutika* of Oppian (second century), the *Deipnosophistai* ("philosophers at dinner") of Athenaios (late second century), *On the Nature of Animals* by Aelian (early third century) and the cookbook that has been transmitted under the name of the first-century gourmet Apicius but probably found its present form in the fourth century. Of these, however, Pliny, Oppian and Aelian focus on fish as a form of marine life rather than as a food, while the work of Athenaios is heavily skewed in favour of the exceptional: the rare, the delicious or the expensive fish. The recipes of "Apicius" give us an impression of the middle- and upper-class menu, but no idea what fish – if any – were consumed by the *polloi.* At first reading, Galen appears more useful, but it is difficult to establish to what degree his statements about the dietary value of specific fish are based on actual observations of their effects. Taken as a whole, however, the literary sources provide some indication of preferences: what fish were particularly sought after, which species were scarce, which were abundant. They also provide us with some scattered evidence for fish prices, though again we find a tendency to focus on the exceptional, such as the exorbitant prices paid by some first-century gourmets (for a detailed survey of this evidence, see Wilkins, 2005, with references).

The value of mosaics as a guide to commercial fish species is often overlooked. Though marine mosaics purport to depict life in the sea, the artist's first-hand knowledge of sea fish is more likely to have been derived from visits to the fish market than from diving expeditions. If this premise is accepted, we may take the species depicted, e.g., on the famous marine mosaic from Pompeii (page 366) as a selection of the fish on sale in the local fish market, though no doubt still with a preference for the colourful and artistically more interesting species.

To get a more objective impression of fish consumption, and for any sort of quantitative indications, we need to look at the archaeological evidence. Human bones are in themselves evidence of the owner's diet when alive; by means of isotope studies it should be possible to assess the proportion of marine food in the person's diet. The method is still in its infancy and the number of published studies is limited. Common to most of them, however, is that the proportion of marine food indicated by isotope analysis is remarkably low, which may indicate that even in coastal communities, only a minority ate fish as part of their everyday diet.

At best, however, isotope studies will only tell us *how much* fish was eaten, not *which fish*. For answers to that question, we must look for remains of the fish themselves in household refuse dumps and waste deposits from processing installations. Minute fish bones are, however, easily overlooked, and recent experimental work by Inge Bødker Enghoff on a mediaeval excavation site in Denmark (Bødker Enghoff, 2004) has shown that unless all excavated material is sieved very carefully, small fish species will be grossly underrepresented in the archaeological record. A cubic metre of earth from which only two fish bones had been identified by the traditional method of visual inspection and hand sorting would produce more than a thousand bones when sieved on a standard 3mm mesh and 25,000 fish remains if sieved on a very fine mesh. Archaeological excavation of fish dumps to the high standards required and the subsequent archaeofaunal analysis of the finds pose great demands in terms of time, resources and trained scientific personnel, but also hold great promise: from dated or stratified deposits, it is possible to make quantitative statements about catch composition and how this changes over time, reflecting either changing faunal composition or changing fishing strategies. A recent study by Alison Locker on fish remains from British sites demonstrates how evidence from individual sites may be combined and, when seen in their geographical and social context, will allow us to "assess whether the cultural effects of the Roman invasion and subsequent occupation discernibly altered patterns of fish consumption" (Locker, 2007, 141; 154-56); patterns that in turn form the background to the fishermens' choice of fishing technology and tackle.

Fishing equipment

This brings us to the question of fishing implements. This category includes, inter alia, creels, jars and tridents, but for the present we shall concentrate on hooks and of course nets. Hooks of bronze or other metals have been found by the thou-

Figures 5-6. Net fishing from a boat using a casting-net, Oman, 1966 (photos Daniel J. Bosch, from Bekker-Nielsen, 2005).

sands, and are susceptible to typological analysis (Bernal, this volume). The size range is very broad while the range of shapes seems rather limited compared with those available to a modern angler. The lack of distinctive features apart from size is reflected in the ancient literary sources, where there are few attempts to describe the hook or distinguish between different types or shapes. One of the rare exceptions is Oppian's detailed instructions on how to string the hook for catching swordfish (*Hal.* 3.531-540), but the description of the hook itself is frustratingly brief – is the "double-barbed" (*diplêisin*) hook a symmetrical double hook with two points, or a conventional single hook with a barb to either side of the point? – and the subsequent description of how the sword-fish uses its sword to carve the bait before eating it sounds rather too fantastic to be trusted.

A difficult question is whether finds of different types of hooks on a site can be used as indicators that specific species were fished offshore of that location. Leaving the problem of survival rates aside – it could be argued that the hook types most often used were most likely to be lost, and therefore underrepresented on a terrestrial site – we still have two problems: first, the number of distinct hook types is far lower than the number of different species, thus one hook type will have been used to target a number of species; second, the choice of hook will depend on the marine environment (turbulence, underwater visibility, seabed structure and vegetation) as well as the target species.

Nets are in some respects easier to deal with. The types are *myrioi* – "innumerable", according to Oppian – and adapted to the size and individual or collective behaviour of the targeted species, as well as the depth of the waters to be fished (Morales, this volume). A selection of these are described detail by Oppian, enabling us to distinguish between various classes and types of nets. Their widespread use is vouched for by the presence, on many archaeological sites, of weights and sinkers for casting or floating nets. The net fabric itself, however, was made from organic materials which are rarely preserved. A carbonized fishing net was found in *Herculaneum*; some nets have been preserved in the dry climate of Egypt (Thomas, this volume), and there will certainly be well-preserved fishing nets in the anoxic levels of the Black Sea, though so far it is not possible to salvage any of these for study.

One problem which deserves further study is the question of continuity from late Roman to early Mediaeval fishing technology. Both as regards technique and organization, there are a number of features in common between the tuna fishery described for us by Aelian in the passage quoted earlier and the modern *almadraba de tiro* (Florido del Corral & Menanteau, 2006, 868-887; García Vargas & Florido del Corral, this volume, 207-209). This is not in itself proof that the ancient *sagênê* is the direct ancestor of the *almadraba de tiro*; the basic technology is not complicated and could have been re-invented, or re-introduced, at a later date, and for the period from the fifth to the tenth centuries, we have neither literary nor pictorial sources for the use of the beach seine.

It is indeed a striking fact that while net fishing is a popular subject in mosaics of the second to fourth centuries AD, it is almost entirely absent from pictorial art – e.g., funerary reliefs and manuscript illustrations – of the fifth to tenth centuries. This is all the more surprising as net fishing plays a role in the Gospels, and the episodes on the lake Genezareth and at lake Tiberias are popular subjects in later Mediaeval art. As late as the early eleventh century, the scribe producing a copy of *De rerum naturis* by Hrabanus Maurus (780-856) placed a vignette of an angler in a boat at the head of the chapter *"De Mediterraneo"* (Amelli, 1896, tav. LXI). Not

Figure 7. Fishing with a seine suspended from cork floats. Mosaic from the "Maison de la Cascade", Utica (Yacoub, 1995, 175, fig. 87).

exactly the image that comes to mind when one thinks of the "high seas"! Could it be that net fishing formed no part of the world-view of our manuscript illustrator because it played no significant role in the world around him?

Let us examine the question from another angle: the societal context of fishing. Coastal seine fishing for tuna or other migratory fish in the manner described by Aelian involves a great number of people, but also produces a vast amount of fish at one point in time; far too much to be consumed by the fishers and their immediate dependents. The surplus must be preserved, processed or marketed at once. Selling the catch as fresh fish would only be possible in a large urban market (such as Constantinople) and even fish processing was, as García Vargas (2006b, 551-555) points out, often closely linked with an urban community which provided a market or a point of contact with interregional trading networks. It is well known that in late Antiquity, urban population numbers declined and trans-Mediterranean trading networks broke down: did this create a situation where sea fishing could no longer find an outlet for its catches and impose a change in organization and technology from the large-scale seine fishing operations of earlier periods to small-scale fishing based on casting nets and angling? The question remains open for the time being – but it serves to illustrate how ancient fishing needs to be viewed in all aspects of its complex context, and how the answers to our questions will only be found through the joint efforts of archaeologists, historians, biologists and anthropologists.

"La pêche du thon", etching by Jean-Pierre Houël, 1782 (detail).

7. The Origin and Development of Tuna Fishing Nets (*Almadrabas*)

Enrique García Vargas & David Florido del Corral

Introduction

The closure of most Mediterranean and Atlantic coastal tuna-fishing sites, or *almadrabas,* during the last thirty years, and the transformation of some into tourist attractions, has triggered an interest in a fishing technique that has been used since ancient times (Ravazza, ed., 1999, 79-90; Ravazza, 2000, 22-24). This paper aims to contribute to the study of *almadraba* and related fishing techniques, especially concerning its origins. Our goal is not a technological reconstruction or an ethnographic description of tuna fishing (for this, see Oliver Narbona, 1982; Regueira & Regueira, 1993, 73-79; Martínez Maganto, 1992, 234-236; Ladero Quesada, 1993, 352), but to study the fishing activity in its context within different historical periods, from the perspective of both archaeology and cultural anthropology, with special attention to the following aspects:

- The physical and biological preconditions for tuna fishing.
- Fishing techniques and available sources of information.
- The historical development of ownership and management models, and the consequent "territorialization" of the coast.
- The general economic context and the commercial implications of coastal tuna fishing with *almadrabas* and related techniques.

Eco-biological requirements

Catching tuna in large coastal nets, which since the Middle Ages have been known to western Mediterranean fishermen as *almadrabas* or *madragues,* depends on two main factors: the presence of fish shoals and access to them by means of socially complex technical resources. Given the technological conditions applicable in Antiquity, offshore fishing was an exceptional activity (Trotta, 1996, 230), only mentioned by pseudo-Aristotle, who reports that fishermen from Cádiz (*Mir.*, 136a) operated in Saharan fishing grounds (Mederos & Escribano, 1999), sailing on vessels which were known as *hippoi* because of their horse-shaped sterns (Luzón Nogué, 1988). The system did not provide large amounts of fish, but enabled the capture of a very specific kind of tuna which, again according to pseudo-Aristotle, was highly appreciated by the Carthaginians, to whom the fish was sold (García Vargas & Ferrer Albelda, 2001; 2006).

Until the European exploitation of the Grand Banks of Newfoundland in the sixteenth century (Cutting, 1955), fishing was, therefore, mainly a coastline activity. Coastal fishing used to require hard labour, and gave poor profits in return. Fishermen could only escape their miserable existence through a "spell of good luck", filling their nets with either a large amount of fish or a smaller amount of very valuable fish (Theocr. *Id.* 21.52). Nonetheless, this spell of good luck could indeed arise in some coastal areas, thanks to the passage of huge numbers of migratory fish a couple of times a year (García Vargas & Muñoz Vicente, 2003).

One such privileged spot is the Strait of Gibraltar (Roselló & Morales, 1988; Serna *et alii*, 2004), where pelagic fish have always been available in plenty, especially tuna (*Thunnus thynnus*). Tuna enters the Mediterranean from the Atlantic in late spring. They travel in search of suitable hatching areas (gametic migration) where they can find nutrients for their offspring and where the temperature and salinity of the water favours the development of young fish (Sella, 1928; 1929; Rodríguez Roda, 1964; 1973). Although most shoals are dispersed in the waters between Algeciras and Ceuta due to anticyclonic turbulence in the sea of Alboran (Compán Vázquez, 1988, 212), they gather again in the area between Sicily and the Balearic Islands, where their main hatching areas are found.

During their outward migration, the tuna follow the currents of the sea, especially the main branch of the Atlantic current. This current enters the Strait of Gibraltar at surface level due to its lower salinity and density compared with the Mediterranean (where, as is typical of a closed sea, the evaporation rate is higher).

On their return from the hatching areas (Serna *et alii*, 2004) in the late summer (trophic migration), the shoals follow the circuits of secondary currents flowing towards the Atlantic, created by the anticyclonic dynamics of the superficial water mass, along the coast of the Maghreb. At this stage, their flesh contains less fat, due to the immaturity of the young fish and the reproductive stress recently suffered by the adults.

Two other biological characteristics of scombrids are crucial for explaining their proximity to beaches during migration (Florido & Ménanteau, 2006): their low tolerance to temperature changes and the fact that they are visual predators. The former forces them to search for warm and salty waters. The latter brings them close to clear water coasts in search of a prey whose location varies, among other factors, with the winds, so fishermen talk about *"tuna winds"* when they are "blown" close to the coast. Finally, attacks by or the proximity of killer whales can also bring tuna very close to the coastline.

Technical aspects of *almadraba* fishing

Types of tuna fishing nets and how they operate

When fishing a shoal of tuna, either on their outgoing or return (*revés*) journey, the net must be adapted to its target. From an ecological perspective, we can distinguish between two basic types of grouping: the shoal and the school. The former is a simple association of individuals created by social attraction, whereas the latter is a group swimming in a synchronised manner. This swimming action may be polarised or not, depending on whether the group consistently moves in a given direction or changes direction in a compulsive and random manner. Both behaviour patterns are adaptative, but the former, apart from protecting the fish from predators, maintains sustained progress towards a reproductive or trophic goal. Generally, tunids progress in schools which, besides a uniform pace, involve organised distribution, with the adults leading the march, and a more or less sustained direction. The shape of the caudal fin of tuna and its oscillating (i.e., not wave-like) motion, the shape of the body and the development of the ossified scales of the chest, which help to reduce water friction, enable tuna to reach a maximum speed of 80km/h (although the average is about 45km/h). A fishing procedure intended to trap one of these large groups, comprising hundreds of fish and following the lead of fully developed individuals, must therefore try to stop the tuna before their capture.

In large static *almadrabas* of the early modern period (figure 1), which were conceived as traps comprising several chambers and fixed to the seabed, this function is performed by the *rabera*, a net curtain stretching from the beach into the sea, perpendicular to the swimming direction of the school. The *rabera* also forces the fish to swim along the walls of the net to the entrance of the succession of chambers which form the central part of a modern *almadraba*. The fish are trapped within these chambers until the fishermen decide to lift the *copo* (terminal section) of the final chamber – into which the fish, in today's practice, are forced by boats as well as frogmen – to the surface (figure 3).

But before the use of these static devices became widespread, tuna fishing was carried out with a technique known in Spain as *vista* or *tiro* (figure 2) which had a much greater heroic or epic component. Descriptions of "lances" or *vista* or *tiro* fi-

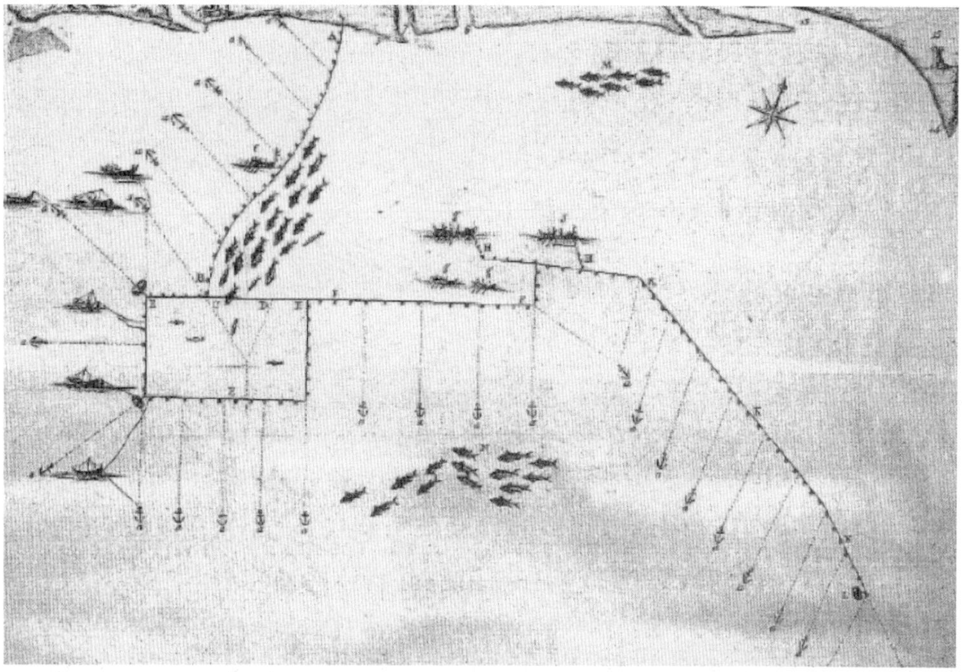

Figure 1. Modern static almadraba of buche type from El Terrón, in Ayamonte, Huelva (Sáñez Reguart, 1791-1795).

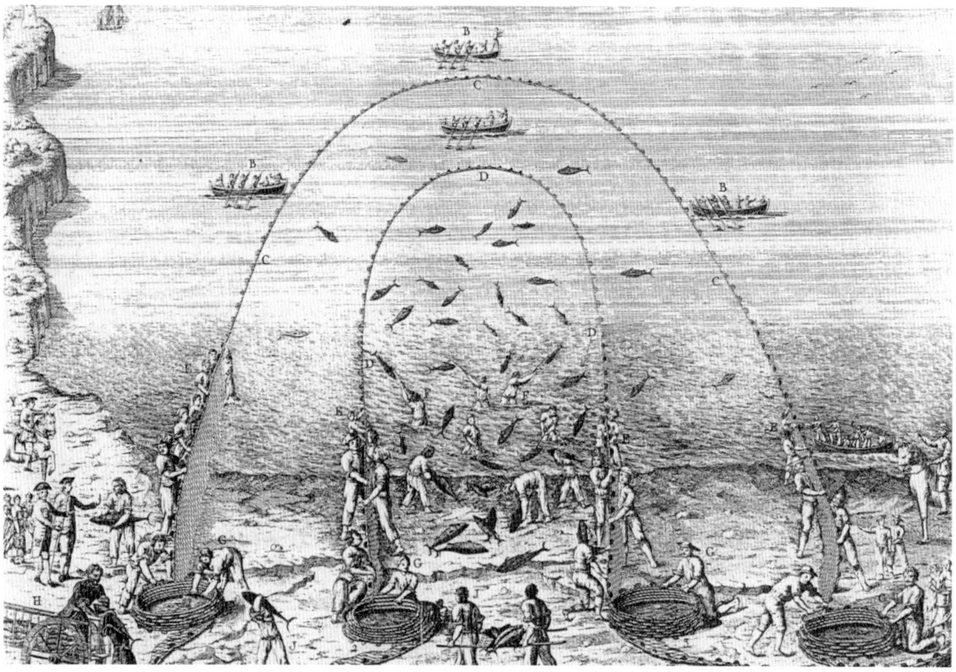

Figure 2. Old fashioned vista or tiro almadraba from Conil de la Frontera, in Cádiz (Sáñez Reguart, 1791-1795).

Figure 3. Tuna fishing in a Sicilian tonnara (Ravazza, 2000).

shing are generally based on those provided by fifteenth- to nineteenth-century historians, topographers and ethnographers (Pedro de Medina, Pérez de Messa, Agustín de Horozco, Suárez de Salazar, brother Jerónimo de la Concepción, Sáñez Reguart and father Miravent being the best known). These descriptions, besides being rather repetitive, are very simplistic. They mention two consecutive operations, which commence at the moment when the *torrero* or *atalayero* (lookout) announces the presence of tuna from a tower (figure 4) on the coast.

Recently, a more detailed description of this fishing system was published by Luisa I. Álvarez de Toledo (2007), after the reconstruction of documents and drawings from the archives of the Dukes of Medina Sidonia (*Explicación de la almadraba de tiro*, 1765), who for decades monopolized *almadraba* fishing. This description describes the vessels and tools, the guilds and categories of fishermen and a detailed sequence of the operations involved. This information is fundamentally consistent with that published in 1791 by Sáñez Reguart (1791-1795), about the organisation of the *almadraba* of Conil. Both documents provide details about the two stages of fishing using the *almadraba de vista* or *tiro* fishing systems.

In the first, several vessels (*boliches, calones, barca delantera, barca de segunda*) with a different number of rowers (from four to ten, depending on the type of boat), were launched to cross the path of tuna coming from the west (because the *almadraba* of Conil is *de derecho*). After sighting the tuna, the *atalayeros* would make signals to the watchmen on the boats with flags. The boats, positioned at the two ends of

Figure 4. "El Puerco" tuna watching tower in Chiclana de la Frontera, Cádiz (Regueira & Regueira, 1993).

the fishing area, the *boliche de levante* and the *boliche de poniente,* would sail and place the nets called *sedales* (Sáñez Reguart, 1796, I, 50). The one at the east end would progress from its position towards the beach, and the one at the west end would sail along a parallel line further out to sea. The first vessel's net or *sedal* would stop the tuna, and the second's net would prevent it from escaping. A third boat, the *barca segunda,* would extend the *sedal de la barca,* equipped with *panda y plomo* (floats and weights) between these two nets, progressing towards the west. Once this net converged with the eastern and western *sedales,* the trap was complete, and the sailors would cease to row. However, there would still be a gap at the ends of the nets (*cabrestreras*) through which the tuna could escape. This was prevented by small boats (*calones*) with crews making noise, hitting the boat and the sea surface with their oars and throwing stones to keep the tuna away from the gaps. In the meantime, the ends of the nets would be provisionally stitched together from the boat called *calón de cabrestrera.*

This signalled the start of the second stage of the process, encircling the tuna with a thicker net and pulling it in towards the land in order to capture it on the shore. A boat called *calón de sirga* would set a rope (the *sirga*) encircling the *boliches* and through which the *cinta* – thicker net, reinforced in the eastern side (Sáñez Reguart, 1791, II, 291) – would be set. The watchmen of the boat performing this operation would be directed by the *atalayeros,* in order to follow the arch required to catch all the tuna. The *ventureros* were responsible for pulling in the *sirga.* Among them, the *cabeceras* would be located on the shoreline, inside the water, to prevent the tuna passing above the floats of the *cinta,* while the *ventureros* pulled the sirga in from the beach. According to most sources, the *cloqueros* or *paralelos,* who had pushed the boats out to the sea at the start of the operation, would then kill the tuna, armed with hooks.

Ancient tuna fishing in literary sources

Ancient literary descriptions of tuna fishing appear to refer to a *vista* or *tiro* system, although is not entirely clear whether the net circles used were double, as in later sources. However, the ancient descriptions available are incomplete or metaphorical, and difficult to understand and interpret. The elder Philostratos from Lemnos, who wrote in the third century AD, describes (*Imagines* 1.13) a painting showing tuna fishing – without giving many details, however, since the text refers to a drawing. The second-century author Oppian (*Hal.* 3.620-648) prefers metaphoric language (doubtless making interpretation more difficult), pointing out that the nets take the shape of a city, with precincts and gates. His near-contemporary Aelian (*N.A.*, 15.5) describes the setting of the nets from the five boats advancing in line (this volume, p. 186), but does not explain how the tuna are encircled. The best way to get an idea of this fishing technique applied to tuna is possibly to combine these descriptions and compare them with similar fishing techniques described by the Spanish chroniclers mentioned earlier.

According to Philostratos, Oppian and Aelian, the manoeuvres always begin with the sighting of the fish, from a high point which literary and epigraphic sources normally refer to using a derivative of the verb *skopeo* (watch, observe or sight): *skopía, thynnoskopeion.* In Aelian, the watchtower is a structure of pine timber, to which the net that was thrown after the sighting of the fish would be attached. This kind of wooden watchtower, in which two converging posts bend towards the sea, as described by Aelianus, is known throughout the Mediterranean in the Byzantine, Medieval and Modern periods, but its existence in Kyzikos, in the Hellenistic period of a watchtower (*skopía*) belonging to the municipality (Robert, 1950, 81-83) and rented by a guild of fishermen, might indicate that this structure was more than a simple wooden frame. In fact, on the Andalusian coasts around the Strait, the *torres almenaras*, conceived in the modern period for the surveillance and defence of the coastline against the incursions of the Barbary pirates also played an important role as watchtowers for tuna fishing. Sometimes, as with the *almadraba* of Torres de Hércules, in Cádiz, these towers were built especially for watching the horizon during fishing operations. The presence of these towers for fishing became so pervasive that the *avistador* (corresponding to the *skopiazon* of ancient sources) was commonly called *torrero*, "tower-man".

Once the tuna-watcher, whose wonderful wisdom is praised by Aelian (*N.A.* 15.5), had estimated the size and direction of the school, fishing operations would begin. Once again, Aelian reports that each of the five boats advancing in line drops a net, but he does not extend his description to the other operations, although he does not forget to point out that the rowers capture the fish as if they were assaulting a city.

The picture of a besieged city is also evoked by Oppian (*Hal.* 3.640-641), who indicates that the fishing technique consisted of encircling the school with the net, consistent with Philostratos (*cf.* also Manilius, *Astron.* 5.667: *circum vallata sagena*); he also adds that the method recalls the image of a city with its gates and guards (*Hal.* 3.643). This last remark has lead to belief in the existence of static nets, similar to the ones used nowadays. The insistence of ancient authors on the encircling operation resulting from throwing the nets from boats, however, seems to point not to a passive device to trap the fish, but to an active system similar to modern *vista* and *tiro almadrabas.* This is also implied by the use of the term *sagena* or sag n to refer to the net itself.

Oppian could be describing a hybrid device, a very primitive sort of *almadraba de buche*, including *sagena* and a death chamber; alternatively, he may be referring to something else, and has been misinterpreted. The latter seems more likely, because when comparing the *almadraba* to a city, the poet seems to be picturing the most characteristic image of a city: its walls. The comparison between *sagena* fishing and the conquest of a city by siege (an image also present in Manilius and Aelian, who refers back to Homer) is an old Greek literary *topos* (Mastromarco, 1998), in the different versions of which citizens are compared to the tuna trapped in the *sagena*; and hence Oppian's comparison, a few lines earlier, comparing tuna with the phalanxes of an emigrating nation which is surrounded and annihilated.

Leaving the explicit reference to gates and guards aside for a moment, although we will return to it shortly, a joint analysis of the texts by Philostratos and Oppian points to a circular manoeuvre encircling the school of tuna. The operation was, according to Aelian, carried out from five boats setting the net with which, thanks to the strength of the rowers, the school was blocked shortly after being sighted. According to Perez de Messa (1595), a similar fishing technique was used in Conil during the Modern Period by six or seven boats, although in this last case, the boats were already forming a circle before the arrival of the tuna.

Nothing in the ancient references points to the existence of nets arranged concentrically, with the possible exception of Oppian's reference to precincts and guards, which recall the connections between the *boliches* (*sedales de boliche*) and the sedal *de segunda* in Conil's *almadraba* (see above; Sáñez Reguart, 1791-1795); the boats called *calones* would be located in the gaps between nets, called *cabestreras*, the rowers beating the water to prevent the fish from escaping through the gaps. In Colibre's *almadraba* (Rosellón), some of the nets were lifted to direct the fish from one chamber to another, so the precincts, gates and guards of Oppian can be understood as a reference to several nets somehow connected. This net structure by sections is also confirmed by Philostratos, who says that sometimes the fishermen would open the net partially to let some of the fish go, preventing the net from breaking under the weight of too many of them.

The conclusion seems clear. In sum, during antiquity, the literary sources are focused on a relatively late period (second-third centuries AD), describing some type of fishing technique *a la vista* (García Vargas, 2001). The possibly earliest mention of a static *almadraba* trap is not found in the ancient "literature" until the provisions of Byzantine emperor Leo VI "the Philosopher" (*Const.* 57), regulating some aspects of these fishing techniques and devices – that in his time (the early tenth century) were known as *epokhai* or *remorae piscatoriae* – especially the minimum distance between nets to prevent them from damaging each other. Leo VI himself believed that the lack of previous legislation on this issue was due to the fact that it was a fishing technique unknown before his own time.

Ancient tuna fishing in epigraphic sources

The ancient epigraphic sources coming from Greek cities, referring to coastal fishing, are also devoted to tuna fishing techniques *a la vista.* Legal inscriptions (Fernández Nieto, 2006) often define public or private ownership of watchtowers or towers (*thynnoskopeia*), in which rental and fees were commonly a matter of dispute between neighbouring communities; all the fishermen's trades and denominations mentioned in the few sources generated by fishermen guilds refer to the typical operations performed in an *almadraba de vista y tiro* (Fernández Nieto, 2002).

Two inscriptions from Parion, dating shortly after the foundation of the Caesarian colony, refer to the organisation and operation of two fishermen's guilds created to rent two watchtowers from which to sight tuna (Robert, 1950, 81-91). Unfortunately, one of them (*CIG* II *add.* 3654b = *IGSK* 25.6) is in a very fragmentary condition, but the preserved text mentions a fishermen's corporation, created in a place called *Phroy*[...]. There was a *skopiazôn* or watchman among them. The other inscription is preserved in full (*IGR* 1.817 = *IGSK* 25.5) and refers to a fishermen's guild organised in a place called *Neilaios* within the colony. Among the members of this corporation, under the leadership of the principal investors (*arkhontes*), there are five net-masters (*diktyarchai*), two of whom also act as watchmen, and five boat-masters (*lembarchai*). This indicates that the fishing involved at least five boats. The inscription portrays the typical labour organisation for *almadrabas de vista y tiro* in the Strait of Gibraltar in the Modern period. Indeed, the trades mentioned clearly reflect those with which we are already familiar. The number of boats is five, corresponding to the five mentioned by Aelian (*N.A.* 15.5) for the operation of setting the net, and the six or seven mentioned by Pérez de Messa for the Conil *almadraba* in the late sixteenth century.

The situation in Parion during the late Republican period perhaps reflects continuity from the Hellenistic period. If we compare this information with that provided, for the same geographical context, by literary sources (above, p. 207), we can conclude that between this period and the middle imperial period, the fishing devices today known as *almadrabas de vista y tiro* were very common in the Aegean. The situation in the west must have been very similar.

Historical models of fisheries management

Ownership of fishing grounds: legal basis and historical development

Gianfranco Purpura is authoritative about the absence of any kind of public monopoly over coastal fishing during Antiquity: "*in nessun caso nel mondo greco ed orientale – e poi nel romano – è attestata la concessione statale del mare o del lido*" (Purpura, 2008). F.J. Fernández Nieto (2006, 207) argues a similar case for the ancient Greek world, based on philosophical texts (Plato, *Leg.* 7.824a; Arist., *Econom.* 2.2.3a [1346b]) and a rich collection of Greek legal inscriptions illustrating many examples of privately owned fishing grounds or of the rental to individuals, not of coastal fishing rights, but of the public watchtowers themselves. P. Ørsted (1998) also asserts that the state did not exert monopoly over or concede rights for coastal fishing, based on the Roman legal concept of the sea as *res communis.*

It seems unlikely that, in normal conditions, private individuals could impose any kind of restriction on coastal fishing. An opinion of Ulpian (*Dig.* 8.4.13.pr.) records a legal dispute over the tuna fishing prohibition imposed by the owner of the Geronian *fundus* over the purchaser. The seller claimed that he or his estate had been damaged because he kept an adjacent estate, called Botriano, dedicated to this activity (*Venditor fundi Geroniani fundo Botriano, quem retinebat, legem dederat, ne contra eum piscatio thynnaria exerceatur*). The judge's decision, in this case, is to uphold the prohibition, but only in consequence of the *bona fides* of the contract. It is clear that, under normal conditions, no private law restrictions could be applied to the sea (*quamvis mari, quod natura omnibus patet, servitus imponi privata lege non potest*). The text is, however, more interesting for what it allows *de facto* than for what it disallows *de iure.* It seems that this is one of the earliest examples of private ownership being imposed on the shore. This would justify the application of private law limitations on fishing on coastal estates, provided that the owners had previously agreed over the issue. If that is the case, the *servitus thynnos non piscandi* would belong to the sort of relationship between provincial estates mentioned by Gaius (*Inst.* 2.31) as susceptible to be agreed upon on the basis of *pactiones et stipulationes.*

This process leading to private ownership over fishing rights which took place in the Severan period, would have its culmination in the *pars orientalis* at dates much later than those witnessing the collapse of the western empire. The Byzantine emperor Leo VI's "fishing legislation" implies an extension of regal rights not just over the shoreline, but also over the sea. The development of new static trap devices (*epokhai*) required new legislation affirming the exclusive ownership of the *dominus* of the coastal waters adjacent to the properties, or at least over a certain section of them. Therefore, Leo VI establishes, through constitution number 56 that "the owner will have exclusive ownership over the shoreline, having the right to prohibit unauthorised fishing on the beaches". Constitution 57 establishes that the minimum distance between neighbouring *almadrabas* shall be 365 steps – *orgyai* – (700m), or 350m (182½ steps) to each side. Constitution number 102 forces owners for whom the adjacent shoreline extension is not large enough to respect the minimum distance, to communally set a single net, calculating costs and profits in proportion to the size of the estates. Finally, constitution number 103 eliminates this proportional division by imposing an equal one, irrespective of the size of the estates, because "fish are not always in the same place waiting for fishermen, and because the bigger part cannot survive without the smaller". With this, the fishing device is in legal terms detached from the shoreline upon which it relies, something to be expected when the fishing system does not depend on the beach as is the case with static or semi-static *epokhai* or *remorae pescatoriae.*

It is interesting to note that legal disputes among owners, which constitution number 104 attempts to regulate, establishing a statute of limitations for infractions related to distance between nets, include churches, monasteries, hospitals, charities and public tax institutions among the owners of *almadrabas.* These were affluent institutions, able to fund capitalisation costs, although the statute of limitations was longer for them (forty years) than for non-institutional owners (ten years, extended to twenty *in absentia*).

There were some attempts to recover public ownership of fishing grounds during the reign of Michael VII (1071-1078), although, during Nikephoros III Botaneiates' (1078-1081) reign, private ownership of the shoreline and the sea itself was definitely sanctioned, "returning what belonged to God to God", returning to the churches and monasteries of Constantinople full ownership over their *skalai* and *epokhai* (docks and fixed nets) as a natural extension of their landed property (Dagron, 1995, 66).

Management of ancient and medieval tuna nets in the Mediterranean. From the *liberum mare* to the privately owned resource. Territorialisation processes

The inscriptions from Parion, cited above, point to a continuity of a lending system originating in the Hellenistic period over the Roman period, in which fishermen's guilds acquired temporary exploitation rights over certain publicly owned estates in exchange for an annual fee or *telos.* The inscription *IGSK* 25.5 reveals the internal organisation of one of such *societas* or *koinôn,* as well as the social position of its members, the relationships among them, and their different trades. The basic criterion for membership is kinship, including both free and dependents (freedmen and serfs). The *manceps* or *archôn* of this association is a free man called *P. Avius Lisimachos,* who shares control over the nets with his son *P. Avius Ponticus* and three other diktyarchs, including a man freed by *Lysimachos (P. Avius Bithys*), a free man (*M. Apicius Quadratus*) and the slave (*Epagathos*) of an Artemidoros who does not seem to be a member. One of the five boat leaders *(lembarchai)* is a slave belonging to the *manceps;* another is a slave of the freedman Bithys; a third belongs to the eponymous magistrate mentioned in the inscription, *L. Flavius*, priest of the Imperial cult, while the two others are slaves belonging to a certain Asklepios, who holds no position in the guild. The helmsmen (of the larger boats?) are a freeman (*Tubellius [L.?]aetus*) and a slave belonging to Lysimachos, whereas the floats-keeper, in charge of the floats on the nets (*Tomgillius Cosmus*) and the secretary and accountant of the association (*Cassius Damassipus*), are free men having no kinship links to the *manceps.*

The complex system of dependent relationships between the members of the guild and the indirect relationship of the guild itself (through a freedman) with the city's magistrates, point to the guild being part of the community's socioeconomic life. We can assume that, at least on the Sea of Marmara, fishing was an "urban" activity, subject to the regulations affecting the renting of real estate belonging to the colony. As usual, the cession of the properties refers to the exploitation of the rented resources and existing buildings, but real rights are not yet considered.

This situation will have changed over time, but it is difficult to admit legal processes leading to private ownership of publicly owned tuna fishing installations prior to the second century AD. In those owned privately, the concept of the *res communis* affecting the sea and the shoreline would, as we mentioned earlier, prevent exclusive ownership claims over the exploitation of the shore. Therefore, we cannot refer to state monopoly (public ownership of certain coastal properties is a different matter) or concessions of such (non-existent) monopolies to private owners, typical phenomena in "aristocratic" processes of territorialization of fishing grounds. The communal organization of labour and the personal nature of the concessions point in the same direction, keeping fisheries within the area of jurisdiction of public and citizen magistrates, and regulating rights and fees due during the term of duration of the *conductio.*

The process leading to private ownership of the shore and the sea began during the middle Imperial period, due to *pactiones et stipulationes inter privatos.* It had its culmination in the east, with the legislation of Leo VI and Nikephorus III, involving profound transformation of the legal consideration of the sea and the shoreline, to which the typical property rights affecting rural estates could therefore be extended. The progressive acquisition of public *almadrabas* by aristocratic houses and religious institutions also played a part in breaking previous management systems, which, at least from the Hellenistic period, were based on the leasing of public watchtowers and their respective fishing grounds.

The privatisation of publicly owned Byzantine fisheries from the tenth century onwards follows a general trend of private acquisition of leading economic activities by the aristocracy and the Church, thus creating a set of economic activities outside the direct control of the state, managed by monasteries and imperial, aristocratic and ecclesiastical foundations: the *oikoi.* They would finally control the entire economic process, owning fisheries, buildings, workshops, harbours and commercial fleets in several cities (Dagron, 2002, 427-428.). The *oikoi* would basically operate in urban contexts, for the houses from which the corporation's properties were managed were also based on cities, although they also owned rural property. In fact, the economic role of the *oikos,* in which imperial tax institutions and the imperial house itself had a part, should not be seen as something different or alien to the state's structure, but as a special private system aimed at guaranteeing the cities' supply and organising local charities. As such, and because of its aims, the *oikos* had privileged taxation status. In these conditions, and despite the quasi-privatisation of property related to economic processes, we cannot yet refer to the territorialisation process that will eventually remove it from the urban control.

If we now focus on a region closer to Spain, we also find examples of concession of fishing rights to bishops and monasteries in Norman Sicily, at a similar date to that for Byzantium (eleventh-twelfth centuries). Nevertheless, it is normally assumed that the Norman state was able to retain most of the state's property accumulated during the Byzantine and Islamic periods (Bresc, 1981, 17), because there is no evidence of clear predominance of a manor system until the reign of Alfonso V of Aragon. The most common management system was, until the mid-fifteenth century, the leasing of public *almadrabas* to private individuals in exchange for a fee and a certain amount of fish as church tax. We are familiar with the social composition of companies of *"gabellotti"* (Costanza, 1999, 33-35), as well as the relationship between these entrepreneurs and those providing the initial capital: members of the local aristocracy or the traders, Sicilian or not, who distributed salted fish. Once again, we find the relationship between those exploiting the fishery and the citizen elite that we could infer in the case of Parion. At least in Sicily, management procedures for *almadrabas* remained fundamentally urban until a very late period, leading us to think that they were part of a pre-established territorial framework, not yet playing *per se* the territorialisation and population role that they would eventually play.

In Sicily, the conditions for this transformation came with the kings of Aragon, especially since the creation by Alfonso V, after the annexation of Naples, of the Kingdom of the Two Sicilies. The abandonment of bush agriculture and the economic retreat to the grain-producing areas of the inland regions during the twelfth century due to the dangerous conditions of the coastline, was, from the beginning of the Aragonian presence, compensated by encouraging for fishing settlements and *almadrabas* that helped to repopulate the region (Ravazza, 2000, 22-23). The general climate of insecurity produced by warfare and piracy led to fortification of the *almadrabas.* They began to attract fishing and peasant population, and contributed to the economic restoration of the coast. From the mid-fifteenth century, this repopulation process came to rely on the aristocratic houses that were enjoying the concession of fishing property rights in a feudal regime, previously in the hands of the state. It is now that we can properly refer to a process in which the aristocrats took over of the *almadrabas*, fundamentally trap-like devices, and the coastal resources associated with the activity (timber, salt, textile plants, animals, floats, labour and services from the inhabitants of the fishery…). This had an impact on their role as population attractors and as elements for the territorialisation of the coastline.

In Castile, the Atlantic *almadrabas*, in this case *de vista* or *tiro*, performed a similar function after the Strait of Gibraltar was conquered by the crown. In a true social and political *limes*, reflected in place names such as "de la Frontera", aristocrats progressively accumulated land, population and assets, including exploitation rights over *almadrabas* and other maritime resources, such as salt-pans. In exchange, the nobles had to commit themselves to the defense of the border which, from the second half of the thirteenth century onwards, separated Castile from the kingdom of Granada. In this case, repopulation difficulties also led to effective reliance on aristocratic control, after a very brief period in which some *almadrabas* were controlled by the military orders.

The process by which the Andalusian *almadrabas* were transferred to the aristocracy goes back to 1299, when Alfonso Pérez de Guzmán the Good received the Guadiconís' *almadraba*, between the capes of Roche and Trafalgar, from Fernando IV, along with authorisation to establish the settlement which later became Conil (Santos, 2003). The monopolistic strategy of the Pérez de Guzmán family in relation to the *almadrabas* originated as their possessions grew from Niebla to the Kingdom of Granada, in the following centuries, thanks to their grant as Dukes of Medina Sidonia. They were not, however, the only nobles to create *almadrabas*. Some villages and cities under aristocratic control also established these net-systems, which became a common issue for litigation over property rights between aristocratic houses and villages, or between two aristocratic houses, in a general environment dominated by the monopolistic pretensions of the Pérez de Guzmán-Medina Sidonia.

Here, the monopolistic issue is of less interest than the ability of some families to impose exclusive rights over fisheries that were to attract population, especially, as in Sicily, from the mid-fifteenth century. The number of legal actions to prevent others from establishing *almadrabas* increased hugely during this century, despite the fact that a royal order signed by Charles I, and dated in 1551 – directly related to the regulation of sardine fishing imposed by the Marchioness of Ayamonte on her own estate – insisting that "the sea was common to all" (Carriazo, 2001, 50). The state would not regain the fishing rights over the western *almadrabas* until the late eighteenth century. It was then that the static nets (*buche*) were introduced in the area, in the context of a public debate dominated by the theses of the productivists and the national nature of coasts, seas and fishing resources.

Economic framework and commercial distribution of *almadraba* tuna fishing

From very ancient times, fishermen specialised in the *almadraba* system were a "privileged" group in poverty-ridden fishing societies, largely because of the commercial significance of salted tuna products. Given their added value, they have traditionally been an important part of the "exports" shipped from urban ports and tuna-fishing settlements.

In the second century BC, Polybius (*Hist.* 4.38.4-5) lists salted tuna among the superfluous things (*pros de periousian*) that Athens received from the Black Sea, showing that it was a luxury. Apart from those coming from the Black Sea (Dumont, 1976-1977, 100; Braund, 1995, 163; Purcell, 1995, 138), Athenian classical literature also mentions the salted products from Phoenician Cádiz (García Vargas & Ferrer Albelda, 2006) in a specific literary context: comedy. It is a literary genre with its own rules, in which food was commonly used as a very effective literary topic, not only because the description of the characters' diets made it easy to portray their social status but because the greed of the rich for luxury foodstuffs, especially fish, provided a large number of comic possibilities with which to make fun of their greed for *tryphê* and hypocritical pretensions of virtuous conduct (Ferrer Albelda & García Vargas, 2001).

The role of fish and fish preserves was not, therefore, just a literary topic, but the result of very specific social developments, which led to the emergence of a social class rich enough to consume increasing amounts of expensive fish (García Vargas, 2001, 24, n. 37). It is clear that the increase in demand was favoured by the development of the market institution, as well as by enhanced trade between Athens and the rest of the Mediterranean.

The situation in the other Greek *poleis* was probably very similar. The development of a literature of the *tryphê* in the western Greek world from the fifth century BC onwards suggests rapid development of this phenomenon in the cities of Sicily and *Magna Graecia* (Degani, 1982; Wilkins & Hill, 1994). This was the source of most of the wheat consumed in the Peloponnese, the traffic of which was controlled by Corinth (Dunbabin, 1948, 214). The scarce amphora remains of these products from Cádiz were found precisely in these regions: Corinth (Williams, 1979, pls. 105-24) and Olympia (Gauer, 1975, pl. 22, no. 3). The role of preserves from Cádiz in places like Athens can perhaps be extended to other Greek cities of the central and eastern Mediterranean, in the context of urban elite consumption.

In the far west, the urbanisation of the Atlantic and the Mediterranean coasts of Iberia generated a series of "markets", closer to the Cádiz-based merchants who, from the late sixth century BC onwards, focused their activity on the marketing and consumption of salted fish. The type T-11.2.1.3. (Ramon, 1995, map 116) map of the distribution of western Phoenician amphorae can be used as evidence of the marketing of western Phoenician salted fish towards the eastern coast of the Iberian Peninsula and the Balearic islands, where its presence is attested in a few coastal settlements of varied ethnic and cultural adscription.

During the time of Cato the Censor, salted fish products from the Black Sea were a luxury on the dining tables of Rome. According to Pliny (*N.H.* 19.57), the taste for *garum* was in this period considered to be a questionable fashion. Nevertheless, ostentatious consumption spread fast among the Roman aristocracies during the second century BC, as a result of the enormous wealth streaming to the *urbs* after the conquest of *Hispania* and the defeat of Macedonia, and also as a result of the Greek fashions adopted by the city's upper classes.

It is likely that the fact that most exported products were made with red tuna and other prestigious fish, such as the sea bream, had an impact on the general perception of salted fish products, and specifically *garum*, as a luxury well into the Republican period. Archaeology provides evidence of the predominance of tuna products among the food remains found inside transport amphorae from the fifth century BC onwards (García Vargas, 2006a; 2006b). The export of products made with minor scombroids, such as different types of mackerel (*Scomber scombrus* and, especially, *Scomber japonicus*), begins to be more common from the first century BC onwards.

We have argued elsewhere (García Vargas, 2008) that the "democratisation" of taste was well under way during the early empire. This relative "gastronomic revolution" enabled mass production of mediocre preserves affordable for wider population groups, urban or rural, civilian or military (García Vargas, 2007). We are referring to *garum* cheaper than *sociorum*, fish sauces of inferior quality than the ordinary *garum* (*hallec*, *muria*, *liquamen*), solid salted products (*salsamenta*) made with mackerel, less exclusive than the *melandryae* or tuna loin steaks.

The growing monetisation of exchanges and the monetary "injection" caused by the euergetic expenditure of urban aristocracies and the imperial policy towards some tax categories increased the circulation of wealth in cities through medium- and long- range merchants. This clearly benefited salted fish producers, along with other products, such as wine, not included in the *annona*.

This "democratisation" of the consumption of salted products not only affected the number of exploited species, but also the way they were captured (García Vargas, 2003). The large *almadrabas* continued to operate, but their exploitation costs and the limited short term profits offered in return, made them viable only for rich rural land owners and corporations which, such as the one we know from Parion, could count on external capitalisation. Simpler fishing systems, for the capture of mackerel, horse mackerel (*Trachurus trachurus*) or similar species, must have proliferated. Although also costly, and profitable only in the medium term, they did not require so much financial effort as the big tuna fishing devices. It is the case of the seine, operated by a limited number of men (from sixteen to twenty-four), including some on land, and the crew of the single boat (García Vargas, 2001). The little seines required for the capture of the sardine, controlled from the land or from a small vessel, were even smaller. All these techniques provided sufficient fish at a low cost, according to Oppian (*Hal.* 3.589-591), who presents a stoic-like anecdote according to which the mackerel which had escaped from the full nets battled to enter it from without, finally trapping themselves by their heads in the meshes of the net. This reminds us that these fisheries must have also used simple net techniques, with much lower labour costs.

Although sardines (*Sardina pilchardus)* and other engraulids (anchovies: *Engraulis encrasicolus*) are present in the archaeological record from a very early date (García Vargas *et alii*, in press: sixth century BC), their role in diet and trade was restricted to local circuits and regional distributive patterns (Amores *et alii*, 2007). From the third century AD, however, these species predominate in the large excavated salting factories, as well as in the amphorae in which the products originating in the Atlantic and the Mediterranean factories were distributed. N. Desse and E. Desse-Berset (Desse, 1993, 341-346; Desse & Desse-Berset, 1993, 327-340) have related this fact to the overexploitation of mackerel fishing grounds, establishing a need for the commercial exploitation of smaller species. Recently, Morales and Roselló (2006, 63-73) have pointed out that the recorded variations in ancient fishing and fish consumption cannot be fully explained by overexploitation of fishing grounds; they may also be the product of environmental changes or socio-economic phenomena.

The socio-economic transformations are clear from the final years of the second century AD, in which the structure of the Roman state goes into a major crisis as a result of the collapse of the distribution networks for coinable metal (Chic García, 2005). We therefore favour, without ruling out the existence of overexploitation, the likelihood of a restructuring of the commercial fishing industry in the Late Empire, favouring the capture of smaller species.

A constitution passed by Honorius and Theodosius, dated to AD 415 (*C.Th.* 14.20) attempts to legislate in favour of the imperial house's fish providers, who argued that taxes did not allow them to purchase the thirty pounds of first-class fish that they had to provide in exchange for one *solidus*. This reminds us that although the majority of the fish marketed in ancient cities was of low to medium quality, there was always a market for first rate products.

In case of exceptionally large captures, the price of regular and white tuna (*Palamides*: *Sarda sarda*) could drop dramatically in fishing communities. This was the case in Constantinople in 582 when, during a severe famine (Dagron, 1995, 73), a miraculously abundant catch provided the markets with a sudden supply of food, and a nine-pound tuna could be bought for only twelve *folles*. It appears that under normal conditions, tuna was not a cheap product, especially not the select parts such as dry cured tuna (*homotarikhos*) or tuna belly (*hypogastrion*). In fact, the concentration of the Byzantine tuna fisheries, or of the taxes extracted from their exploitation, into the hands of bishops and religious orders, is clearly related to the ecclesiastical need for quality food, especially for Lent and the common penitential periods of the liturgical calendar.

An anecdote about the life of saint Luke the Stylite (died AD 979) shows how the saint provides an abundant catch to the members of a fishing collective through a miracle performed from his column in Constantinople (*Life of St. Luke* 38-40 ed. Vanderstuyf, quoted in Dagron, 1995, 61). After blessing the nets, and after at least two catches, the saint angrily demands the proportion due from the second catch in church taxes, for the fishermen are trying to cheat about the number of fish caught. This "protection" offered by Luke to the Chalkedonian fishermen can be interpreted, on the basis of later Byzantine fishing legislation, as a claim to fishing rights on the part of monasteries and bishops.

The bishops and monasteries that owned fisheries *(epochai)* also had the means to market the product: urban ports (*skalai*), warehouses and shops, fleets and "commercial agents" in other cities (Dagron, 2002, 458). During the term of office of the patriarch John the Almoner (early seventh century), the church of Alexandria owned a fleet of twelve merchant ships (Jones, 1964, vol. 2, 867). All were hit by a storm on the Adriatic Sea during a commercial voyage and had to jettison their cargo. The loss was estimated at 25,000 gold *solidi.* On another occasion, a captain of John's fleet was sent to *Britannia* with a cargo of Egyptian wheat to relieve a famine in the island (Whittaker, 1983, 168). Some monasteries, such as the Pantokrator in Constantinople or the Amonites of Thessalonike and Hierissos, sent fleets, either their own or chartered, from headquarters to their *emporia* or *aulai* in other cities to trade and to collect the rents they were due (Dagron, 2002, 422). Large secular landowners did likewise; their involvement in these economic urban transactions often kept them apart from an autarchic life style, which was more of a literary *topos* than a social reality.

In the western world, the relationships among these same aristocratics included bonds of interdependence, often reinforced by exchange and gift-giving dynamics in luxury goods, among which salted fish. This is the case of the gift of *muria Barcinonensis* that Paulinus (future bishop of Nola) had prepared in one of his possessions and given to his friend Ausonius of Bordeaux. The gift is answered with a friendly and learned reprimand from his friend (Ausonius, *Ep.* 25*)*, for the use of the word *muria* to refer to what was nothing but a *garum* that Ausonius called *liquor sociorum.* The literary resonance of the name suggests a very exclusive product, *hematitou* (made of blood), prepared in the Spanish fashion, probably with mackerel, *bonito* or bullet tuna. All this means that major amounts of fish were still moving through the trade routes, although after the fall of the western empire the commercialisation and distribution patterns for sea products had changed considerably, in Europe as well as in Byzantium. With regards to Vandal Africa, we should note the case of Carthage, where (in the circular harbour, although not in the city) tuna remains were found, apparently testifying to the export of quality salted products as late as the sixth century AD (Hurst, 1994, 319).

The situation described above includes more modest trade circuits which moved tuna sub-products of poorer quality and lower price, made from the discarded parts of the fish: the heads and the spine. A Baetic amphora, rescued from the Chiessi shipwreck (Bruschi & Wilkens, 1996, 165-69) lost in the second half of the first century AD off the isle of Elba, contained pieces of tuna heads, a part of the fish which, according to Athenaeus (*Deipn.* 4.135e), was typical of the poor classes.

In the cities of fifteenth-century Sicily, almost all good quality fish were consumed by a very exclusive elite, as the price of tuna was higher than that of all other fishes except eels (Bresc, 1981, 15). The discarded parts from the tuna salting process: heads, bones and bloody parts, were sold to the Sicilian salting factories at a lower price than the flesh from the back (loin), the belly and the trunk, for local consumption (Bresc, 1981, 15; *cf.* Ladero Quesada, 1993, 354 for a similar situation in Spain). Nevertheless, luxury products became the main factor in the commercial revival of the Sicilian fisheries. The growth of the Sicilian coastal cities coincided with the emergence of new "markets" for luxury products in Liguria and peninsular Italy, including the courts of the popes first in Avignon, then in Rome, which became the main destinations for the tuna prepared in Mediterranean fisheries. The *almadrabas,* therefore, benefited from the economic growth of the Late Middle Ages and the early Renaissance, which created the conditions for the restoration of the trade in luxury products, also promoted by the financial "injection" derived from the exploitation of new metal resources, first in Europe and later in America.

During the Middle Ages, the manorial system was essential to the economic structure of *almadraba* fisheries. It provided basic resources: textile fibres for ropes and nets, timber for boats, casks, wagons and oxen, salt for salting, etc. From the sixteenth century on, the services due to the lords became regulated through contracts, which established reciprocal relationships. Until the mid-sixteenth century, the Dukes hired a "company of the *almadrabas*", the owners of which would provide the fishing vessels. From that time on, the Duke himself became the boat owner, eliminating all competition. Personnel was paid in kind according to professional category, their labour supplemented with that of slaves and serfs, who took care of the jobs on land and the hardest tasks on the beaches.

A central aspect of the activity of the *almadraba* was the commercialisation of the catch. Commercial dynamics, indeed, affected the performance of fisheries. During most of the sixteenth century, both demand and prices were high. Salted tuna products were highly appreciated on the eastern Spanish coast and in the Mediterranean, hence the exploitation system called *a la valenciana*, "Valencian style". Merchants from Florence were often seen in the *almadrabas*, although they were not as common as Catalans, Majorcans and traders from Valencia.

In some years, salted fish products, especially tuna in casks, were packed in Zahara and shipped from the Salado river (Conil). This has been confirmed by studies of the commercial activity of the Dukes of Puerto de Santa María (Franco Moreno & Moreno

Figure 5. Zahara's almadraba in a sketch by Wyngaerde (sixteenth century).

Ollero, 1981). In the best period, the catch was sold in a single operation, but this became increasingly difficult. This is confirmed by some contemporary descriptions of *almadrabas* such as those by Santibáñez or Pedro de León (Antón Solé, 1965).

During the sixteenth century, the first symptoms of crisis begin to appear. The factory of Zahara (an *almadraba* located between Vejer de la Frontera and Tarifa) called "Mesón del Sol", which did not have a stable year-round population until the nineteenth century, was promoted. It was visited by buyers from Andalusia, eastern Spain, Castile, Italy and Flanders, all of whom were seeking quality products ("sweet" and "refined" tuna), but at the end of the sixteenth century, demand declined due to the competition of other salted fish, cod or hake, and dumping strategies were used by traders to keep prices down.

The *Extractos de la Pesquería de Atunes, de sus Productos y sus gastos* and the annual *Libros de Almadravas* (Bohórquez Jiménez, 1999; Guillaume-Alonso, 2006; Santos, 2007) reveal something of the economic organisation and productivity of

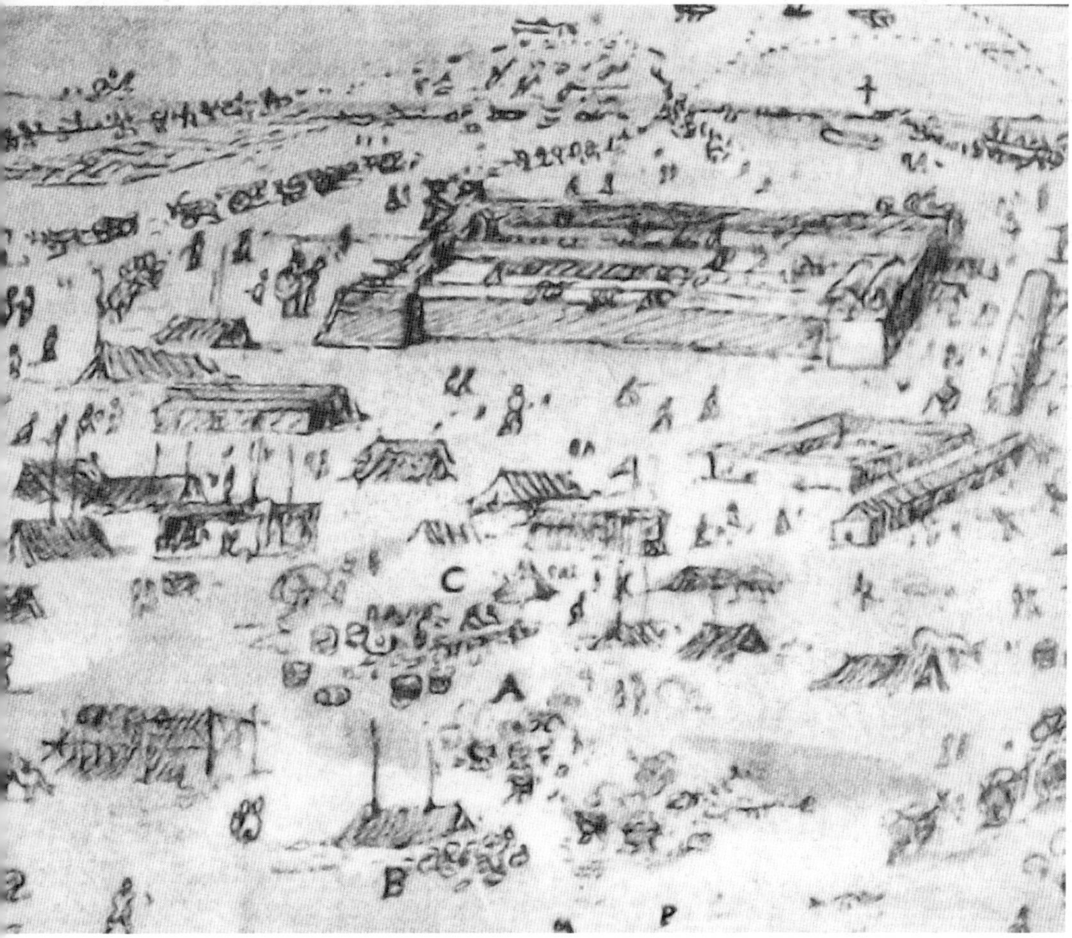

fishing for the Duchy of Medina Sidonia in the sixteenth century (with Conil and Zahara as the most important centres of production). The importance of these documents lies in the fact that they not only contain the total production of the *almadrabas* in long time series, but they distinguish income and expenditure. The former include the fresh tuna sold in the retail market, the tuna sold "because of its small size" (Guillaume-Alonso, 2006, n.10), the salt sold, the barrels, and the fish sold wholesale to traders (from twenty to thirty a year). The latter include the wages paid to the land and sea workers, the cost of the barrels, the salt, the flour used for baking bread, and other sundry expenses. Ultimately, we can distinguish between two major commodity flows: those sold through different agents and those who took the high quality tuna in barrels to different parts of the Mediterranean, and those sold to smaller traders, including those from Andalusia. This was salted and fresh fish consumed among the lower classes, thanks to the dietary constraints and taboos regarding meat in the society of the Ancien Regime.

Wyngaerde's sketch (figure 5; Kagan ed., 1989) show us that there were pans everywhere, overflowing with heads and bones, cooking the remains of tuna fish: this was a way to obtain the fat that was used to treat the boats. Álvarez de Toledo (2007) describes another production system in which fat was collected from tuna during the salt pressing process. The production of the *almadrabas* was thus not limited to fresh and salted tuna fish, but also included heads, bones, bloody parts and fat.

We particularly have to consider the role of the tuna economy in the Duchy's economic and financial fabric (Salas Almela, 2006): tuna sales provided cash which was essential to the Duke's finances. Payments were made in Madrid, where the Duchy had an agent, saving the transactional expenses involved in making payments over long distances. In other words, the income from tuna fishing was enjoyed in Madrid, which was already an important economic centre during the reign of Philip II. The financial importance of the tuna economy may also explain why this activity retained its prestige, even though catches after 1580 were considerably smaller than before.

Final considerations

The first point to consider is the possibility of drawing general, diachronic conclusions concerning tuna fishing in the Mediterranean from ancient times to the early Modern period. Although the *almadraba* is historically a relevant analytical unit and can be seen as a unitary social system in itself, it does contain elements of both continuity and discontinuity.

1. The most relevant elements of historical continuity are: i) trade-oriented economic specialisation, even at a long distance, ii) the constant relationship between technical and social systems and forms of territorial, political and social organisation, with different models of territorialisation at each point in time, and iii) the contextualisation of *almadraba* fishing in the dominant social and economic system in each period.
2. The most important transformation processes include i) going from an urban to a rural socioeconomic context, where repopulation and military functions are more important, ii) going from an urban system of ancient social relations (not excluding personal dependence) to another pyramidal system based on patronage, iii) going from a management model based on the idea of *mare liberum* to another in which the ownership of the sea and coast is predominant, tending towards a feudal organisation controlled by nobles and military and religious orders.

The second point is to consider processes of both continuity and change, together with a need to view the historic evolution of fishing techniques in their socio-cultural context. Economics, political systems, product trade methods, legal concepts regarding ownership, etc., are all important for understanding the specific characteristics of fishing in a specific period. A global/local perspective of this kind, capable of articulating local dynamics against the background of overall continuities and transformations, requires interdisciplinary research and different sources for long historical periods. The conclusions will only be relevant if the data are viewed as part of a global perspective. Following Morán, we agree that "a research strategy to understand transforming societies needs to be concerned with process, with history, with the role of political and economic power as it influences social relations in time and space on a number of scales from local to global" (Morán, 1996, 9).

Lead tube with double flange inside the hull of the Grado wreck, near the keel.

8. Fishing from Ships. Fishing Techniques in the Light of Nautical Archaeology

Carlo Beltrame

Introduction

Tønnes Bekker-Nielsen (2005, 84), in his work about ancient fishing, points out that the archaeological evidence for fishing activity at sea, that is to say fishing from boats or ships, is still very rare. Considering that this kind of fishing is better documented by iconographical and written sources, the author hopes that, in the future, nautical archaeology will be able to offer artefactual evidence to review Gallant's theory about fishing in ancient times. Gallant (1985), in fact, was convinced that fishing in the ancient world was quite primitive; he thought that fishing was not done from boats, but from the seashore and that this kind of activity could be only just enough for survival. That is to say fishing activity could complement the fishermen's diet, but could not be intended for the market.

Bekker-Nielsen does not agree with Gallant's theory, but he admits that the study of the iconographical documents would indicate that fishing boats were quite small, that they had no sails and were moved simply by oars. He presents rare exceptions such as the African mosaic of the third century AD representing Bacchus fighting against the pirates (Bekker-Nielsen, 2005, 87).

In his publication, Nielsen also reminds the reader that the limit to the fish trade in ancient times was the apparent impossibility of keeping fish fresh or alive for many hours (Bekker-Nielsen, 2005, 88).

Fishing boats

Here, we want to present the archaeological evidence for sea fishing which we have collected in the course of our past studies about life aboard Roman ships (Beltrame, 2002,

65-69). This evidence is not able to radically challenge neither Gallant's nor Bekker-Nielsen's views, but it allows us to add some "food for thought" to the subject.

Evidence of boats used to fish at sea (*piscatoria scapha*) is not completely lacking, but it is not always easy to interpret. Excluding the well-documented fishing boat from the *portus Claudius*, dated to the second century AD (Boetto, 2006a) – which, in our opinion, because of the small size and the presence of a box full of water inside the hull, could be used only on inland waters – and the so-called "Jesus boat", recovered from the lake of Kinneret, Israel, dated to the first century AD (Wachsmann, 1995), we can mention a few other boats which could have a marine use: the Roman boats from *Herculaneum* (Steffy, 1985), Toulon (Borreani *et alii*, 1988), Naples (Boetto, 2005), and the Greek boat from Place Jules-Verne in Marseille (Pomey, 2000).

Only the last offers clear evidence of its use for fishing. In fact, coral traces were found in its hull. The use of the others for fishing is only probable and the definition of "fishing boat" applied to the rowing boat found in 1982 on the shore at *Herculaneum* is only hypothetical. The interpretation is based entirely on fishing gear and nets found next to the wreck.

Finally, little information is available about the more recent discovery of a second, very small boat near the Villa dei Papiri at *Herculaneum* (Tuccinardi, 1998).

Archaeological evidence for fishing from ships

Our analytical study of all the shipwrecks in the Roman age, concluded in 2001 (Beltrame, 2002), allowed us to recognize objects related to fishing activity aboard 52 of the 177 vessels considered. This datum allows us to say that about one third of the Roman vessels would be carrying fishing gear. But is this equipment proof that fishing was the primary activity of these ships, or is it not more prudent to assume that the crew used the fishing gear only either to complement the onboard diet or for recreation?

Let us consider the different categories of objects to answer this question.

On the *Cavallo 1* Roman wreck (Bebko, 1971, pl. 11), an impression left by an iron trident-head on a concretion was found. This tool could be used to fish octopus, sharks, sword-fish and tunny. Its appearance has not changed over the centuries, as evidenced by a similar find recovered in the seventh century wreck from Dor in Israel (Galili & Rosen, 2008, 70) (figure 1) and on the Serçe Limani eleventh century wreck (Bass, 2004, 429-430) and on the *Glass wreck* of Venice (D'Agostino, 1995-96, 44-45).

Although the find from the *Cavallo 1* wreck seems unique, at least for the Roman world, its use is well documented both by the ancient authors, such as Oppian (*Hal.* 3.552-54; 4.252-53) and Petronius in the *Satyricon* (109.1.6), and by

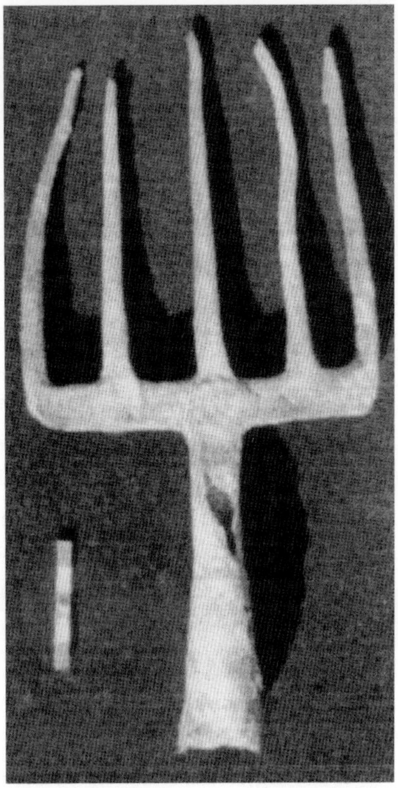

Figure 1. Trident head from a Byzantine wreck of Dor, Israel (after Galili &, Rosen, 2008).

iconography. These sources would testify to its use not only aboard small boats but also on ships. See, for example, the mosaics from Dougga now in the Bardo Museum, from Uadi ez Zgaia (Foucher, 1965, figure 23) and from the Maison de l'Arsenal at Sousse (Gauckler, 1897) as well as the child sarcophagus in the Vaccari-Bacchettini palace of Rome (Lawrence, 1962, 290, tab. 77.2).

The harpoon could be used from the ships for more or less the same kind of fish as the trident. Although no evidence of its use from ships is available from iconography, where only small boats are represented, a rare archaeological testimony is the iron head found in the Dramont G Roman wreck (Joncheray, 1987, figure 10) – figure 2.

Among the fishing tools found in wrecks of ancient ships, the gears for the line fishing and the nets are much more diffuse than other finds.

Both Oppian (*Hal.* 3.72) and ancient iconography describe line fishing aboard ships. From the iconography – as an example, we can cite the mosaic from La Chebba in the Bardo Museum – we learn that it was common to fish with rods from ships. Since rods had to be of wood or cane, they have never been preserved in a wreck. Instead of the use of rods, we could have evidence of the practice of line fishing with reels. The only iconographical evidence we know comes from the painting made by a Greek craftsman in the "Hunting and fishing tomb" at Tarquinia, where a man is fishing from a small boat (Steingräber ed., 1984, 299-300).

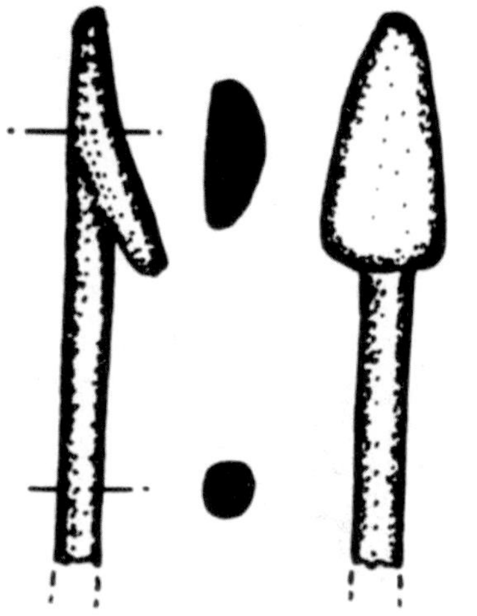

Figure 2. Harpoon head from the Roman wreck Dramont G (after Joncheray, 1987, figure 10).

This kind of object could perhaps be identified both among the finds from the Greek wrecks of Porticello – Italy – (Eiseman & Ridgway, 1987, 16) and Kyrenia – Cyprus – (Swiny & Katzev, 1973) and in the Roman wreck off Colonia de Sant Jordi in Majorca (Cerdà Juan, 1980, 92-93) – figure 3 . Also, thirty-one similar objects were found during the excavation of the ancient harbour of Marseille (Hesnard *et alii*, 1999, 64) which is a multi-stratified context of the Greek and Roman periods.

We must stress that the interpretation of these finds as reels is only hypothetical – also because we have never had the opportunity to analyse them *de visu*. Other authors have suggested that they were toggles for sails (Eiseman & Ridgway, 1987, 16; Hesnard *et alii*, 1999, 64).

Loose fish hooks were recovered in wrecks of the Archaic period, such as the Isle of Giglio (Bound, 1991, 26) and Gela 1 (Panvini, 2001, 63) wrecks, and, more often, in wrecks of the Roman age such as those of Grand Congloué (Benoit, 1961), Port-Vendres 1 (Chevalier & Santamaría, 1971), Titan (Benoît, 1958; Tailliez, 1961) and Punta Patedda (Jurlaro, 1972). But hooks were also found inside wooden boxes: in the Port-Vendres II (Colls *et alii*, 1977, 123-128) wreck, they were inside a cylindrical box, while in the Grado and Cap Camarat B (Carre, 1994) wrecks they were found inside rectangular boxes with sliding lids. In the Comacchio ship, numerous hooks were contained in a wicker basket (Berti, 1990, 271) – figure 4a.

In this last wreck, a grapnel hook was also found (Berti, 1990, 271) – figure 4b. This type of fishing gear, an example of which was also recovered from the Medieval Serçe Limani wreck (Piercy & Bass, 2004, 400), is still used today for catching octopus and squid.

The practice of line fishing aboard can also be documented by lead weights. These can be of various shapes and weights. The shape is often a truncated cone with a hole for the line (figure 5). This type of weight was used as early as the Greek period, as documented by finds in the Isle of Giglio wreck (Bound, 1991, 26-27).

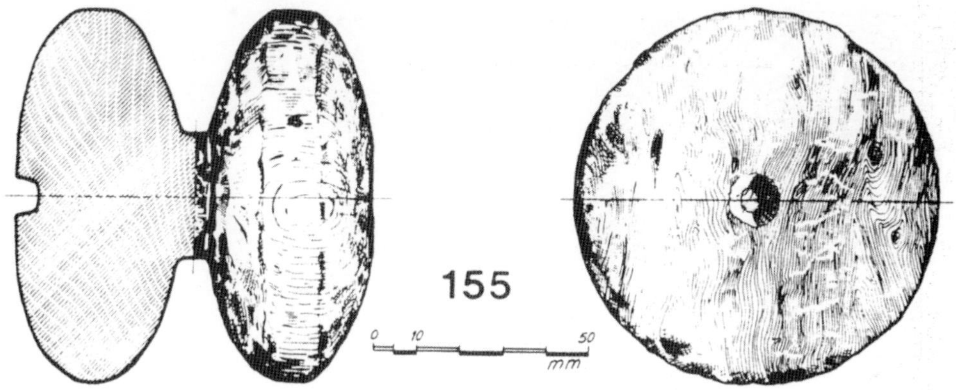

Figure 3. Possible reel from the Colonia de Sant Jordi Roman wreck (after Cerdà Juan, 1980, 92-93).

Indirect evidence of the use of line could also be the finding of mussels in the Comacchio Roman wreck (Corazza, 1990) and in the Byzantine wreck of Yassi Ada (Katzev, 1982b, 294). The mussels could, in fact, have been fished to be used as bait for hook-and-line fishing.

Net fishing is well documented both by ancient authors (especially Oppian) and by the iconography of the Roman period. These sources would indicate that these nets were used aboard boats rather than ships (Bekker-Nielsen, 2002b, 18-23). Because of the perishable nature of the organic material of which they were made, there are no traces of nets in the wrecks, but these objects are indirectly documented by some inorganic finds. We are speaking of lead sinkers, clay weights, net needles and lead rings.

Lead sinkers can have various shapes and weights and can be confused with weights for line fishing. In the Byzantine wreck of Yassi Ada, the weights had sphendonoidal, crescentic and triangular shapes, but folded-over strips of lead were also found (Kuniholm, 1982) – figure 6. This last type of weight appears for the first time in the Archaic Isle of Giglio wreck (Bound, 1991, 26-27) – figure 7. These sinkers could be cast on board as the pieces of raw lead and, especially, the lead-coated spoon bowl found in the Yassi Ada wreck would testify (Katzev, 1982b, 280-281).

Sometimes, in Greek and Roman wrecks, clay weights both of pyramidal (figure 8) and lentoid shape (figure 9) can be present. Since they are very similar to loom weights, some scholars have not interpreted them as fishing gear, although they had been found in wrecks. This interpretation, curious at least, has been given to the objects recovered in both the Archaic wrecks of Gela (Panvini, 2001, 62; 95) and the Pisa San Rossore river harbour excavation (Bruni ed., 2000, 103).

Evidence that could support the interpretation of these finds as fishing sinkers, at least when they are found in a wreck or in a harbour, is provided by the discovery of a cluster of pyramidal terra-cotta objects in the excavation of a site at Eliza-

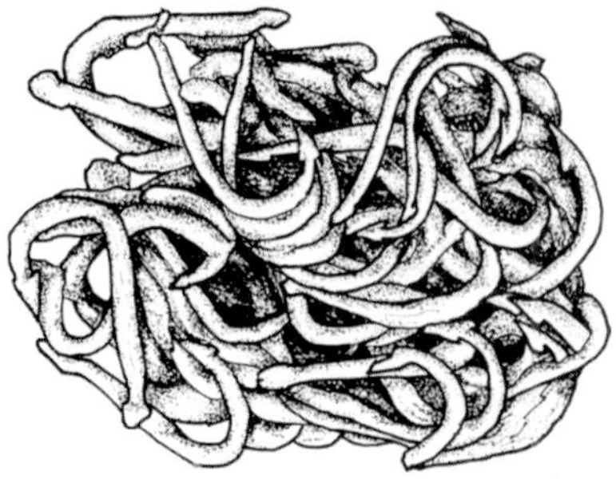

A

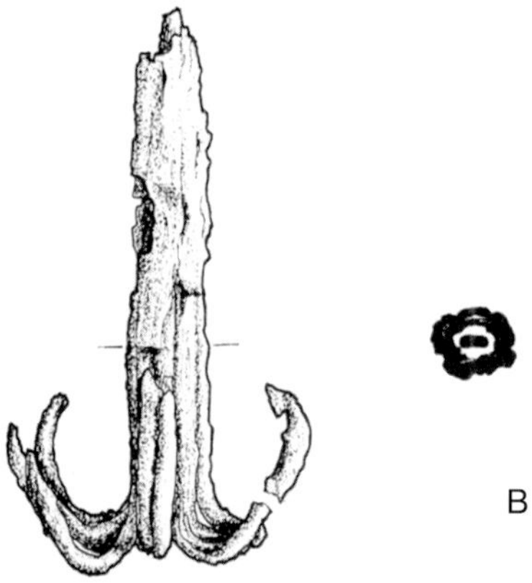

B

Figure 4. A. Hooks contained in a wicker basket of the Roman ship of Comacchio (after Berti, 1990, 271). B. A grapnel hook from the Roman ship of Comacchio (after Berti, 1990, 271).

vetovka on the norther coast of the Sea of Azov (Højte, 2005, 135). Another indirect testimony to the use of nets aboard is the discovery of net needles in some wrecks, such as the Byzantine wrecks of Yassi Ada (Katzev, 1982b, 262) and Serçe Limani (Piercy & Bass, 2004, 421-22). These objects have preserved the same shape from the ancient times to the modern age (figure 10).

The small lead rings that were found in the Archaic wreck off the Isle of Giglio (Bound, 1991, 26-27) – figure 11 – and in the Roman wrecks of Grand Congloué (Benoît, 1961, pl. 30) and Mahdia (Päffgen & Zanier, 1994, 127), but more often at the sites used for anchorage, have been interpreted as net sinkers, too, in contrast to their past interpretation as sail rings (figure 12, lower part). In our study of life aboard Roman ships (Beltrame, 2002, 67) we have reviewed the previous theory about their function as parts of the sails. Here, we have, in fact, suggested that it seems more logical that the rings, which are represented in iconography on the square sails of ancient ships, were made of wood, therefore of a less heavy material. A possible proof for this assumption comes from the recent discovery of a piece of sail with wooden rings in Egypt (Whitewright, 2007). In addition, 25 lead rings were found in the galley area of the Cap Lardier 4 Roman wreck (Beltrame, 2002, 67-68) – figure 13. Since they were found near a hearth used to cook aboard the ship, it seems quite obvious that they must have been attached to a net rather than

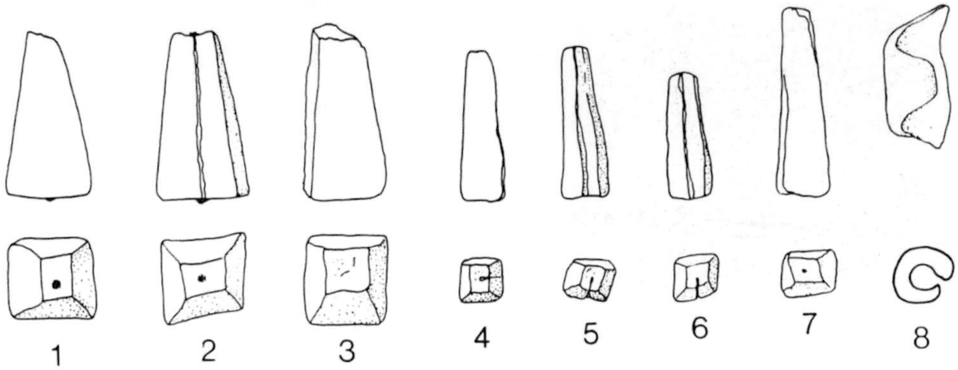

Figure 5. Fishing line weights from the Cavaliere Roman wreck (after Charlin, Gassend &, Lequément, 1978, 52).

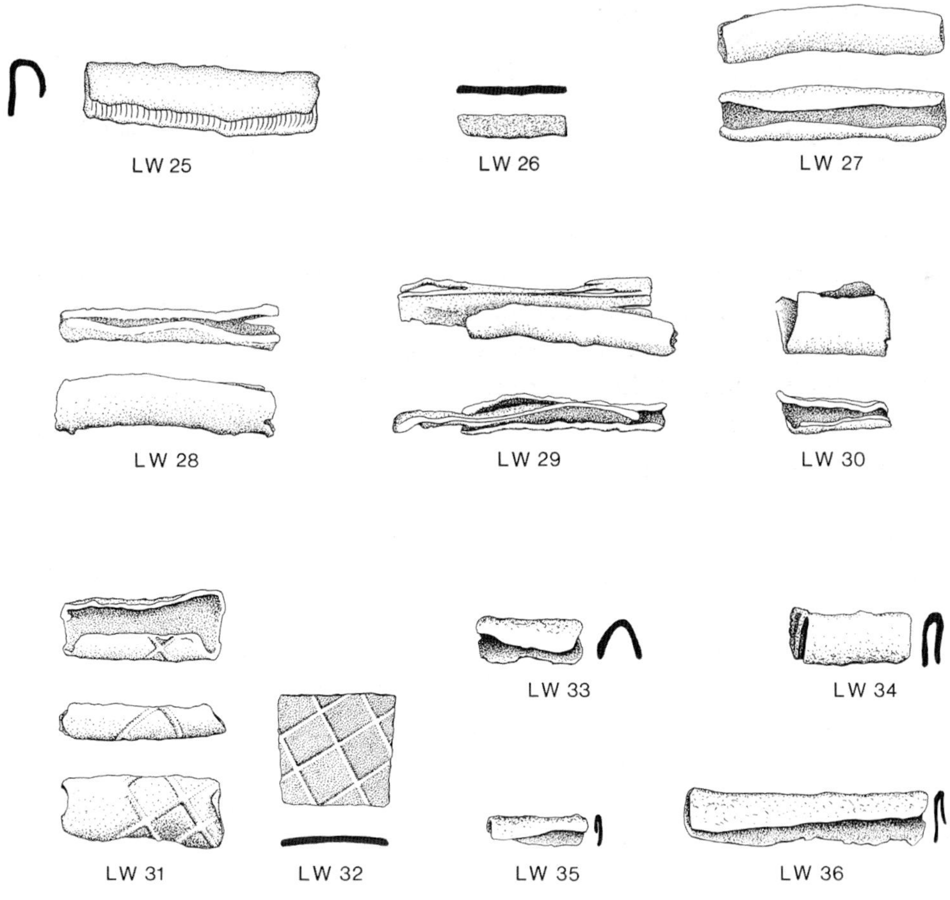

Figure 6. Net weights from the Byzantine wreck of Yassi Ada (after Kuniholm, 1982, figure 13.5).

Figure 7. Net weights from the Archaic wreck of the Isle of Giglio (after Bound, 1991, 27).

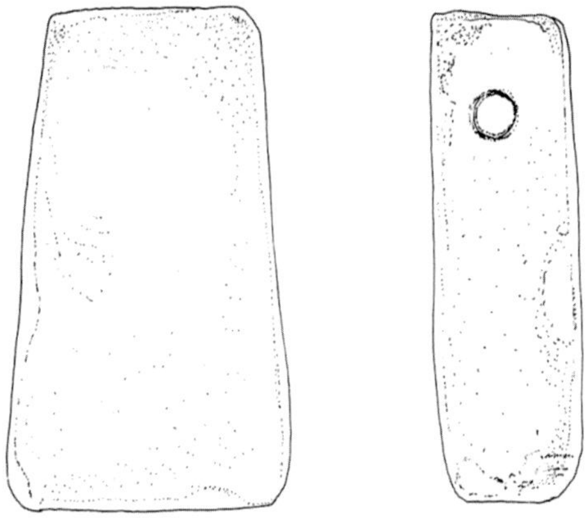

Figure 8. Terracotta net weight from the Roman ship of Comacchio (after Berti, 1990, 271).

Figure 9. Terracotta net weight from the Archaic wreck of the Isle of Giglio (after Bound, 1991, 27).

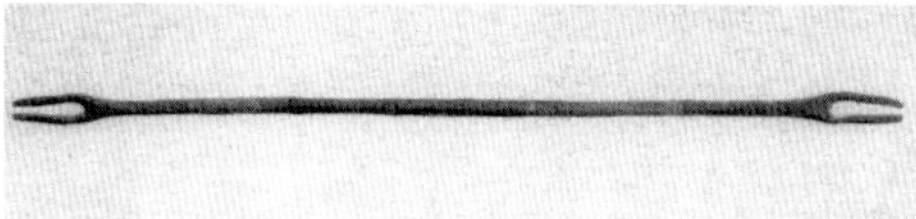

Figure 10. Modern netting needle.

Figure 11. Lead rings from the Archaic wreck of the Isle of Giglio (after Bound, 1991, 27).

to a sail. Finally, we should note that such lead rings were also recovered on fishing-related land sites such as the fisherman's hut at Casa Brancazzi near Orbetello, Italy (Cavallo, Ciampoltrini & Shepherd, 1992, 104 and figure 1).

The biggest rings, with an external pierced protuberance, found both on Greek and Roman ships (figure 12, top), are likely to have had a different use. An interesting interpretation suggests that the holes could be used to fasten a rope, which could be used to free nets or lines. The rings perhaps were sent down, along the line or the net, by a working line, to pull the hooked line down from the snag which held it, freeing it and giving the fisherman the chance to recover his gear (Foerster Laures, 1989).

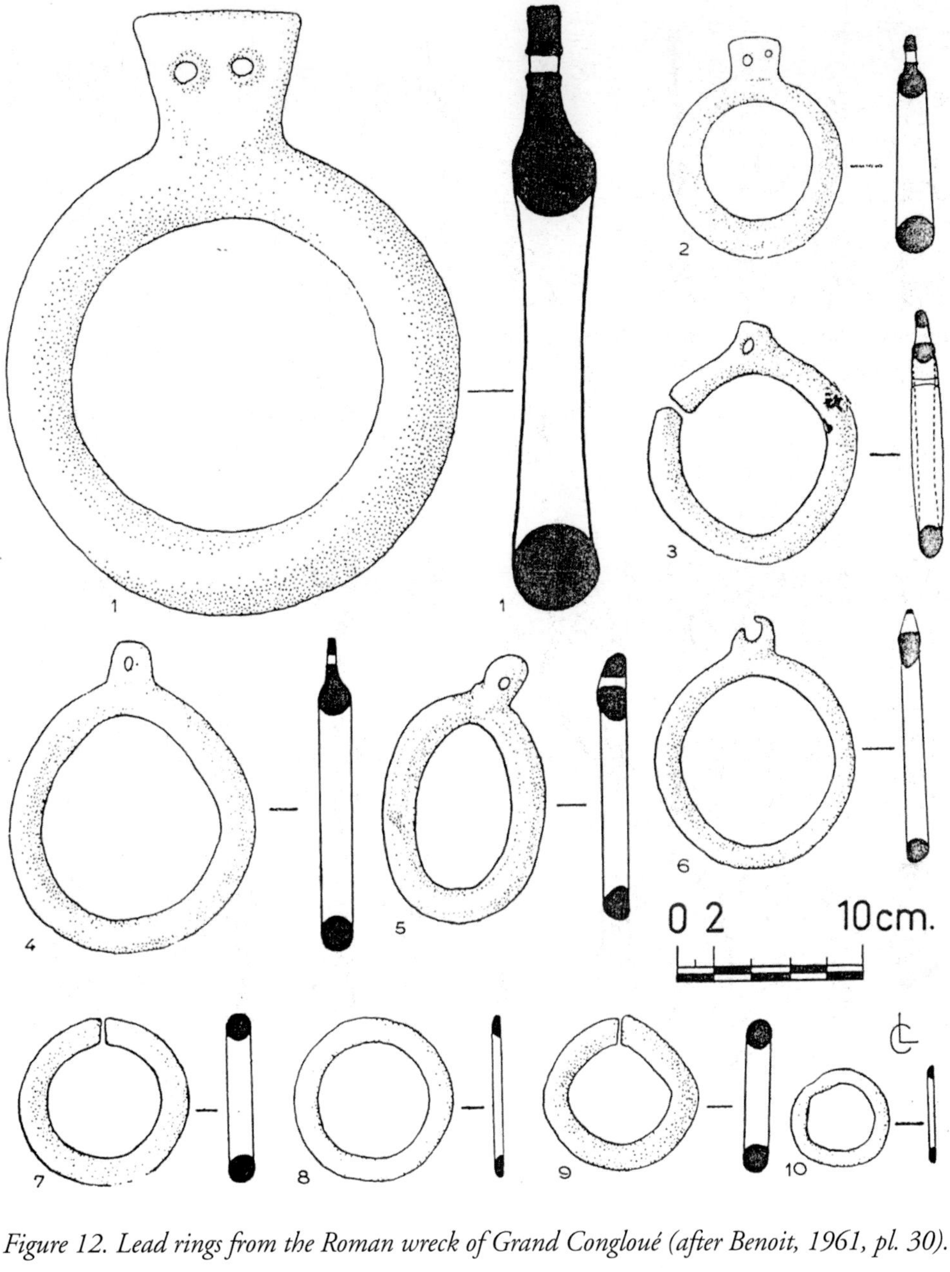

Figure 12. Lead rings from the Roman wreck of Grand Congloué (after Benoit, 1961, pl. 30).

The quite small number and size of lead sinkers occurring in wrecks, could indicate that they belonged to rather small nets. We can then presume that only casting nets (amphiblêstra), well illustrated in the African mosaics, were aboard ships, which is to say that nets were used only when the ship was at rest, in port or at anchor.

No evidence of floats for seines has been found in wrecks of ancient ships. In this regard, the only finds we are able to mention are the objects, made both of

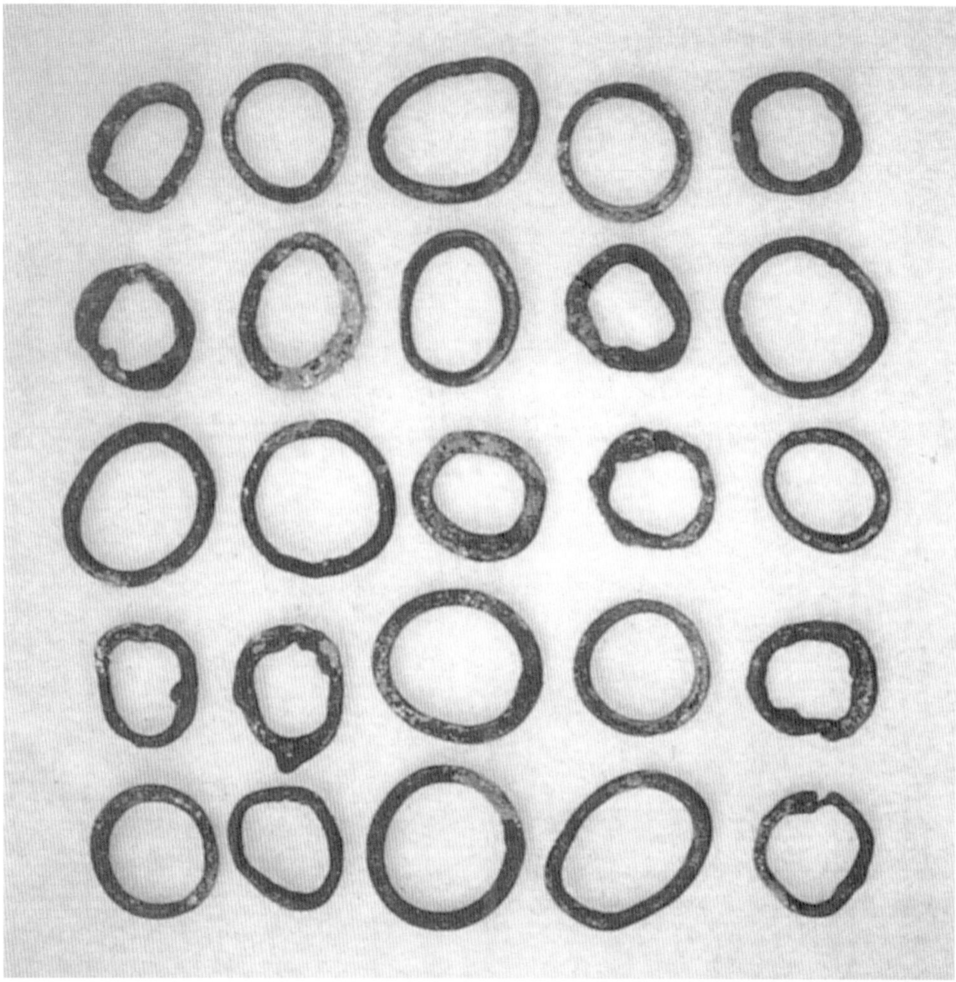

Figure 13. Lead rings from the galley area of the Cap Lardier 4 Roman wreck (photo J.-P. Joncheray).

bark and wood, discovered in the eleventh-century Serçe Limani (Piercy & Bass, 2004, 411-14) and perhaps in the ninth-century Bozburun ship (Hocker, 2005, 104). The perishable nature of wood and cork may explain why so few objects of this type survive on the shipwrecks, in contrast to the hundreds of lead sinkers.

It is important to note that the dating of fishing gear from wrecks and the association of fishing gear with an underwater site is not easy. Wrecks, in fact, are exposed to contamination by fishing activity in periods subsequent to the foundering of the vessel. The fishing implements that are lost over the site during these later fishing activities cannot be easily recognized, due to the strong conservatism in the design and use of this same kind of objects. Only a careful excavation will permit us to distinguish finds belonging to the wreck from contaminations.

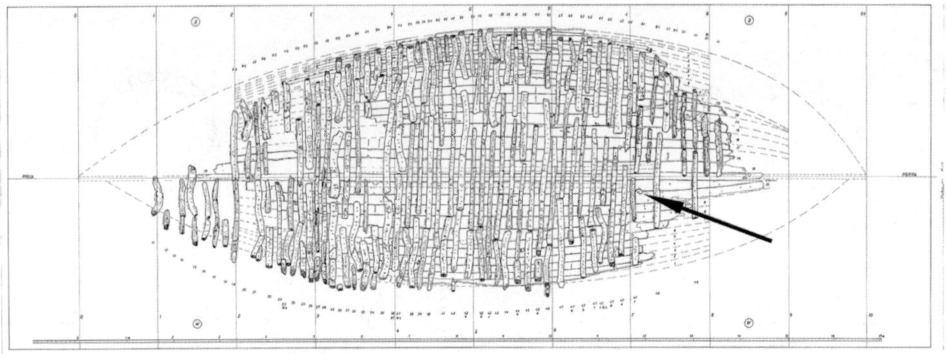

Figure 14. Plan of the hull of the Roman wreck of Grado. The arrow marks the hole in the hull where a double flange, connected to a lead tube, was located (drawing G. Penzo).

Conclusion

In short, the evidence from wrecks of ancient maritime ships indicates that, aboard vessels from the Greek period to the Middle Ages, fishing activity was not rare. However, a limited number of finds related to fishing in each ship could indicate a subsistence or recreational activity rather than a professional one. That is to say that we still have no archaeological proof of offshore fishing for the market in ancient times. The earliest evidence would be the Serçe Limani Byzantine wreck where it was calculated that three nets, 40 metres long, were aboard (Piercy & Bass, 2004, 424). The evidence of fishing from sporadic finds, because difficult to identify, is also quite poor (Gianfrotta, 1999, 14-26). In any case, a greater attention to this kind of isolated traces could give us very important information about aother aspects of fishing from ships as the technique of coral fishing (Galasso, 1997, 122-124; Gianfrotta, 1999, 16-18).

Then, it is also clear that this activity could have limits because of the well-known difficulty keeping the fish fresh or alive.

The Fiumicino boat no. 5 would offer the earliest evidence of the practice of keeping fish alive in well-boxes aboard (Boetto, 2006a). Here, a wooden box, in the middle of the boat, was filled directly from the sea through some holes in the bottom of the hull. A similar solution, used in many fishing vessels of modern times, could be present aboard ancient maritime ships too.

Another solution could be the use of a tank filled with water by pumps. Although there is no evidence for any kind of fish tanks on ancient marine ships, the second-century wreck from Grado offers very important news. On the floor of the hull, near the keel, a lead tube was connected to the strake planking, exiting through the bottom of the ship (figures 14, 15, 16). Piero Dell'Amico, the naval archaeologist who excavated the wreck, has proposed that a piston pump was used to recover water from the bilge. This pump would have discharged water not from the deck, which was the normal arrangement on an ancient vessel, but from the bottom in a very modern

Figure 15. Lead tube with double flange inside the hull of the Grado wreck, near the keel (after Beltrame & Gaddi, 2005, figure 13).

Figure 16. The hole in the planking of the hull of the Grado wreck after the removal of the double flange with the lead tube (photo the author).

way. If true, this would be the earliest evidence of the use of a piston bilge pump aboard ancient ships where, normally, simpler chain-pumps are attested.

Recently, Dario Gaddi and I (Beltrame & Gaddi, 2005; Beltrame, Gaddi & Pavizzi, in press) have demonstrated that a bilge pump of this kind has no sense and a more logical explanation was that the presumed piston pump was used to suck water from the sea. But at this point the question is: what was the purpose of this system? The only explanation we can propose is that the water was used to fill fish tanks in order to change the water frequently, which was necessary to keep fish alive.

Considering that the lead tubes are the only evidence of this system, we must assume that the tanks had been dismantled in order to convert it to a freighter, to carry the cargo of amphoras found over the hull of the wreck. On the other hand, we should not forget that the amphorae carried in this ship were filled with pieces of fish, which is to say the *nauclerus* of the vessel was engaged in the fish trade.

Obviously, because of the poor evidence, this interpretation of the Grado ship requires caution (Oleson & Stein, 2007; Beltrame & Gaddi, 2008) but it seems clear that it could open a very important window on the problem of fishing in ancient times.

Fiumicino 5 wreck during excavation (photo Soprintendenza Archeologica per i Beni Archeologici di Ostia).

9. Fishing vessels in Antiquity: the archaeological evidence from Ostia

Giulia Boetto

Introduction

A large number of ancient Mediterranean merchant ships have been examined on the seabed, but very few of them can be attributed to activities other than trade (Gianfrotta & Pomey, 1981; Parker 1992; for fishing activity on merchant vessels, see Beltrame, 2002 and this volume). The study of ancient Mediterranean warships, for instance, is mainly based on iconography (Basch, 1987). We have only two archaeological wrecks that can be identified as warships: the Punic ship of Marsala, Sicily (Frost, 1976) and the ship driven ashore at Athlit, Israel (Steffy, 1994, 59-61).

Since the late 1950's and early 1960's, when the seven Fiumicino wrecks were discovered within the silted maritime harbour of Rome (Boetto, 2001 and 2006b), the number of known non-merchant vessels is growing. Rescue archaeology finds in urban areas previously occupied by ports, channels or rivers have provided a new body of evidence (Pomey, 2009).

However, within the large category of non-merchant vessels, it is very difficult to distinguish between a fishing boat and a craft used for other activities (for harbour services, as tug, as tender for cargo ships, etc.). This is the case, for example, for the *horeiae* of Toulon (first century AD; Borreani *et alii*, 1988; Brun, 1999; Boetto, 2009) and Naples (end of the first century AD; Boetto, 2005 and 2009) the *Herculaneum* boat (first century AD; Steffy, 1985 and 1994; Bockius, 2002), the *Pisa C* vessel (first century AD; Camilli, 2002 and 2004; Barbagli, 2005; Bonino, 2006) and the *Pisa F, G* and *H* small boats (second century AD; Camilli, 2004 and 2005), the Lake Kinneret ship (Israel, first century BC to late first century AD; Wachsman, 1990a, 1990b and 1995; Steffy, 1994) and the shipwreck *Jules-Verne 8* from the Place Jules-Verne in Marseille (*c.* third century AD; Pomey 1995, 462-463; *idem,* in press).

In any case, the small dimensions of these craft (no more than 10m in length), their shape or their propulsion (oars with, sometimes, a complementary sail) could not be considered as unequivocal indications of fishing activity. Nevertheless, at least two shipwrecks, *Jules-Verne 9* and *Fiumicino 5*, have been securely associated with fishing.

The Greek archaic (last quarter of the sixth century BC) shipwreck *Jules-Verne 9*, excavated in 1990 within the ancient harbour of Marseille (Pomey, 1995, 471-75; *idem*, 1998), was a light and fast boat 9.50m in length propelled by with oars and sail (Pomey, 2003). This sewn boat has been associated with fishing activities because of the discovery of fragments of coral within the vessel inside the hull (Pomey, 2000).

The *Fiumicino 5* boat dating from the late second century AD (figure 1) and discovered in 1959 within the basin of the harbour built by the emperor *Claudius* three kilometres north of Ostia in the first century AD, was a fishing boat equipped with an internal wet well for keeping live fish (Boetto, 2002b, 2006a and 2006b; on the port of Rome, see Keay *et alii*, 2005).

We should also mention another shipwreck, discovered in 1965 not far from *Fiumicino 5*, that has been tentatively identified as a fishing boat (Boetto, in press): the third century AD *Fiumicino 4* sailing vessel. This craft was about 10m long with a load capacity of 3.79 tons; it had an elegant shape and was finely built. Its structural characteristics are typical of the largest maritime cargo vessels (Boetto, 2002a and 2006b). It could have been used for a variety of assignments, such as transport, travel for pleasure or other purposes.

The fish-well boat *Fiumicino 5*: architecture and function

Fiumicino 5 was about 5.60m long and had a beam of 1.50m (figures 2 and 3). Amidships, the bottom was rounded. The keel was rectangular in cross-section and had no planking rabbet except in the extremities. These were joined in hook scarfs forward (stem) and aft (stern). The garboards were attached to the keel with pegged tenons, to the stern and the stem with copper nails. Planking, about 2cm thick, was edge-to-edge joined by pegged mortise-and-tenon joints. The frames were not arranged in the standard alternation of floor timbers and paired half-frames. Their interval varied from 5cm to 23cm. Wooden treenails, copper and iron nails held the frames to the planks.

According to the xylological analyses, the keel extremities were fashioned from oak, the planking was made of cypress and the frames were mostly juniper. Other tree species were also identified, such as stone pine, cypress and oak.

The fish-well occupied a central position (figures 1 and 2). Its planking was fashioned from elm, oak, stone pine and cypress. It was 89cm long, 85cm wide and 56cm high. Its maximum capacity was about 300 litres of water. Nineteen holes, from 2.5cm to 3cm in diameter, assured the continuous flow of water into

Figure 1. Fiumicino 5 wreck during excavation (photo Soprintendenza Archeologica per i Beni Archeologici di Ostia).

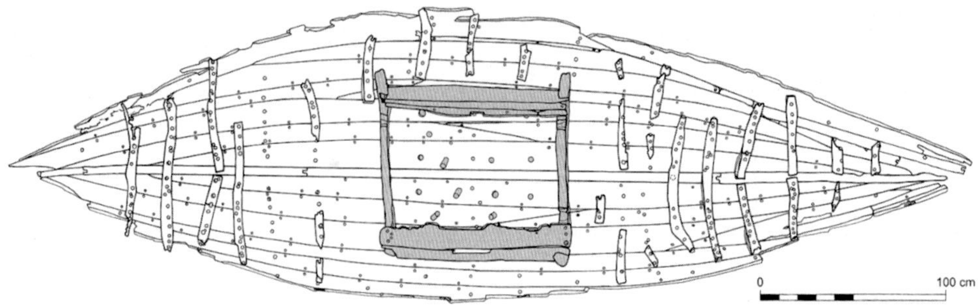

Figure 2. Plan of Fiumicino 5 wreck (drawing G. Boetto).

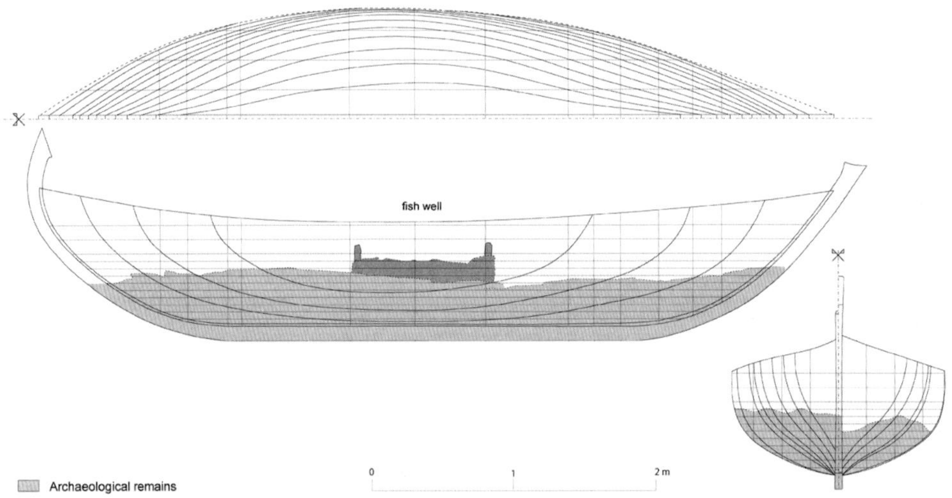

Figure 3. Reconstruction plan of Fiumicino 5 boat (drawing G. Boetto).

the fish-well (figure 4). Several of these holes had been closed with plugs carved from stone pine branches and measuring from 10cm to 14cm in length.

The constructional characteristics observed in *Fiumicino 5* allowed us to identify its function and its typology. Since fish could be transported alive in the central fish-well, this craft can be identified as a fish-well vessel.

Fish-well vessels in written sources

We know from Athenaios (*Deipn.* 5.208a) that the *Syrakousia,* a giant vessel built for the tyrant Hieron II of Syracuse in the third century BC, possessed a fish-tank filled with live sea fish. This fish-tank, built with wooden boards and coated with lead, was an integral element of the ship's structure and it was filled with sea water.

Figure 4. Fiumicino 5: the bottom of the fish well with the holes and the plugs (photo G. Boetto).

In the first century AD, *Optatus*, Commander of the Fleet of *Misenum*, imported live parrot wrasses *(scari)* from the Carpathian sea (the southern part of the Aegean). He scattered them offshore between Ostia, at the mouth of the Tiber, and the coast of Campania (Pliny, *N.H.* 9.62-63). Macrobius (*Sat.* 3.16.10) named the ships used for the transport of the live fish as *naves vivariae.*

Wet-well vessels have probably been used to restock some lakes of Latium with marine fish and to transport a special kind of oysters from Brindisi to *Baiae* near Naples (Gianfrotta, 1999, 24).

We can imagine that, thanks to the *naves vivariae*, fish in particular demand, such as red mullets from Corsica and Taormina or moray eels from Sicily (Juvenal, *Sat.* 5.92-99), could reach the market of Rome. Moreover, it was possible to supply directly to the imposing banquets of the wealthy Roman aristocracy (Suet. *Vitellius* 13).

In conclusion, we can infer that different types of fish-well vessels (*naves vivariae*) existed in Antiquity. The *navis vivaria* used by *Optatus* was bigger than the small boat *Fiumicino 5*. In the *Syrakousia*, the giant ship built for the transport of grain and then transformed in a fabulous floating palace (Zevi, 2001; Pomey & Tchernia, 2006), the internal wet-well was only one element of its equipment.

These examples already reveal the large variety of types and sizes of fish-well crafts. A similar range of variation can be found in nautical archaeology and ethnography.

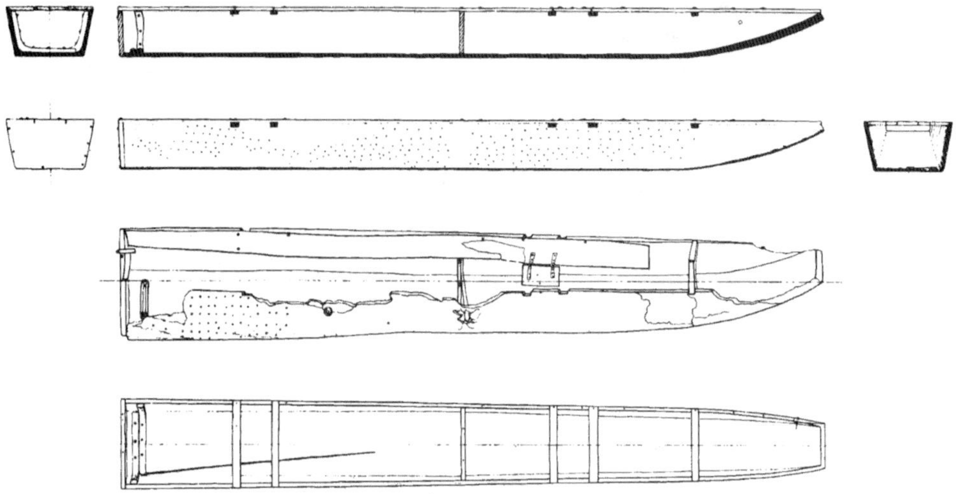

Figure 5. Zwammerdam 5 wreck, Netherlands (drawing after de Weerd, 1988).

Fish-well vessels in archaeological and ethnographical sources

The *Zwammerdam 1* and *5* pirogues, found in the Netherlands and dating from the second or third century AD, are *naves vivariae* (figure 5). These pirogues were re-used as floating fish-tanks after drilling a large number of holes through their hull planking (de Weerd, 1988; Arnold, 1995, 118-119).

In Central Europe, we know of some pirogues equipped with a fish-well, normally made from one log (*monoxyloi*), where the bottom is perforated by holes. They date from the Medieval period and were used in rivers and lakes (Arnold, 1995, 138; 180; Ossowsky, 1999a, 46, 114-115; *idem,* 1999b, 65).

During a recent excavation along the bank of the Saône river at the Parc Saint-George in Lyon, seven fish-well vessels were found. These vessels are dated to the sixteenth century (Ayala *et alii,* 2005). Finally, we could mention the two wrecks of *waterschepen* excavated in the Zuidersee, Netherlands, and likewise dated to the sixteenth century (figures 6 and 7). The excavators estimated that one of these vessels had a carrying capacity of *c.* 10 tons of water and fish (Reinders *et alii,* 1986; Pedersen, 1996).

The ethnographical sources for fish-well ships are especially abundant for the nineteenth century (Boetto, 2006a). In Europe, the fish-well ships have a wide geographical distribution and were used inland, on lakes and rivers, along the coasts and on the open sea. Their form and dimensions vary according to their transport zone and to their fishing activity (see also Duhamel de Monceau, 1771).

However, fish-well craft can be used for four principal functions:

a) transport of fishing gear: the boat is used to reach the fishing sites and to transport the fishing equipment;

Figure 6. The waterschip ZN42, Netherlands (photo after Pedersen, 1996).

b) conservation: the wet-well is supplied with an ever-changing flow of water from the sea or the river penetrating through holes bored in the bottom of the hull at the position of the storage well;
c) transport: the fish, conserved in the wet-well, could be transported alive from the fishing sites to storage infrastructure or directly to the market;
d) long term storage (or stabulation): the fish could be kept alive in the wet-well for a longer period for different purposes. For example, in order to increase their size or wait for their selling price to go up.

In the Venice and Comacchio lagoons in the northern Adriatic region, the fish-well vessels known as *marote* could reach 14m in length (figure 8). The *marote* were punted or towed to the fishing areas, and were generally used for the storage and transport of eels. The eels could be held within these ships for weeks. The capacity of the fish-tanks varied from 500kg to more than 2 tons of fish. These special floating tanks formed part of a complex system of exploitation of brackish water lagoons for fishing and pisciculture called *lavoriere*. In the second half of the nineteenth century, convoys of *marote* could reach Naples through the strait of Messina or Milan by the Po river. The *burciello*, a vessel of more limited dimensions, was used to supply the fishing vessels called *bragagne* (Boetto, 2006a, with bibliography).

Figure 7. Waterschip ZN42: the bottom of the fish-well (photo after Pedersen, 1996).

Figure 8. Small marota from the Comacchio lagoon, Italy (photo after Cecchini, 1990).

From the nineteenth century, a flourishing lobster fishing activity developed in Sardinia. Small fish-well boats, 5m to 10m long, similar to vessels like the *gozzo* of Alghero and the *carlofortina* or *schifetto* of the island of S. Pietro near Carloforte, could hold 20kg to 40kg of spiny lobsters (figure 9). In these vessels, 6 or 12 drilled holes, each 4cm in diameter, were drilled through the second strake of the hull bottom. Plugs were used to block the holes of the wet-well. Once the holes were plugged and the water was evacuated, the fishermen could easily catch the lobsters. The well was empty when the boat was moored, during navigation in a calm sea and when a ballast was not needed to stabilize the boat.

Such small fish-well boats were commonly used to fish crustaceans along the rocky coast. Thanks to the wet-well, the lobsters could stay alive to reach the wooden fish-tanks or other special coastal installations situated in the fishing port. From there, bigger fish-well vessels called *ragostiere*, transported lobsters to the markets in Spain, France and Italy (Pomata & Sanna, 1998; Boetto, 2006a).

In these traditional vessels the apertures have different forms and dimensions according to the fish species stored in the wet-wells. The *marote* and the *burcielli*, used for eels, had small round holes or small rectangular carved openings (figure 8; on the dimensions of holes, see also Duhamel de Monceau, 1771, 27). Similarly, the Dutch *waterschepen*, used for sardines, had very small holes (figure 7). On the other hand, the wet-wells for lobsters had holes up to 4cm in diameter. As a result, eels could be transported within these fish-wells only after placing a grid or a net on their bottom (Boetto, 2006b).

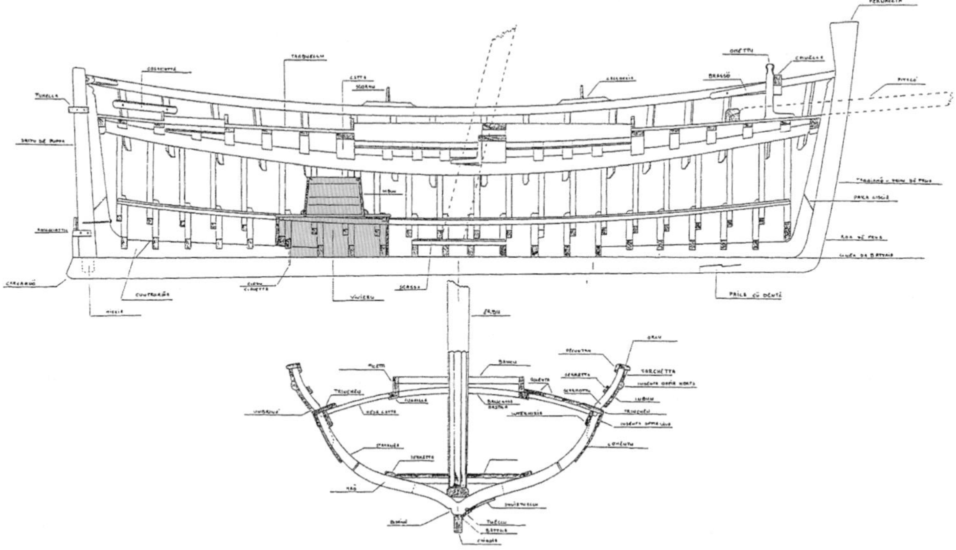

Figure 9. Schifetto from Carloforte, S. Pietro island – Sardinia, Italy – (drawing after Pomata & Sanna, 1998).

These considerations on ethnographical examples can help us towards a better understanding of the function and use of the *Fiumicino 5* fish-well.

First of all, the wet-well of *Fiumicino 5* is placed amidships and does not occupy the full length of the boat, unlike the floating fish-tanks used for storage in the northern Adriatic region. In *Fiumicino 5*, alongside the wet-well it was possible to store propulsion equipment (oars) and fishing gear.

In addition, if we consider the limited capacity of the wet well, we can suppose that the catch was conserved during the time required to return from the fishing grounds, and for the live fish to reach the temporary storage basins and/or the market. Therefore, a small craft such as *Fiumicino 5* could actually be part of a composite transport system of live fish, similar to the one ethnographically documented for lobster transport in Sardinia.

Thirdly, based on the dimensions of the holes in the well (from 2.5cm to 3cm in diameter), *Fiumicino 5* could hardly contain small fish or small crustaceans (figure 4). The transport of eels would probably only be possible after having covered the bottom of the well with a metallic grid or a net.

Finally, plugs could be removed in *Fiumicino 5* to evacuate the water when the boat was moored, as was the practice in the Sardinian *schifetto*. We can also imagine that these plugs could be used to improve the capacity of the container. In fact, in all fish-well ships the water fills the container up to the boat's floating level. If the quantity of the water was insufficient for preserving the catch, the fisherman had to put in more water after having plugged the holes. But, in this way, the wet-well was not kept supplied with an ever-changing flow of water. We may also imagine that plugs could be used to keep marine water inside the container during riverine navigation or *vice versa.*

The transport and fishing zone of *Fiumicino 5*

From our architectural analysis we infer that *Fiumicino 5* was locally constructed. Based on dimensions and propulsion, its navigation and fishing zone was restricted to a limited navigation zone, notably the Tiber delta with its coastal lagoons, the river mouth and, probably, also the stretch of the river up to Rome (Boetto, 2006a and 2006b) – figure 10.

The younger Pliny (*Ep.* 2.17) informs us that due to the sandy seabed, Ostian waters were not abundant in valuable fish, but yielded excellent soles and prawns. If we look at a modern fishing chart, we can observe that the nearest rocky sea floor is found at Tor Paterno, south of Ostia. These shoals are well known to local fishermen (Fusco, 1941). In any case, the Ostian waters were being over-fished, as Juvenal (*Sat.* 5.92-96) asserts, and, as we have seen, *Optatus* had added variety to the local catch by introducing parrot wrasses from the Aegean transported in *naves vivariae* (Pliny the elder, *N.H.* 9.62; Macrobius, *Sat.* 3.16.10; see above, p. 247).

During the Roman period, the brackish areas of the Tiber delta were exploited as salt marshes (Giovannini, 1985 and 2001; Morelli *et alii*, 2004). Fishing structures (fixed traps composed of wooden posts), dated to the Medieval period, have been recently identified north of *Portus* (Morelli *et alii*, in press). In that period, fishing communities were active on the western lake of Maccarese (*stagnum maius*), the eastern lake of Ostia (*stagnum hostiense*) and the hexagonal basin of the port of Trajan (*lacus Traiani*) (Boetto, 2006a, with bibliography).

We can assume that brackish water fishing was also important in Antiquity, as attested in the lagoon of Cosa on the coast of Tuscany (Gadza & McCann, 1987) or in the lagoon of Lattes, France (Sternberg, 1995, 2002, 2005a, 8-11 and 2005b). Brackish water fishing is based on the migratory habits of some fish species. Fish are captured by means of a complex system of screens formed by bundles of cane positioned in such a way so as to lead the fish into a terminal trapping chamber (Sternberg 1998 and 2000b).

In Antiquity, the Tiber river was also very important for fishing (De Grossi Mazzorin, 2000). The most prized of the river fish was *lupus*, bass (*Dicentrarchus labrax* L.), and Rome's fishermen were organized in a corporation (Le Gall, 1953, 268-270). Tiber fishing also flourished in the Medieval period when fishermen built, along the riverbank, fixed traps named *piscariae* (D'Onofrio 1980, 48, figure 21; Motta, 1986, 124). Here, the catch could also be stored and kept alive: this kind of con-

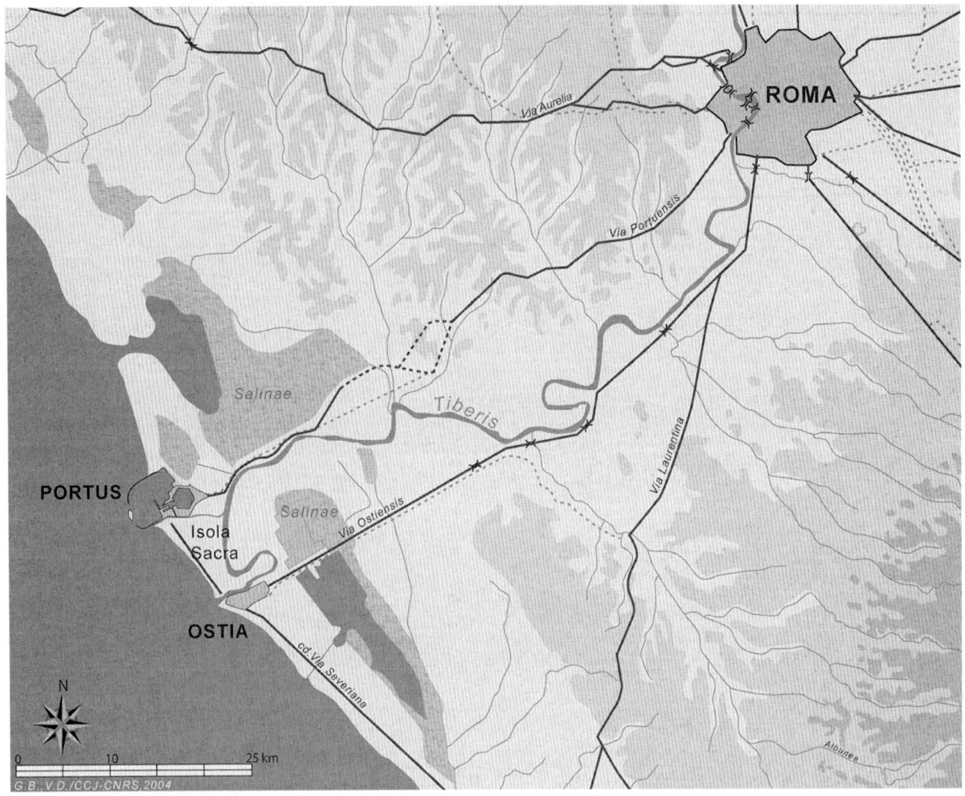

Figure 10. Transport and fishing zone of Fiumicino 5 boat.

servation was advantageous for consumers – who could have fresh fish at their disposal – as well as for fishermen – who could demand higher prices (Vendittelli, 1992, 392-97).

Finally, recent archaeological studies of maritime villas situated to the north of *Portus* near Ladispoli and Sta. Severa (Giacopini *et alii,* 1994; Rustico, 1999; Pellandra, 1997) have established a functional relationship between ponds *(piscinae)* and private harbour structures. These ponds could have been used for the temporary storage of fish before its transport to the market or to private tables (Rustico, 1999, 54). A craft similar to the *navis vivaria Fiumicino 5* could reasonably have been used in this kind of operation (see also Higginbotham, 1997, 34).

Conclusion

The use of fish-well ships, well attested from Classical Antiquity until recent times, is related to the demand for fresh fish to supply coastal and more distant markets.

A live fish is a living guarantee of its own freshness. To keep the fish alive for as long as possible by means, for example, of *naves vivariae*, was one of the *optimum* ways to avoid the quick decay of this highly perishable foodstuff.

Moreover, fish has a better taste immediately after the catch. This was well known in Antiquity and Seneca (*N.Q.* 3.18) tells us that the Roman gourmets "cannot taste a fish unless they see it swimming and palpitating in the very dining-room". The famous cook *Marcus Apicius* thought it especially desirable for red mullets to be killed in a sauce made with the entrails of other similar fish (Pliny, *N.H.* 9.66). Pliny the Elder (*N.H.* 9.95) asserted that "the spiny lobster is the only animal whose flesh is of a yielding texture with no hardness, unless it is boiled alive in hot water". In Rome, this crustacean was in great demand, and *Apicius* (412-416 [9.1.1-9.1.6]) introduced several preparation techniques and cooking recipes.[1]

In the light of the archaeological evidence and the written and ethnographical sources, our study of the *navis vivaria Fiumicino 5* confirms the efficiency of the transport and conservation technique of fish within fish-well ships and fish-well boats. This system was used in order to supply not only coastal markets but more distant markets as well, with fresh fish.

Consequently, the preservation of fish after the catch cannot be considered as a technological constraint on the development of Greco-Roman fisheries as has been previously asserted (Bekker-Nielsen, 2002b and 2005, 93). Beside the *naves vivariae* used for live fish transport, the Graeco-Roman world developed efficient systems to preserve large amount of fish. As such, we could mention the *piscinae* discovered along the Tyrrhenian coast in Italy (Giacopini *et alii*, 1994; Rustico, 1999) or the perforated *dolia* discovered along the Dalmatian coast, probably connected with the conservation of live fish (Radic Rossi, 2006). Perforated amphorae were also used in Roman times to conserve fish or fishing bait (Gianfrotta, 1999).

In conclusion, our study of the *navis vivaria, Fiumicino 5* emphasizes the importance of fresh fish as a valuable foodstuff in great demand among ancient populations. This boat, of local construction and use, is evidence of an important fishing activity in Rome and confirms the role of Ostia and *Portus* as fishing harbours (Meiggs, 1973, 267).

1 On the other hand, the sources do not mention the consumption of live eels. However, live eel seems to guarantee a better taste and an easier digestion after cooking (Coste, 1855).

The Uitgeest logboat (600 BC) during excavation.

10. Fish and "Chips of Knowledge": Some Thoughts on the Biases of the Archaeological Record

THIJS J. MAARLEVELD

Introduction

Maritime archaeology and fishing, one could hardly think of a better combination. Isn't the economic basis one of the strong points of archaeology and isn't fishing the major economical activity in the maritime sphere? There is ample reason to suppose that archaeology could provide important information. In fact, for extensive periods of human history and Prehistory, direct information from the archaeological record is readily available. But what of Classical Antiquity? Do we have extensive material remains of nets and fishing gears? Does the archaeological record provide us with ample direct evidence? Alas, no. In fact, direct information is scarce. So we mostly have to infer from indirect sources. How can that be? Would it have been otherwise if other factors had influenced the way the archaeological record has come about? That may be so. A possibilistic approach, however, will not bring us any further. Instead, this contribution will try to reflect on the way archaeological information is secured and the way archaeological knowledge is produced.

Such reflections can be useful in any discipline. It is certainly a useful exercise in the young discipline of maritime archaeology, where contingencies have so far determined the agenda. It is not just contingencies, however, but also a sequence of systematic processes that filter "what originally was" from what we get to know about it. Not all activities are equally likely to leave traces that will be embedded in deposits.

What happened *originally* is, however, only partly responsible for what we know. What happened *in the meantime* is another filter. Some materials stand better chances of survival than others. And finally: what happens on discovery? Management traditions, research traditions and research agendas are a strong filter on discovery. So is the present use of a particular area. But even more important, perhaps, are the biases in the perception of those who are most liable to see a piece of evidence first, and recognize it for what it is or not. It is on their observations that we are largely dependent: the present-day users of the maritime spheres, the modern successors of ancient and prehistoric fishermen.

Fishing and archaeology

Fishing was quite an industry in the ancient world. Epigraphic, literary and iconographic evidence provide an important background for the interpretation of techniques involved. But archaeological information is relatively scarce. Potentially, the archaeological record is very strong. It can provide very detailed and undeniably direct information on the gear, the material culture that was deployed in the fishing industry. The way in which archaeological data are acquired is, however, subject to all sorts of determining factors. The same evidently applies to the way in which the data are interpreted and transformed into chips of knowledge. So what I want to look at are fish and "chips of knowledge". How do we obtain information? How do we analyse it? How do we interpret its strengths and its weaknesses? How can we influence the processes that lead to information in the first place? I think it is useful to reflect on those issues and I am sure that every researcher does so, occasionally or on a regular basis. If one works with archaeological material, there is a lot to be said for doing so consistently, continuously and systematically.

The ways in which the "chips of knowledge" are approached apply in a very similar way to the archaeological material from different areas of the world, from different periods of the past or relating to different aspects of human activities. That is why I dared to accept the invitation to contribute to this workshop. I am certainly not a specialist on the classical world, nor can I claim to be a specialist on fishing. Occasionally and intermittently, I have dealt with the archaeology of fishing (Louwe Kooijmans, 1987, 238-242; Dean *et alii,* 1992, 35; Brinkhuizen, 1994). At the very start of my career I was employed as technical assistant at the National Museum of Antiquities and experienced a wonderful excavation (figure 1). I had to make sure that the pumps kept the excavation area at 8 meters below sea-level reasonably dry, but more interestingly, I spent day after day lying on scaffolding in order to excavate some incredibly well-preserved fish-traps. The traps are Neolithic (*c.* 4200 BC) and made from cornel-wood (dogwood). The growth of the individual thin sprouts show that the cornel-trees have been pollarded and cultivated for the particular purpose. The thin branches are bound together without knots in two-stranded rope (Louwe Kooijmans *et alii,* 2005, 263, 269; Out, 2009). What struck me then, and continued to strike me is how fragile our archaeological knowledge base actually is; how dependent we are on contingencies. This magnificent, but very small and deeply buried site at Bergschenhoek could easily have gone unnoticed.

Figure 1. Excavating the Bergschenhoek fishtraps of 4200 BC in 1978. The traps are made of pollarded cornel-wood (dogwood), bound together without knots by two-stranded rope (photo Rijksmuseum van Oudheden, Leiden).

On the other hand, an excavation like this can revolutionize our view on technology and culture in a particular area at a particular time (Fokkens *et alii*, 2008). This is what makes "wetland archaeology" and by extension "underwater archaeology" so fascinating. The body of knowledge that we may extract from a well-considered approach to the underwater world – previously unavailable for study – is simply stunning. But we can only see if we look. At present, what we have to go by are "chips of knowledge" – be it in European Prehistory, in Medieval and early Modern times, or in Classical Antiquity.

Against this background I want to examine how these "chips of knowledge" come about. How do we generate archaeological knowledge? In how far is that process dependent on contingencies? How is it biased in favour of one aspect over another? In what way does that influence what we know or can know about fishing technology in Classical Antiquity? In other words I want to approach the subject on the basis of archaeological theory. The reason for doing so is that if we understand such processes, we may be able to influence them.

Archaeological theory or theoretical archaeology has the reputation for producing very complicated schemes and texts. I will try to avoid doing so; I am convinced that in order to be useful, theory can and should be expressed in very simple terms, even at the risk of stating the obvious. My assertions will not be surprising at all, just common sense.

Rules of correspondence

The questions I posed above can be addressed with the aid of "rules of correspondence": how does what we know correspond with what originally was or what we want to know about it. How is our knowledge contingent, how is it biased. In unravelling the processes that influence the way things can correspond, we need to distinguish between three time frames in the history of an archaeological deposit. The first is what happened originally. The second is what happened since deposition and the third is what happened upon discovery.

What happened originally is of course what we want to know if we want to know or construct "how it was" or "wie es eigentlich gewesen" (Ranke, 1885). In fact, however, it is only a very specific part of historic reality that contributes to the "archaeological record", the body of direct, primary archaeological information. What we learn about what happened, is determined by that specific behaviour and those specific circumstances that defined what items, what vestiges came to be embedded in the earth or under the sea. It is a simple "rule of correspondence". It is a rule that is certainly important for what we know about fishing in Classical Antiquity.

What happened since deposition encompasses a whole body of processes affecting the integrity of the original deposit. Many of these processes relate to the natural environment and can easily be described more or less according to laws of natural science. If the salinity of the surrounding water is so and so and if the oxygen level is this, iron objects will degrade at such and such a rate. If organic materials are exposed and bacteria, fungi or marine creatures have access to them, they will be eaten away. With the very practical theoretician on the "formation processes of the archaeological record", Michael Schiffer (1987), we can call these processes Nature-transforms ("N-transforms").[1]

In this intermediate time frame, not only natural processes of protection and degradation determine what will remain, but also a range of human activities. Were spe-

1 Schiffer, in fact, distinguished these factors from "behavioural" rules which determined what came to be the deposited in the first place. Like all archaeology, his project is actually focused on interpreting what happened originally, although to many he has become associated with what happened since deposition (LaMotta & Schiffer, 2001).

cific items dumped in certain places? Were specific items picked up again? Have later strata been dumped on the same site? Has trawling, dredging or whatever affected the survival or integrity of the deposit? Analogously, such processes are called Culture-transforms ("C-transforms").

Now it is evident that these two time frames of formation processes with which Michael Schiffer has been concerned, and on which a lot of research has been done since his seminal book of 1987, are very important to understanding the archaeological component in our study of nets and fishing gear. There is every reason, however, also to look at the third time frame.

What happens upon discovery is a completely different matter again. It seems to be highly coincidental, but in fact it is not: definite patterns can be recognized which influence what is discovered and what is not. Such patterns are partly related to natural processes. At least as important, however, are cultural preferences, cultural traditions and human activities, not to mention the legal framework. For instance, the fact that by 1950 more wrecks from the classical period had been discovered in Greek waters than in Turkish waters, and that the balance has dramatically changed since then, is definitely a function of legislation. The law has promoted reporting of wrecks in Turkey, it has discouraged this in Greece. In an essay dealing with the general assessment of archaeological heritage in the United Kingdom, John Carman has stressed this particular type of influence by citing it as a Law- or "L-transform" (Carman, 1996).

But there is more. There are, in fact, many more factors that influence what happens upon discovery. The interests of discoverers are, for instance, mightily varied, but also mightily important. So are the interests of researchers and even the way in which the interests of researchers are interpreted in the popular image of archaeology (Russell, 2002; Holtorf, 2007). Let us look at that in a bit more detail.

Discovery and action

The theme of the present volume gives a very good illustration of the importance of interest and awareness. In fact, it is very strange indeed that we have quite a lot of direct information on the technology of fishing in a prehistoric period like the Mesolithic or the beginning of the Neolithic, more than in the much more extensive body of archaeological data relating to Classical Antiquity. This is even more striking since archaeology of the submerged environment began in the Mediterranean, was targeted towards Classical Antiquity, and has addressed vast amounts of wreck sites. Very few of these inform us directly about the fishing industry. Indirect information, although hard to extract, is dealt with elsewhere in this volume by Carlo Beltrame.

The seven steps of archaeological discovery		
Exposed	Earth moving or natural erosion	Short or prolonged
Provisionally recognized	Non-specialist observation	Many do not notice
Reported	Telephone call to presumed specialist or authority	Often goes astray
Recognized as relevant	Receiver decides on follow-up	Interest defines priority
Assessed	Observation by presumed specialist	Keep or discard
Described and inscribed	Note in official or specialist register	Registers vary
The archaeological find exists		

Table 1

We can hardly take it that this is the result of *what happened originally.* We cannot assume that more fishing went on between 5000 and 3000 BC than during the centuries of Classical Antiquity. Nor can it be a function of *what happened since deposition.* The fishing gear of the Mesolithic and Neolithic is not in itself sturdier than that of the Classical period (figure 2). In other words, the explanation must lie in the processes that led to discovery and what we consider to be primary archaeological information.

Actually, it is quite positive that the variation between the amounts of data regarding different periods of the past must be explained at this level. We cannot retrospectively alter the processes in the first two time frames, but we can certainly try to influence what will happen to future discoveries.

For material evidence to be accepted as archaeological information it needs to pass through several stages. Each of these steps is likely to fail (see table 1). This is especially true for the discovery of previously unexpected sites and for discoveries outside an archaeologically organized excavation. It is not enough that a site in which archaeological objects and information are contained is exposed. It is equally essential that traces or artefacts are recognized as of potential interest, that they are appreciated as significant enough to be inscribed in the "record", through a reporting system or otherwise. We only see when we look and we only look for what we want, or expect, to see.

Each and of the steps is completely dependent on a level of awareness. Quite obvious objects easily go unnoticed. Potential finds may lie disclosed for a considerable length of time, without anyone seeing or looking. Conversely, the window of opportunity may be very short. In large earthmoving operations exposure may be very limited. Few potential observers may be present, whereas someone needs to get the feeling that something special is exposed before even thinking about reporting

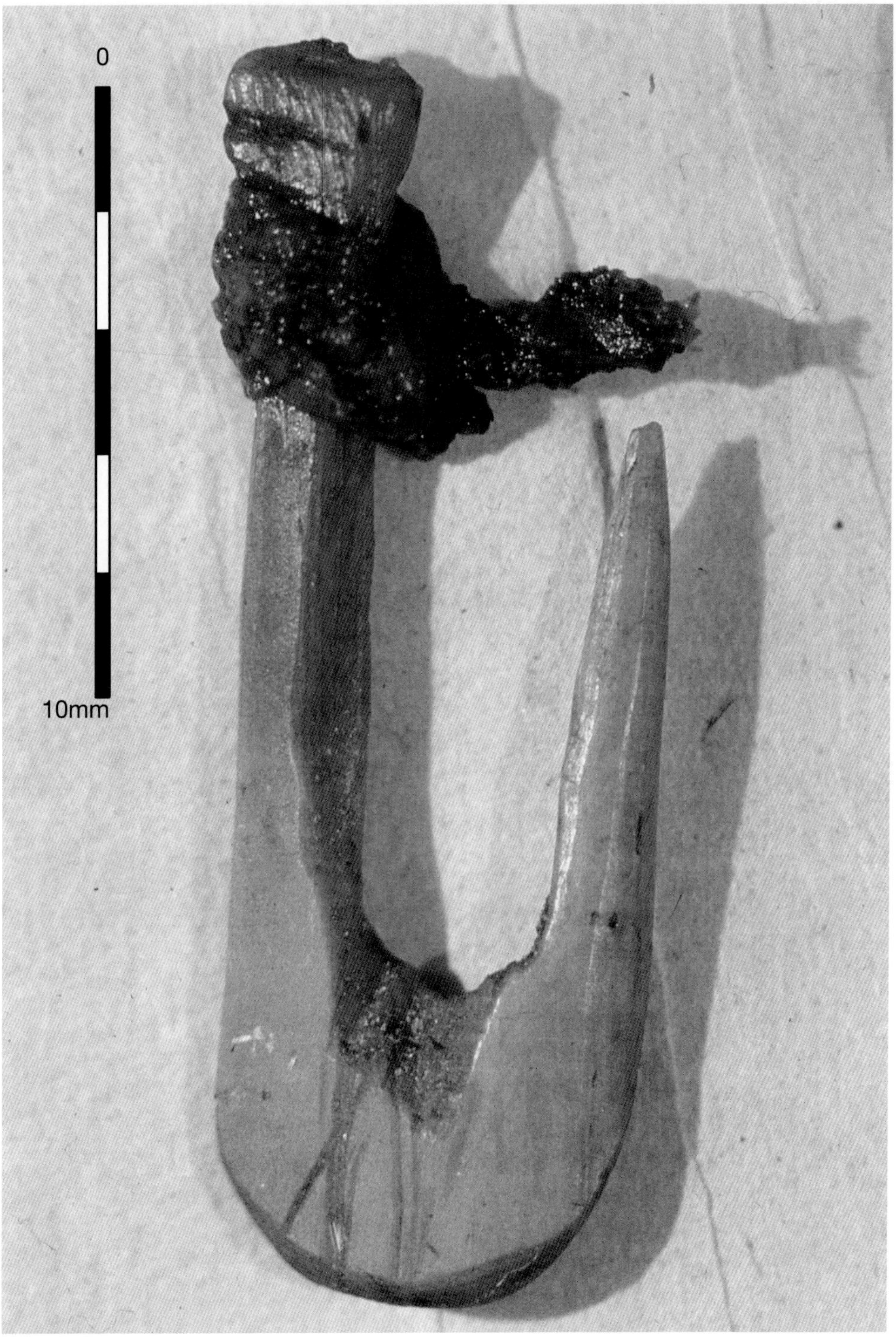

Figure 2. The fishing gear of the Mesolithic and Neolithic is no more robust than the fishing gear of Classical Antiquity. Fish hook from Tybrind Vig (4500-4000 BC). The hook is 2cm high (photo Søren H. Andersen, University of Aarhus).

Figure 3. A large fifteenth century river barge as found in the river Waal near Nijmegen in early May 2003. Everyone present during the clearing operation assumed the find was insignificant on an a priori *basis. In fact very little is known about fifteenth century shipping on the Rhine, of which the Waal is the major branch, and this vessel was more or less complete on discovery (photo J. van den Besselaar).*

it. More often than not "discoveries" are *a priori* recognized as insignificant by whoever is present when the site is revealed (figure 3). As such they will never enter our archaeological record, whatever their actual information potential: they simply will not exist.

When something is provisionally recognized as reportable, there may be many reasons why a report is postponed or ignored. Especially if it concerns something unconventional, people tend to be uneasy. They do not want to make fools of themselves or they may think "this will not interest archaeologists" – of whose interests they may have a very specific view. Sometimes they are right, even though the observations might be of interest for archaeologists in another archaeological tradition. And then, there is always the question of whom to report to. Official authorities? Someone one knows to be interested in "such matters"? Anyone who is consulted in the process may guide the reporter onwards, but is as likely to discourage further action as irrelevant.

But if, indeed, the report ends up on the table of a presumed specialist and is recognized as relevant enough for a follow-up, then note will be taken of the original observation. The findings will be assessed and if considered worthwhile, they will be described and inscribed in our common memory. Only then will they exist in the "archaeological record" But it is somewhat disquieting that this only happens after 7 steps, each of which has a very uncertain outcome. Inscription in official registers of archaeological finds and observations may be simultaneous, or may follow much later.

Some salient examples

To underline the statements made above about on discovery and action, ships and ship finds can serve as a good illustration. They certainly do not fall within the category of "small finds"; in fact they are often quite bulky. The discovery of the Bronze Age Dover boat illustrates how many contingencies were involved in that particular case (Maarleveld, 2004). Just like the Bergschenhoek find (figure 1), it could easily have gone unnoticed. No action would have been taken, no-one's interest would have been aroused. The Dover boat would simply not exist.

Another example is the Iron Age logboat of Uitgeest, discovered in November 2003 (Eerden, 2004). I happened to be at the receiving end of a chain of phone calls that led to an intricate rescue operation. Under strong pressure, the most unlikely partners could be convinced to cooperate (figure 4). It was the sheer attention and interest of the driver of a mechanical excavator and his boss that led to the existence of the find. Evoking latent interest elsewhere made it possible to rescue and preserve the find, and to raise sufficient funds at short notice. But what if that latent interest had not been strengthened by the explicit interest of scholars? Both the Dover boat and the Uitgeest logboat would not exist if they had not appealed to a curiosity and research interest that had implicitly or explicitly been expressed in a research agenda. Even the Nijmegen ship (figure 3) has become inscribed in the archaeological record, although too late to avoid major damage. It appealed to a research agenda. But what if no such research interest had been present? Then these finds would simply have been reduced to splinters without further notice being taken. It is again a simple question of correspondence, or non-correspondence perhaps.

Take a good look at the maritime finds in figure 3. Would they exist in the archaeological record if you had been present at the scene? Perhaps they would, but perhaps you would not have thought of them in archaeological terms. There are many similar examples of the latter from Northern Europe. Both maritime archaeology and historical archaeology were relatively late to develop. Archaeology did not focus on large pieces of wood and complicated constructions. Later, when the interest emerged and inventories of sites were to be built up, many stories emerged. Ma-

Figure 4. The Uitgeest logboat (600 BC) being excavated after a block lift. The evocation of latent interest made it possible to rescue and preserve the find and to raise sufficient funds at short notice (photo Th.J. Maarleveld).

terial that later came to be looked upon as significant had not been taken note of. The discoveries only started to exist after they were registered systematically. The recognition of the significance of the *Vasa* or the Skuldelev ships was the archaeological exception rather than the rule (Cederlund, 2006; Crumlin-Pedersen & Olsen, 2002).

Another good example is what happened with three fifth- to seventh-century vessels that were discovered during dredging of the river Scheldt in the course of the twentieth century. In retrospect, these were hugely important finds. A reconstruction of what happened on discovery indicates that the ships had been more or less intact. It was not the ships themselves, however, but their decorated figureheads that attracted the attention of successive dredge masters. After a detour through the art market all three ended up in the British Museum (figure 5). It was only subsequently that the finds were inscribed in the archaeological record (Bruce-Mitford, 1970; Ellmers, 1972). Evidently in this case, we only know the figureheads and not the ships to which these decorations belong. On hindsight, those would make excellent research material, but apparently the interest was limited at the time (the thirties, forties and fifties of the twentieth century).

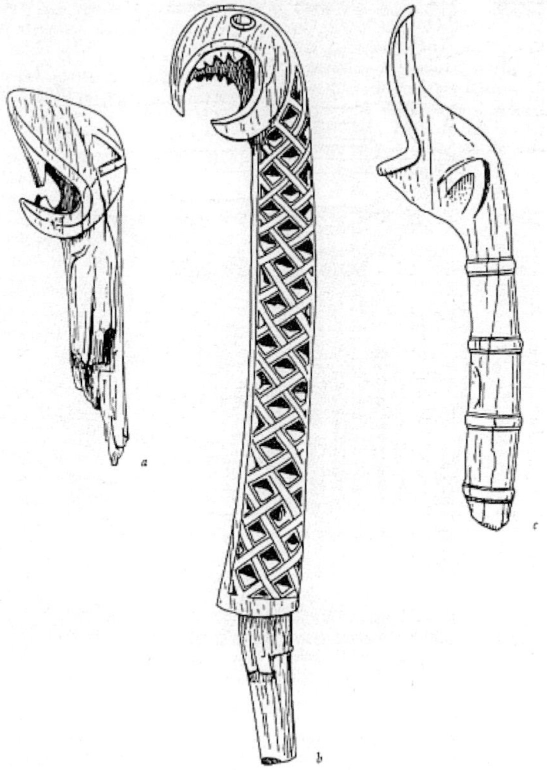

Figure 5. The early medieval stem decorations that ended up in the British Museum as loose objects through the art market. Each of them was taken off a wreck that was demolished when dredging the Scheldt river, respectively at Zele in 1951 (a), Appels in 1934 (b) and Moerzeke-Mariekerke in 1939 or 1940 (c). On hindsight those ships would have made excellent research material, but apparently the interest was limited at the time (after Bruce-Mitford, 1970).

Mediterranean maritime research

Research traditions differ and so do traditions of cultural heritage management (Maarleveld, 1998, ch. 3; 2008). If we consider archaeological discoveries relating to fishing in Classical Antiquity, we must be aware that, as elsewhere, traditions of finds and discoveries in and around the Mediterranean tend to reinforce themselves. By now, Classical Archaeology has incorporated environmental approaches, but for the general public, archaeology in the Mediterranean region is still primarily the study of works of art and architecture. Where underwater archaeology is concerned this was reinforced by the finds of major pieces of classical art early in the twentieth century (Merlin, 1930). Apart from that, let us look at the 1,200 wrecks or so that Toby Parker inscribed in the "archaeological record" or inventory that he prepared around 1990 (Parker, 1992). It is based on all sorts of notes, registers and publications and reflects the discoveries that were made in the years between the introduction of the aqualung (SCUBA) for recreational diving short after the Second World War and up to *c.* 1990. Their distribution is presented in figure 6. What do we see? The histogram is usually interpreted as a definite proof of the rise, apex, decline and fall of shipping in Classical Antiquity. In fact, it may reflect some of this "*what happened originally*". However, it also strongly reflects "*what happened*

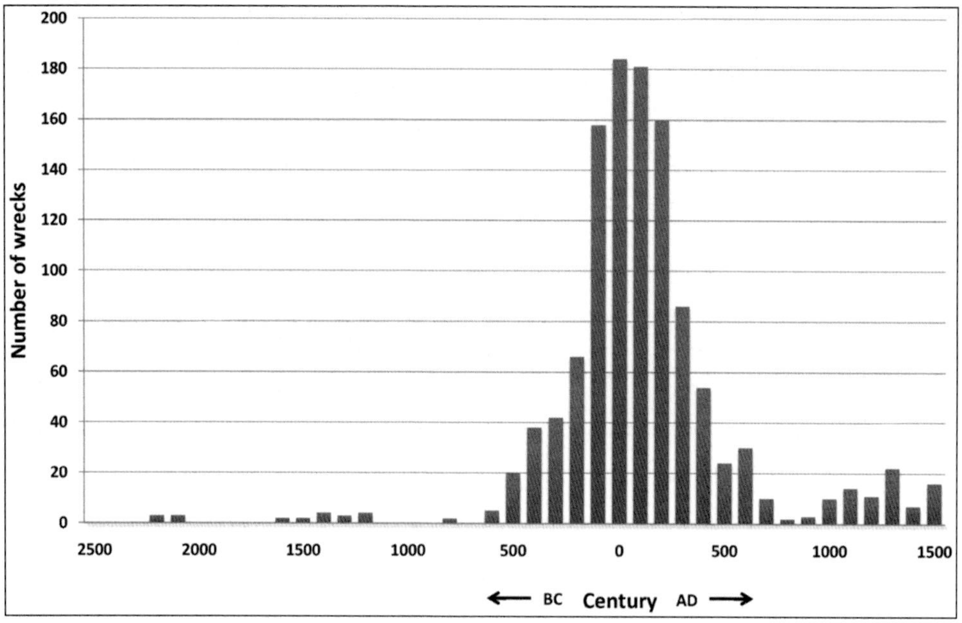

Figure 6. The distribution in time of around 1,200 wrecks that were discovered in the Mediterranean. The histogram is usually interpreted as a definite proof of the rise, apex, decline and fall of shipping in Classical Antiquity. In fact, it may reflect the rise, apex and decline of the amphora as the most common container. In fisheries, baskets, wooden cases and barrels are the preferred container. As a result the fisheries sector is underrepresented in the record, and probably seriously so (after Parker, 1992).

in the meantime" and "*what happened upon discovery*", what was recognized as reportable and inscribable in the archaeological record.

In the Mediterranean of Jacques-Yves Cousteau, the amphora quickly became the icon *par excellence* of recreational diving and of underwater archaeology alike (Cousteau & Dumas, 1953). There is hardly a dive shop in the region that does not show Dressel's classic table of amphora typology, whereas in underwater archaeology the amphora-cargo has become the classical example of "*what happened in the meantime*" since the classical wreck-mound in the sediment-lean Mediterranean was first described (Frost, 1962). This emphasis on the amphora reinforced itself. It was there that heritage was at stake and control and survey should be put in place. The French government for instance launched the "Archéonaute", a patrol and research vessel. Their research may be wider, but their control activities concentrate on those activities that are targeted at amphora mounds and the collection of individual amphorae (Chabert, 1972; Le Gurun, 2006). Other Mediterranean states have a similarly targeted approach and restrictive legislation.

The processes that lead to amphora discovery continue unabatedly, bringing surveyors into ever deeper waters (Long, 1987; Ballard, 2008). Evidently the search and findings are much biased towards cargos with amphorae as containers. This applies to the depth range of recreational air diving that are included in Parker's survey. But this also applies to the depth ranges of recreational tech-diving, to the depth ranges of commercial diving operations and to the depth ranges of Remotely Operated Vehicles or ROV's. As a result, those sectors of the maritime industry and those periods in which baskets, wooden cases and barrels are the preferred container receive less attention. It is a simple rule of correspondence.

Presumably, the association of fish and basket is a very strong one. It is so practical, after all. It is evident in the Neolithic deposits of La Draga (Bosch i Lloret *et alii*, 2000, 42), it is evident in the Mesolithic of northern Europe (Andersen, 1987) and it is (still?) evident in modern fishing. This simple rule of correspondence immediately means that despite attention to maritime matters in the classical Mediterranean, the fisheries sector is underrepresented in the archaeological record, and probably seriously so.

There are exceptions, of course. And those exceptions are perhaps the proof of the biases discussed above. They can also guide future attention. The Mediterranean is sediment-lean, but the estuaries and harbour in-fills are not. With regard to "*what happened in the meantime*" those are interesting areas, as well as for "*what happened originally*". But they have not received much attention in heritage management or research; not by the land-based archaeologists nor by the underwater archaeologists. In a perfect parallel to what happened in northern Europe, we may assume that numbers of fishing vessels will have been dredged or dug away from natural sediments and historical landfills under estuarine harbours in the Mediterranean. It is here, after all that ships stand the best chance of being perfectly embedded, surviving "*what happened in the meantime*". It is here, too, that many vessels may have been discarded after use and dismantling, reflecting "*what happened originally*". In most cases, no one would have thought of such mundane finds in terms of archaeology. They would *a priori* assume that the find was without significance. No report would have followed, and accordingly the finds do not exist. If we can influence "*what happens upon discovery*", directing more attention towards such specific environments, I am convinced that research on nets and fishing gear will benefit enormously. The little boat of Fiumicino *(Fiumicino 5)* is a case in point (Boetto, this volume).

On balance

The comparative wealth of archaeological information on fishing in the Mesolithic and Neolithic periods is related to the fact that a strong research tradition focusing on coastal settlement and its economic basis has existed ever since Prehistoric archaeology was first defined (e.g., Bibby, 1958; Clark, 1965). That is for more than a century. Accordingly, it has long been recognized that information from settlement sites could usefully be supplemented by information from the waste-deposits, often found either adjoining the settlement site or in the water offshore. The water's edge is where maritime artefacts and information are collected.

Such a research tradition has a strong self-reinforcing effect. What is considered interesting by one generation of archaeologists is also deemed interesting by the next. Models are built, based on the existing approach (Fischer, 2004; Deeben *et alii*, 2002) and using such models, predictions are mapped.

Models and predictions inform spatial planning, heritage protection and interested sports divers alike. As a result, both southern England and the Baltic boast groups of recreational divers who specifically look for prehistoric artefacts, and who have been finding one fishing site after another (Momber, 2004; Andersen, 1987). This is exceptional. Worldwide, most recreational divers with an interest in the past are "wreckers", though not necessarily in the destructive sense.

Conclusion

Traditionally, Classical Archaeology has had a strong emphasis on art and architecture. Since the 1950's the archaeology of the Mediterranean has incorporated maritime aspects. At first, this was not a trend that seemed self-evident or natural to all practitioners (Bass, 1975). Moreover, the emphasis on Mediterranean maritime archaeology has been on trading ships featuring amphorae. This may not be completely true for professional archaeological research and management, which has also dealt with harbours, submerged structures and other finds. But it is certainly true for the archaeology of the silent world in its popular image. As a result, most people have never thought of simple artefacts connected with fishing as relevant to erudite study. This view of course reinforces itself. What is not explicitly deemed interesting by one generation of archaeologists is not necessarily found interesting by the next. Moreover, even if the interest is there, there is reluctance to report. One does not want to make a fool of oneself and it is not common knowledge that there is in fact a research interest in such things.

Let me once more state the obvious. If we want more primary archaeological information on nets and fishing gears in Classical Antiquity, key issues need to be addressed. One key issue is awareness. Another is soft sediments. These are not the most characteristic of the Mediterranean basin, but they are not absent either: river plains, estuarine plains and deltas, those are the areas that deserve more attention. To bring the two issues to come together, we should ensure that archaeologists in government and other civil servants devote proper attention to archaeological considerations during planning and development, especially in areas with soft sediments. Also we should make sure that preliminary assessments are not made on the basis of "nothing was found previously so nothing is there", but on the basis of preservation potential.

Finally, it is essential that we make scientific interests explicit. That means drawing up a research agenda that specifically addresses fishing, nets and fishing gear in Classical Antiquity. The present volume will provide an excellent basis. If those three things, a research agenda, attention to filled-in basins and soft sediments and above all awareness and consciousness are catered for, then more discoveries will certainly follow. They may radically change our present interpretations of fishing in Classical Antiquity.

Detail of a Tunisian mosaic in Bardo Museum.

Posters

Blades from La Esparragosa.

11. The Lithic Tools of the La Esparragosa Site (Chiclana de la Frontera, Cádiz, Spain, fourth Millennium BC): A Methodological Contribution to the Study of Lithic Tools for the Consumption of Fish

IGNACIO CLEMENTE, VIRGINIA GARCÍA, JOSÉ RAMOS, SALVADOR DOMÍNGUEZ-BELLA, MANUELA PÉREZ, EDUARDO VIJANDE, JUAN JESÚS CANTILLO, MILAGROSA SORIGUER, CRISTINA ZABALA & JOSÉ HERNANDO

Introduction

In the course of the fourth millennium BC, a form of settlement characterized by large groups of storage pits ("campos de silos") established itself in western Andalusia (Arteaga, 2002). It represents a way of life based on agriculture (Martín de la Cruz, 1994; Montañés *et alii,* 1999; Nocete, 2004), but in the coastal settlements of the Bay of Cádiz and the nearby countryside, traditional habits of collecting molluscs and fishing were retained as an economic complement (Pérez, 2004 and 2005; Ramos, 2004 and 2008). In this paper, we will study the functional aspects of the lithic tools of this culture in relation to the local malacofauna and ichthyofauna.

The occupation of the fourth millennium BC is contemporary with a very clear territorial organization in the Atlantic Band of Cádiz and in the adjacent countrysides. At this time, important social and economic transformations are taking place in the Neolithic tribal societies.

We basically intend to find new approaches to the study of these Neolithic communities, the function of their tools and the different forms of work for which they were used. Therefore, we try to establish the association of the lithic tools with the working of vegetable and animal resources, but above all for the cleaning, scaling and filleting of fish.

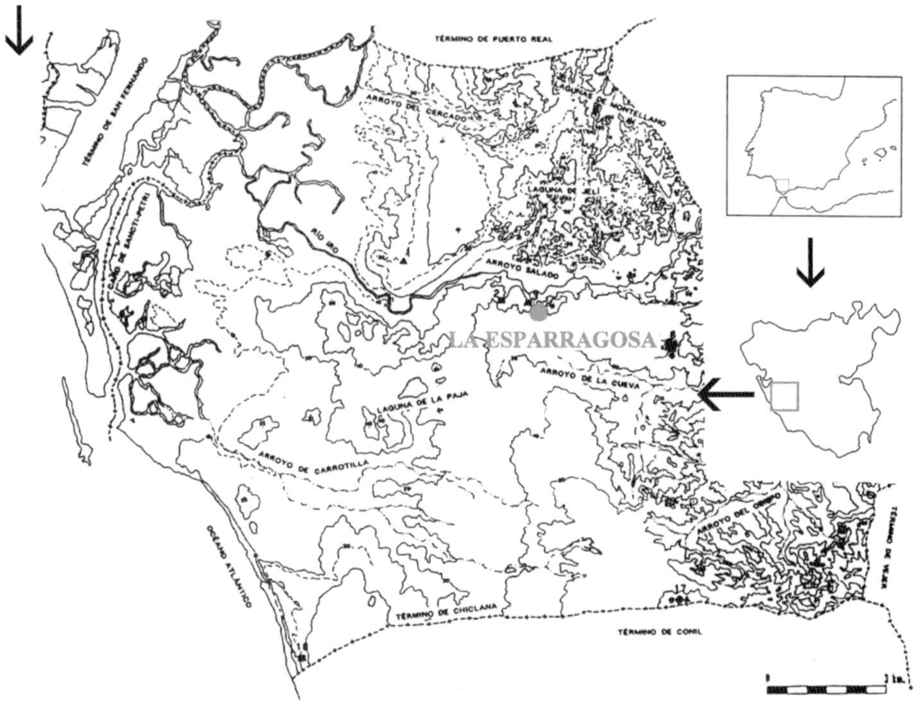

Figure 1. Geographical situation of La Esparragosa.

La Esparragosa

The site of La Esparragosa is located in the Atlantic zone of Cádiz (figure 1). It occupies a projecting plateau (figure 2) beside the River Iro, in Chiclana de la Frontera (Cádiz). Here, we have excavated a surface of 40 × 10m (Pérez *et alii*, 2005) during the 2002 and 2003 campaigns (figure 3). The excavations have documented 8 storage silos and 1 burial (figure 4).

The silos are semicircular in shape with different sections (figure 5), both bell-bottomed and cylindrical types, whose diameter varies between 1 and 1.20m and with a depth of 1 to 1.40m. We have two dates obtained by TL from ceramic samples associated with the burial. They have contributed the following chronology (Laboratory of Dating and Radiochemistry. University of Madrid): MAD-3961: 5255 ±433 BP, MAD-3962: 5129 ±476 BP.

Use-wear analysis of Prehistoric tools

The use-wear analysis of Prehistoric tools allows us to identify the materials that were worked and the activities that took place (Clemente, 1997). This analytical method,

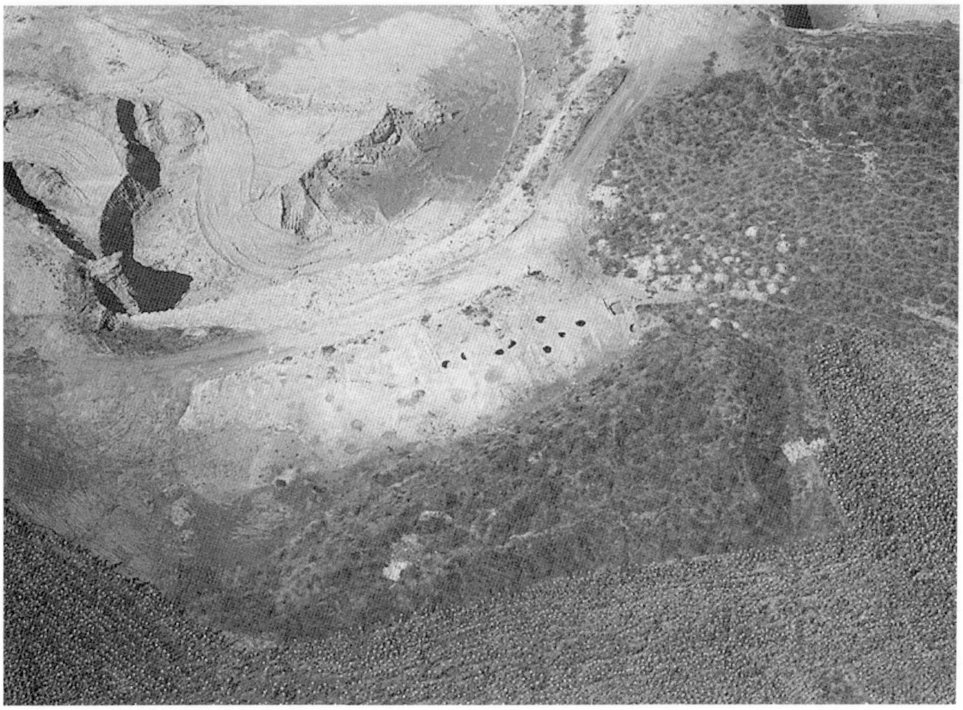

Figure 2. Field of silos of La Esparragosa. Aerial perspective.

Figure 3. Detail of field at La Esparragosa.

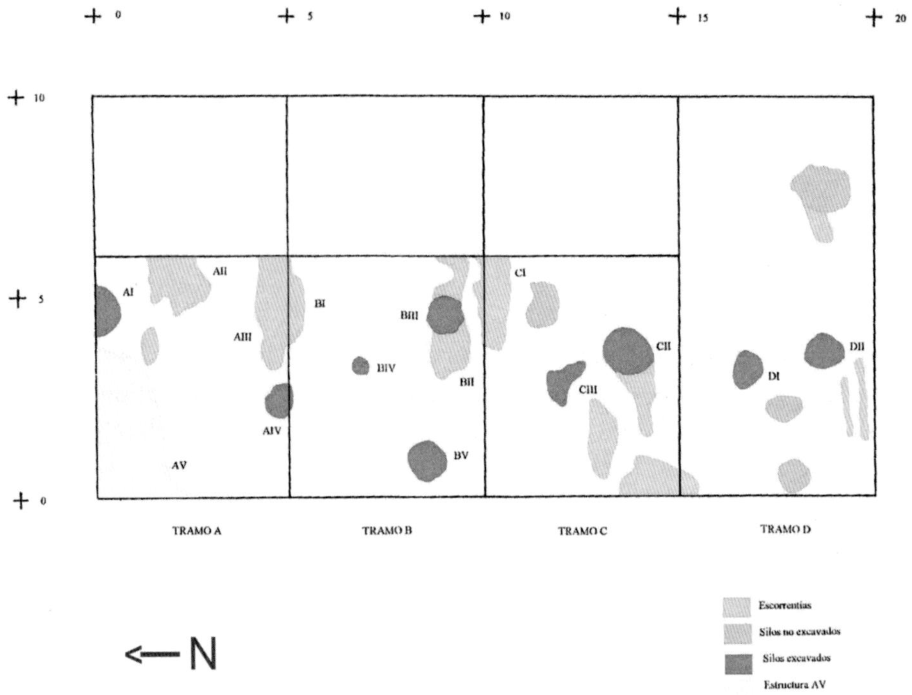

Figure 4. Surface of field work in 2002 and 2003.

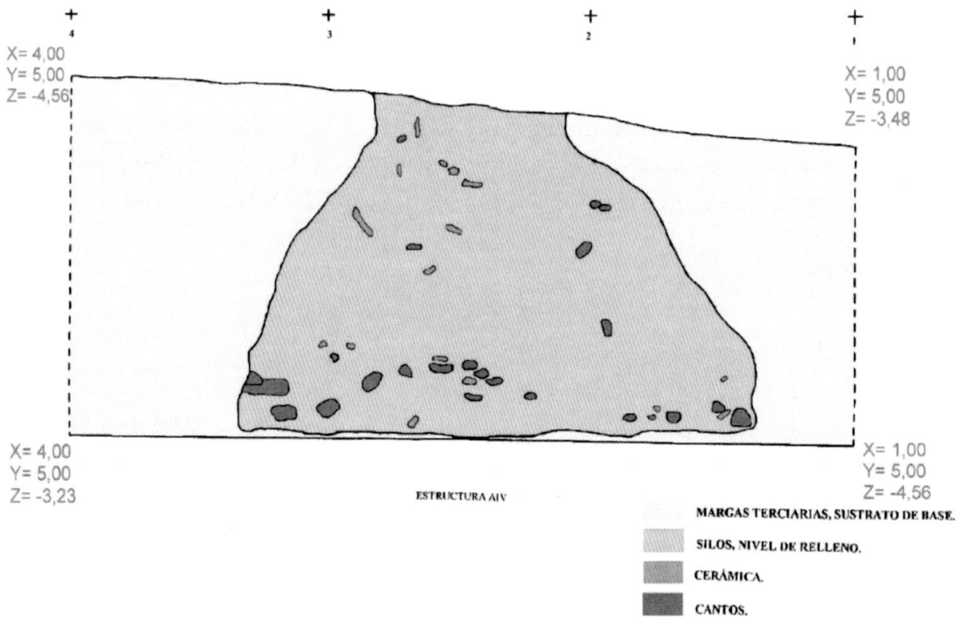

Figure 5. Section of silo.

proposed by S.A. Semenov (1964), provides essential information about productive process in prehistoric societies. For our case study, this method does not allow us to determine the techniques used for fishing; we can, however, extract information on how the fish was processed before consumption.

Fish scaling and cleaning with lithic tools before consumption has been documented on some hunter-gatherer sites of the nineteenth century AD in Tierra de Fuego (Clemente, 1997). We have also established that bone tools made with elk ribs were used for scaling fish on some Mesolithic and Neolithic sites in the Russian plain (Clemente *et alii,* 2002). At the site of La Esparragosa, we think that cleaning and scaling was carried out before filleting the fish for preservation (by smoking or by salting). Preservation of the catch will permit a gradual consumption of fish resources and, perhaps, a long-distance exchange of these products with other groups. The proportion of lithic tools from La Esparragosa showing use-wear attributed to this productive process is very striking, especially compared it with other Neolithic sites. These results suggest the important role of fish resources in the society of La Esparragosa.

Experimental program

An experimental program was carried out to corroborate the use wear interpretation (figure 6). Small or medium size marine species were processed: striped red mullet (*Mullus surmuletus*), common sole (*Solea vulgaris*), european sea bass (*Dicentrarchus labrax*), gilthead seabream (*Sparus aurata*) and atlantic pomfret (*Brama brama*). The activities carried out were filleting, scaling, gutting and head cutting. Each activity was carried out separately to document the different wear patterns related to particular tasks. However, we also used one tool to carry out the whole process. Our hypothesis is that La Esparragosa settlers used the same knives for the whole process.

Tool remains

During the 2002 and 2003 fieldworks on the La Esparragosa site, 388 knapped flint remains were recovered. All the objects have been analyzed macro- and microscopically to determine their function. Less than 2% are macrolithic tools, e.g., adzes, pebbles and hammers manufactured from local raw materials. The rest are lithic tools manufactured from different kinds of flint (93.8%), jasper (1.5%), quartzite (2.5%) and quartz (0.25%). Blades (around 50%) (figure 7) and flakes (26.6%) were made on the site. Although some cores were recovered (5%), we have limited information on the technical processes of tool production.

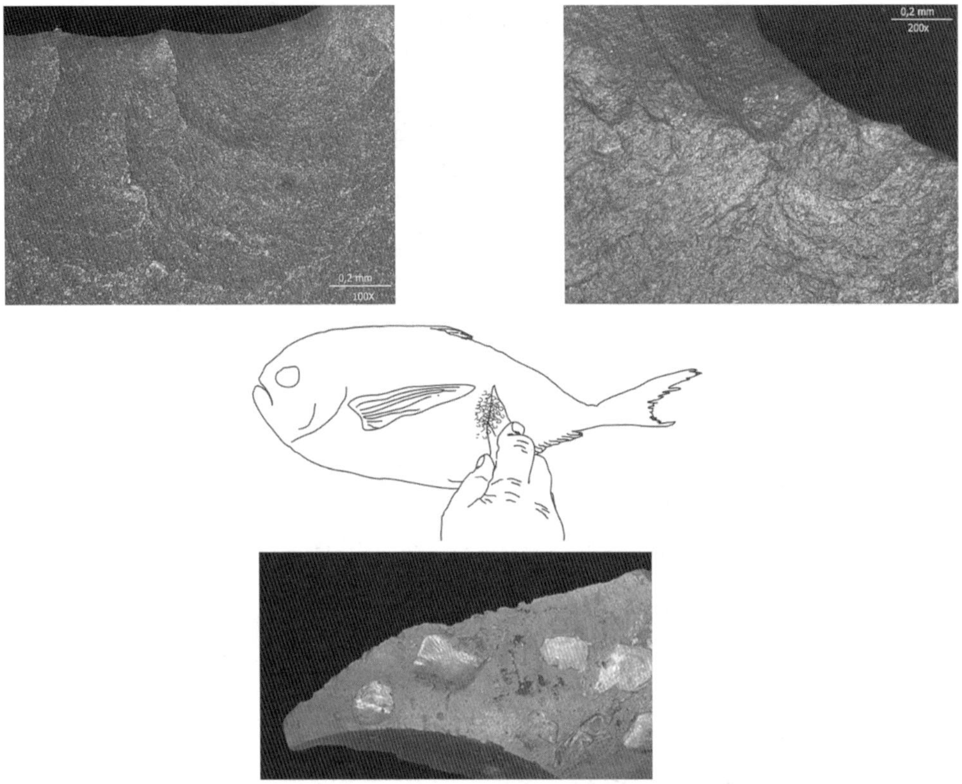

Figure 6. Experimental use-wear, scars, micropolish and residues.

Almost 30% of the lithic remains show evidence of being used as tools. Among these, 180 edges present certain or probable traces of use-wear. 2 edges had been used for working some mineral material; 28 edges for working plant resources; 6 show a double use for plant as well as animal resources and 132 had been used to work on animal resources (figure 8).

Among these 132, 94 edges had been used for filleting, scaling and cleaning fishes. They were mainly constructed as fragmented blades (94%). These fractures were intentionally made for eliminating the distal end that was usually curved. Remains of mastic can be observed on several tools. Sometimes mastic is present on a used edge, probably indicating that the position of the hafted blades was changed after primary use, using as the new active zone the edge that had previously been inserted into the haft.

The use wear related to fish processing is characterized by a combination of wear patterns formed by the contact with different kind of materials (soft, medium and hard materials). The formation of fish use-wear is the consequence of cutting fish meat, scaling and gutting. The contact of the instruments with bones is also revealed by use-wear. As we have observed at the La Esparragosa site, blades were probably used in all stages of fish processing.

Figure 7. Blades from La Esparragosa.

The high percentage of these tools on La Esparragosa seems to be an indicator of the importance of fishing in the economical system of the site. Probably this resource was intensely exploited not only for direct consumption but for its an exchange value after conservation by smoking or salting (Clemente & García, 2008).

Faunal remains: ichtyofauna and malacofauna

To estimate the degree of importance of the various species, the indices most frequently cited in the bibliography (Roselló & Morales, 1994; Moreno, 1994) were utilised the abundance, which is expressed as the minimum number of individuals (MNI). Given the degree of fracturing of many of the remains found, the MNI of bivalves was determined from the hinge fragments, right or left.

For the identification, characterisation of the life cycles and habitats, general reference works (for example, Hayward & Rayland, 1995; Whitehead *et alii*, 1984-1989) have been consulted.

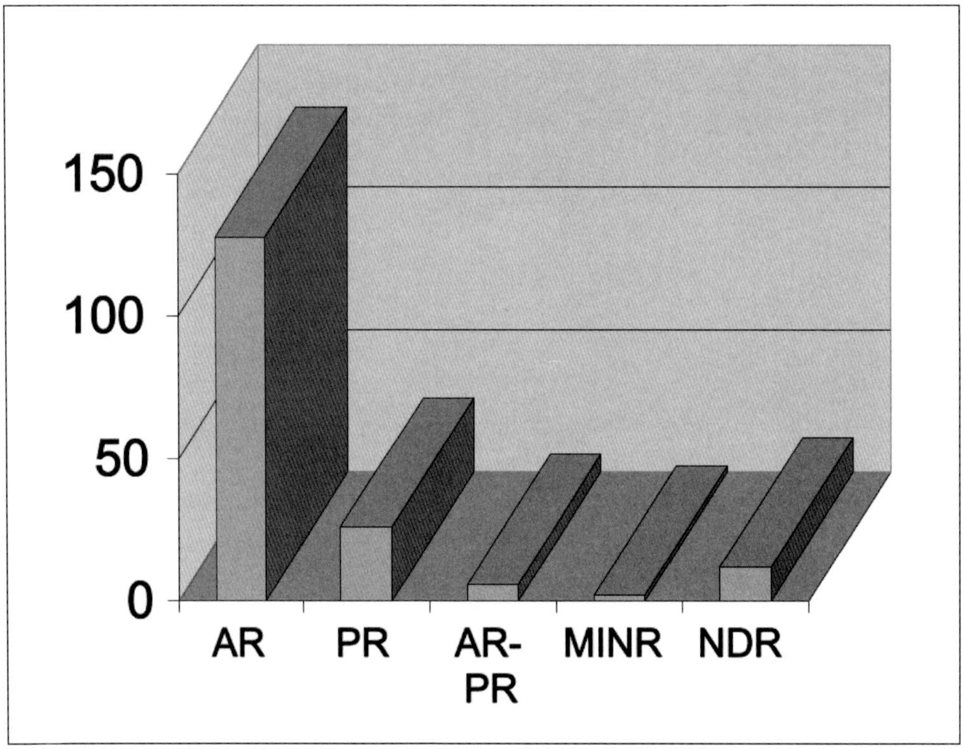

Figure 8. AR-Animal Resources; PR-Plant Resources; MINR-Mineral Resources and NDR-Not Determined Resources.

Results and Discussion

La Esparragosa belongs to a prehistoric site formed by an authentic field of silos. In each of the 4 sections (A, B, C and D) silos have been excavated, in which remains of a minimum number of 2,235 individuals have been found, belonging to a total of 29 taxons (table 1).

Of the 29 taxons, 16 are marine bivalves, constituting the dominant group; 6 are marine gastropods; 4 are terrestrial gastropods found very frequently in the zone, with the exception of *Pseudotachea litturata*, which is characteristic of rocky zones like cliffs; one species of freshwater bivalve and remains of other two invertebrates, shells of crabs and quills of a sea urchin have also been found.

Of all of these, the predominant species is *Tapes decussatus*, which accounts for almost half of the specimens found and is present in 92% of the samples analysed. By sections, it was found in 100% of the samples in both B and D, and in around 85% of the samples in the other two sections excavated. This bivalve, of up to 75mm in length, is characterised by presenting a notably oval- or square-shaped shell, with a distinctly anterior umbo. They present a cardinal central tooth in the left valve, and in the right, central and posterior bifids. The paleal cavity is U-shaped and does not

	Dominance (D)	Constancy (C)	Global Importance (C×D)
Marine bivalves			
Tapes decussatus	48.37	92.08	4453.58
Scrobicularia plana	1.16	5.94	6.91
Glycimeris sp	0.36	6.93	2.48
Chlamys sp	2.33	33.66	78.32
Pecten maximus	6.44	48.51	312.58
Cardiidae	0.27	5.94	1.59
Solen marginatus	0.27	5.94	1.59
Acanthocardia sp	0.63	5.94	3.72
Cerastoderma edulis	0.04	0.99	0.04
Anomia ephippium	1.12	5.94	6.64
Panopea glycimeris	0.22	3.96	0.89
Ostraeidae	1.92	22.77	43.81
Ostrea edulis	0.27	3.96	1.07
Crassostrea sp	2.24	20.79	46.51
Pholas dactylus	1.03	6.93	7.13
Mactridae	0.04	0.99	0.04
Marine gastropods			
Nassarius reticulatus	0.04	0.99	0.04
Cymbium olla	0.18	2.97	0.53
Hydrobia ulvae	0.04	0.99	0.04
Osilis lineatus	0.04	0.99	0.04
Charonia lampas	0.04	0.99	0.04
Turritella comunis	0.04	0.99	0.04
Terrestrial gastropods			
Rumina decollata	0.81	12.87	10.37
Theba pisana	29.22	34.65	1012.47
Otala lactea	1.30	2.97	3.85
Pseudotachea litturata	0.18	3.96	0.71
Freshwater bivalve			
Potomida littoralis	0.45	9.90	4.43
Others			
Unidentified crab	0.54	9.90	5.32
Unidentified sea urchin	0.40	4.95	1.99

Table 1. List of the malacofauna found in the prehistoric site of La Esparragosa, indices of constancy and dominance, and global importance (C×D).

Figure 9. Illustration of the specimens of T. decussatus found in the AV burial structure. Detail of the size of the specimens (left) and of their state of conservation (right).

extend beyond the mid line of the shell. It inhabits the lowest zones of beaches and sublitoral shallows, on sandy, clayey or muddy bottoms, and is also abundant in bays and lagoons, in zones of calm water. It is found on the southwestern coasts of the Atlantic, including West Africa, and the southern shores of the Mediterranean.

With respect to constancy, it follows another marine bivalve *Pecten*, and the terrestrial gastropod *Theba pisana*, present in more than 40% of the samples from all the zones, with the exception of section C, where the value is less than 5%. This species is the second in terms of dominance, although it is distributed very irregularly, being very abundant in section B, and within B, in the silo denominated BIII.

In the case of *T. decussatus*, its dominant presence is striking in the burial structure, where 477 specimens of great size have been found, many of them complete, with the two valves (figure 9). The relative dominance of this species in this structure is notably the highest of the entire site, since the only other findings made together with this species are the remains of 1 specimen of *Pecten maximus*, another of *Chlamys* sp, and 3 of *Theba pissana*. In none of the structures, mainly silos, excavated does such a clear pattern of dominance by a single species appear so clearly.

We are faced with a very diverse type of malacofauna in the silos where, for example, in the AIV, more than 73% of the total species discovered in the site have been found. Generally, the malacofauna of the silos is more diverse than in other

Figure 10. Detail of the size and conservation of a specimen of Pecten maximus.

structures associated with fast-flowing outflow waters, generally also with a lower numerical representation of remains. This relationship however does not apply in the burial zone, where the diversity of remains is very low despite containing almost 22% of the specimens determined.

The role played as a food resource is clear for many of the species found, such as *T. decussatus, S. plana, S. marginatus,* etc. Many other species may have a double function, as is the case of *Pecten maximus,* of which very well-preserved remains of great size have been found (figure 10). In sediments in study evidences of fish have also been documented.

Conclusions

The societies of the fourth millennium BC had an economic base of farming and cattle raising. The investigation of the La Esparragosa site has demonstrated the importance of marine resources among the archaeological evidence, especially the relation of lithic tools to fish consumption. The experimental application produced use-wear patterns similar to those observed on tools found in the settlement.

Acknowledgements

This work is related to the following Research Projects: "La implantación de las especies domésticas en la Europa Atlántica. Sociedades neolíticas y uso de las materias primas en la región del Estrecho de Gibraltar" (HAR2008-06477-C03-C02). Ministerio de Ciencia e Innovación (MCI); and "Forgotten resources in Prehistory: the case of fishing among the Meso-Neolithic communities in the Russian plain" (HAR2008-04461/Hist). Ministerio de Ciencia e Innovación (MCI).

Scenes linked to fishing by the potter GERMANVS from La Graufesenque.

12. *Terra Sigillata* as a Source for Fishing Gear of the Early Imperial Period

MACARENA BUSTAMANTE ÁLVAREZ

It is generally accepted that *terra sigillata* was an essential element in the luxury tableware of the Roman age period. Several explanations have been proposed why *terra sigillata* came to play such an important role. Firstly, its physical similarity with vessels and objects made of precious metals. Secondly, as part of the process of Romanization, provincials actively strove to acquire and use objects associated with the Roman way of life (Bustamante, 2008).

During the late Republic and the early Principate, *terra sigillata* established itself as an essential component of Roman lifestyle. Its wide dissemination was a result of standardised methods of mass production. Scenes from daily life formed an indispensable part of the iconographic programme of *terra sigillata* and were disseminated to the distant limits of the Empire. Depictions of marine scenes and fauna were a part of Roman material culture and *terra sigillata* is an example of this.

In this paper, I shall attempt to study the role of marine resources in the iconographic register of *terra sigillata*, the Roman pottery *par excellence*. The study will be focused on Italian, Gaulish and Spanish production centres during the early imperial period. I shall try to find the meaning and the function of the figures of marine resources in the *terra sigillata*.

Terra sigillata as a source for the study of fishing in Antiquity

The historiography of ancient fishing nets has traditionally focused on the classical literary sources and the magnificent pictorial sources of high intrinsic value (e.g., roman mosaics). In contrast, the iconography of vascular objects has been assigned to a category commonly known as "minor arts", without taking into account that this category includes a range of consumer objects that could be described as basic necessities and objects of commerce, reflecting the direct reality and everyday life of the Roman world.

Among scenes of Roman daily life, marine fishing motifs played interesting and multifaceted roles that have to be taken into account when studying this traditional economic activity.

The value of the *sigillata* as a source of information for the study of ancient fishing derives from the innumerable life-like representations of marine life, depicted in various ways:

- First, fish in isolation. Fish or marine life in general – dolphins or seals – are represented in a generic composition where the main figure appears independent and it is an element that completes the decorative programme.
- Second, as part of a larger composition. That is, a fishing composition with one or more characters in the same act, a fishing composition with a moving group of generic fishes or with dead fishes once they have been caught.
- Third, representations of fantastic animals related to marine life as part of pagan mythological scenes covering a broad spectrum. For instance newts, isolated sea horses, sea horses forming part of the court of Neptune (Elvira, 1981, 63) or the same animals led by *erotes*. As we shall see below, this last decorative element is frequently used by the Roman elite.
- In fourth place, fish scales stylized into a geometric form, mainly used as dividing friezes in the decorative metope of the *terra sigillata* from southern Gaul.

The level of precision of the picture itself an of the engraving of the punch allows visualization of different fishing tools (nets, fishing rods and harpoons, inshore or on the open seas).

Contrary to these positive characteristics of *terra sigillata* as a source, the study of its decorative programme also involves some inevitable problems. First, the decoration is constricted to a background of small dimensions, making it difficult to depict and appreciate minor details. Second, stylistic characteristics play their part, not least a tendency towards schematization of the motifs, making it difficult to visualize an entire fishing scene. Thirdly, the technical discourse of *terra sigillata* is based on punches that are used communally. In spite of that, the potters that mainly or exclusively use these marine motifs are minimal in number: L. COSI and GERMANVS of La Graufesenque and PATERNVS, BANNVS and IVSTVS from the workshops of Central Gaul situated in L'Argonne. Nonetheless, studying the iconographic inventory of the *sigillata* potters could be a good avenue of exploration in the search for more information on ancient fishing, and consistent with the interdisciplinary orientation of modern research. We have at our disposal an important source of information that even with a minimal amount of analysis will be able to complement the discourse of ancient fishing that is based on Roman mosaic art, sculpture or metalwork.

Italian *Terra Sigillata:* beyond a mere iconographic discourse

Turning to the question of chronology, it is well known that Italy was the first region to produce *terra sigillata.* The peak of Italian *terra sigillata* production was reached around the beginning of our era. At this point in time, a decline in the main production centre, Arezzo, and a growing production in the centres of Southern Gaul can be observed. This expansion takes place in a specific political period and context – the transition from the Republic to the early Empire, the process of conquest, Romanization and establishment of the *Pax Augusta.*

The end of the nineteenth century marks another milestone in the history of *terra sigillata.* In the year 1895, Dragendorff for the first time classified these products, establishing a comprehensive typology embracing 55 shapes. This was a prelude to other works that were soon to appear. In 1908, Chase produced the catalogue of potters for the Loeb collection, but an even greater work of the time was that of Loeschke (1909) based on the stratigraphic evidence from the Roman army camps on the Rhine and Danube. For the present study, the iconographic studies of Dragendorff and Watzinger (1948) are also of great interest.

Generally speaking, Italic pieces, represented mainly by type Ritterling 1-9 (type XI according to the Dragendorff terminology) rarely show fishing scenes. Instead, they tend to focus on scenes emphasizing great human strength (e.g., gladiator scenes of MISSVS or hunting scenes of M. PERENNIVS TIGRANVS, where a character dressed in a short tunic is holding a lance, threatening something or someone).

At present, we have only one piece of *terra sigillata* showing fishing nets. This was found in *Septem Fratres* (Ceuta) (Posac, 1998, figure 6, no. 30). On this piece, a person of short stature is represented who could symbolize a Cupid or Eros. He appears to be carrying an object on his left shoulder that seems to be a rolled fishing net. This fits in with the nature of Ceuta as a big fishing port and adjacent to the sea.

In contrast to the absence of fishing activities, these pieces do depict mythical marine animals. One of the best examples is the image of a Gorgon from the workshop of *Rasinys.* In the middle of the scene we see the face of Medusa with her hair flowing in curly strands to all sides, in accordance with the descriptions found in classical mythology. Her face appears to be inserted in circular cartouche, held at the ends by a phytomorphic garland with projecting leaves, as well as a small rose with numerous petals in its centre part. All these elements are limited by a border of double ovolo with a short tab. A piece with similar characteristics appears in the work of Dragendorff and Watzinger (1948, figure 29, no. 405-406).

The forms Consp. 19-20 with *appliqué* decoration show more of these iconographic elements (figure 1). The first group consists of dishes – Goudineau 30 in classical terminology – with a vertical, concave wall, sometimes lightly flared with a projection to the inside of the dish. It is interesting to observe concentric fluting

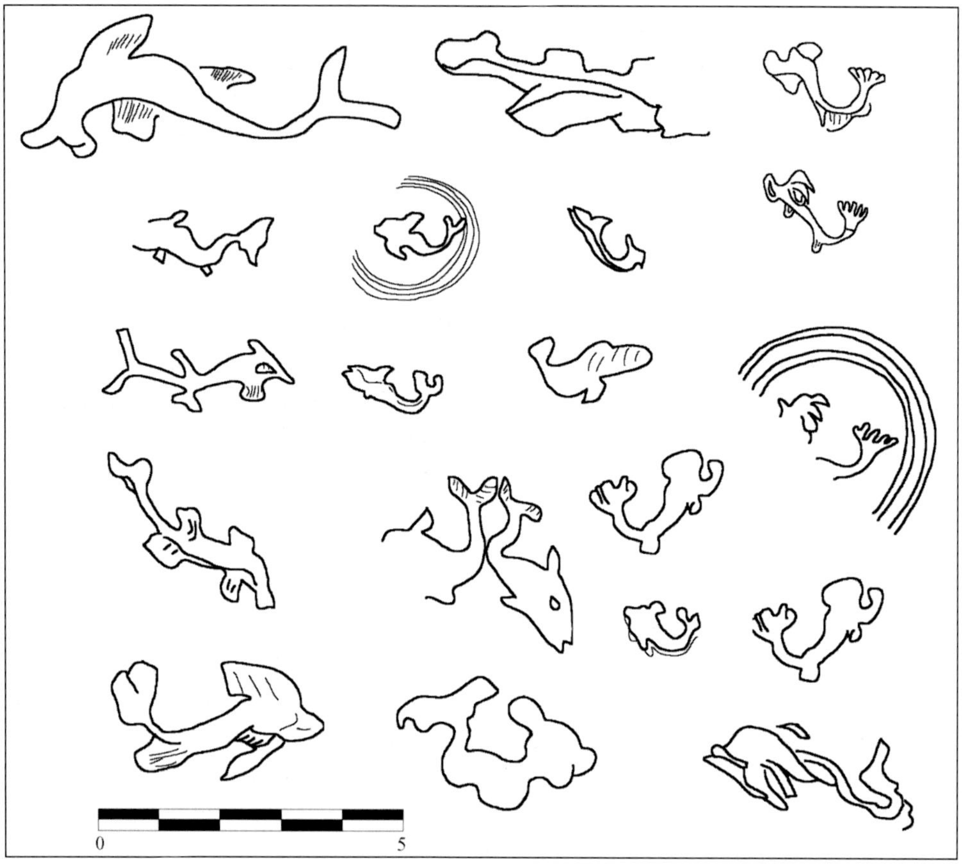

Figure 1. Marine motifs applied in Italic or Late Italic Sigillata (Medri, 1993, 233-234).

on the bottom, which implies a certain dynamism of the forms. Consp. 19 derived from the Consp. 18 form previously mentioned, but with a difference: form 19 has concentric fluting on the interior as well as a marked step.

Consp. 20, a *catinus* with vertical or lightly flared wall, can be plain or can have mouldings on the exterior. The production area of this type encompasses all the Italic production centres, including the region of Padua and early Imperial production centres. Typologically, it matches with Barocelli 5, Berenice B209 and B213, 1-3, Goudineau 39 a-b, Pucci X, 9, 25 and 27-28, Hayes 8 and Ohlenroth 1.

Although estimates for the earliest production date of this type varies, according to whether or not it has *appliqué* decoration, in general terms, we can place it in the period from Augustus to 20 AD. To judge from the chronostratigraphy of the pieces studied here, the type appears at a moment when Italic production is well advanced, in fact already on the decline due to competition from La Graufesenque production. In this context, a promotion of an ideological discourse by means of *sigillata* pottery can be observed (Bustamante, 2008). In this sense, fish and dolphin motifs were articulated as a symbol of initial Julio-Claudian power thanks to the use of dolphins on the *spina* of the Circus Maximus during public games associated with the candidature of Octavian and sponsored by Agrippa. These motifs were used in order to ingratiate with Roman power; the same can be observed in the diffusion of *erotes* riding on dolphins. These icons attempted to link mythology and power at the beginning of the Empire. Mainly they tried to give a divine origin to the Julian dynasty, as descended from Venus, in a moment of crisis and a changing political situation. We have before us a group of objects that tries to bring the tastes, interests and feelings of the privileged class to the people.

According to Jerez Linde (Jerez, 2000, 95) artists and craftsmen strove to underline the "official" character of the new principate through their work, destined for a clientele that demanded everything associated with the tastes of the Imperial court. The decorative elements could have been derived from a distinct series of products that was favoured by Imperial power for dissemination through military distributive circuits. At the same time, Italic pottery production was threatened by competition from ruthen products. So this situation of power caused a tendency to favouritism because of friendly relationships, whether by establishing a fashion to be imitated by the common people, creating monopolies within state commercial circuits (the *annona militaris*), or contributing to its intrusion in others (Bustamante, 2008). Examples of such families are the *gens Ateia*, whose members were politicians, *negotiatores* and *mercatores rei cretariae* (Sangriso, 1998) and the *Cilnii*, with an inversion in the mentioned production because of the friendship with the imperator.

Consequently, in the Italic case we observe design considerations going beyond aestheticism to an inherent political symbolism. In this context, iconography and fishes go together in order to save an abundant source of income in Italic territory, not only for the imperial family but for also for the imperator's closest circle.

The Gallic *figlinarii* L. COSIVS and GERMANUS

At about 15 BC we start to see a migration of potters from the leading Italic producing centre of the moment, Arezzo, to the south-eastern part of Gaul. Here, they began to establish branch workshops producing a type of ware characterized by Passelac as *presigillata sudgallica* (Passelac, 1993). Arrival of this ware is associated with the arrival of *Cn. Ateius*, who moved to Lyon-la Muette about 10 BC. This created a new trend in pottery that was one of the main factors behind the decline of Italic production. Potters from Narbonne, Bram, La Graufesenque, Loyasse or Montans on some occasions made pieces with moulds whose archaeometric analysis indicate that they are native to Arezzo.

From this, we can observe the initial configuration of Gaulish pottery production around the beginning of our era, contemporaneously with a hiatus and a de-

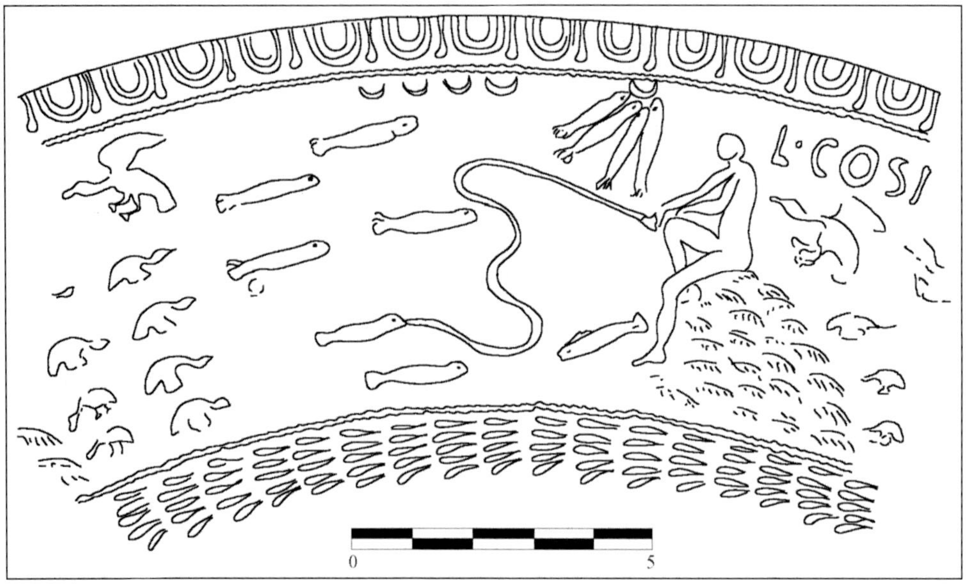

Figure 2. Fishing scene of figulus L. COSIVS *from La Graufesenque (Knorr, 1912, pl. XV and Oswald & Pryce, 1920, 72 and 73, no. 6).*

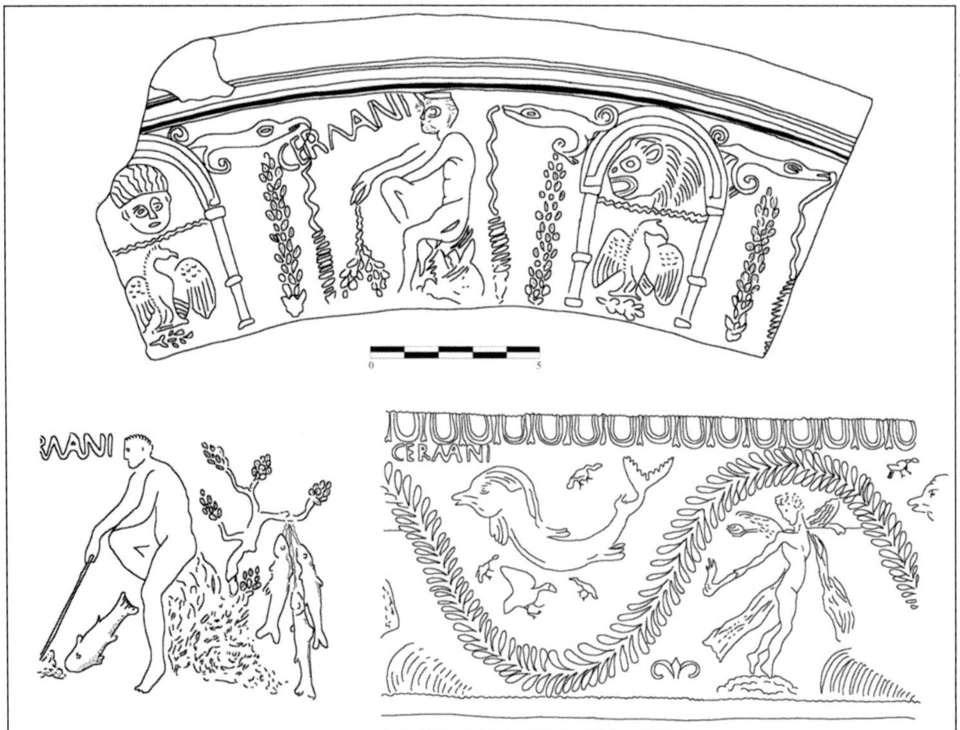

Figure 3. Scenes linked to fishing by the potter GERMANVS *from La Graufesenque (Knorr, 1912, pl. VL, 1; Oswald & Pryce, 1920, 52 and 53, no. 6; Vázquez de la Parga, 1934, 110 and 112).*

cline in Italic production. As regards volume, Vernhet (1991) estimated figures of around 15 million pieces a year in a geographical zone that traditionally did not have a strong tradition of pottery. This vast volume had to be distributed to markets, with the sea as the most convenient means of transport. If we take into account the finds of shipwrecks with Ruthen productions, e.g., Culip IV (Nieto & Puig, 2001), we can observe that most pottery arrived by sea from *Gallia* to *Hispania*. This fact could make the South Gaulish producers reproduce marine scenes.

In terms of marine elements, the decorative richness of south Gaulish production is vastly greater than those of Italic and Hispanic pottery. One of the particularities of the Gaulish *sigillata* production is the possibility, thanks to epigraphy, of linking names of potters with particular iconographic schemes. Potters of the Flavian period, whose decorative schemes shaped a particular style, used this iconographic resource. In general terms, in the decorative register of Gaulish *sigillata* we doesn't observe boats, since these pieces show river fishing scenes, not deep sea fishing. This could be explained by the inland position of the production centres.

Within this production, I should like to highlight two *figuli*:

- L. COSIVS VIRILIS: this potter from la Graufesenque is active between the reigns of Vespasian and Hadrian. Decorative patterns linked to fishing activities are characterized by individuals sitting down on some sort of rock, angling with a fishing rod. This motif is found in the central stripe of the decorative scheme of Drag. 37's forms. Characters have an active schematization and a clear inexpressiveness (figure 2).
- GERMANVS: Ruthen potter whose productive life was from Claudius to Trajan, covering about fifty years. Pieces from this *figulus* are characterized by their rigour, thoroughness and fineness of the representation. In these scenes individuals fishers appear with their instruments and with their catch. We can also observe the potter's name in an intra decorative way. The most common forms are Drag. 29 and 27, but mainly Knorr 78 (figure 3).

Both cases could be considered exceptional ones as they are the only potters in the Gaulish, Spanish and Italian repertoires to show fishing scenes. This could be a symptom of fishing origins, that is, a family descended from sailors, fishermen or people living in a coastal territory, since it has been established that the most of the potters at La Graufesenque are immigrants.

Besides these examples from the production centre of La Graufesenque we have observed others that can be described as anonymous, in so far as they are not associated directly with potters' marks. In some cases we see isolated fishes, fishes grouped in shoals, or simplified fishing scenes (figure 4). We also see dolphins, seals (figure 5) and mythological elements.

With these scenes, geometrical ornamental borders are used to simulate fish scales (figure 6).

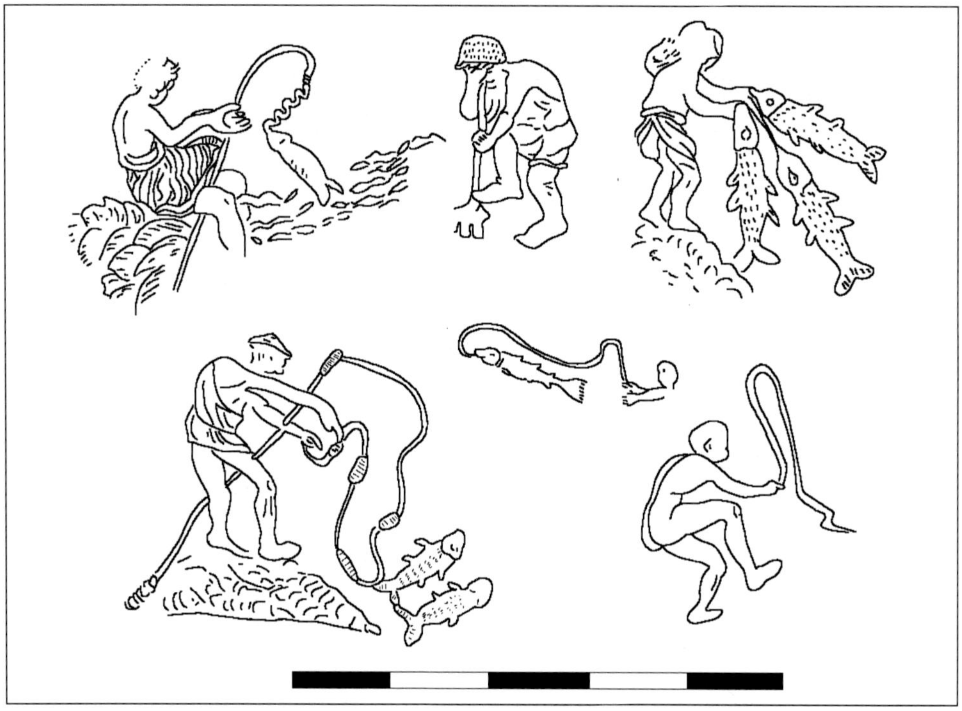

Figure 4. Fishing representations in decorative repertoire of Gallic pottery (Hermet, 1934, pl. 28).

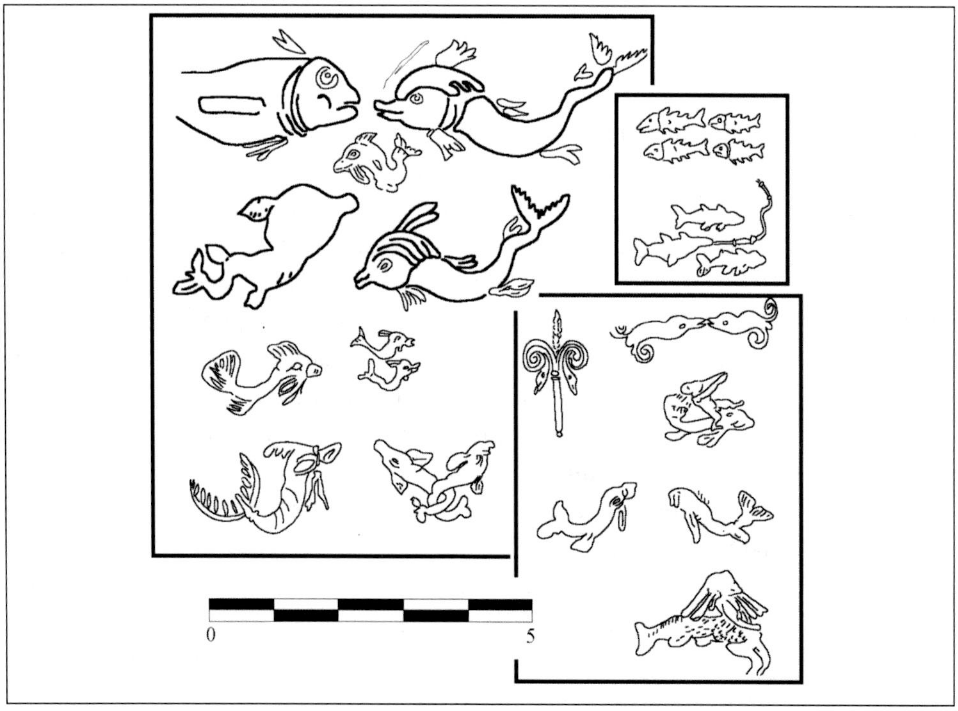

Figure 5. Marine fauna in decorative Gallic repertoire (Hermet, 1934, pl. 28).

Figure 6. Decorative Gallic pieces: styling with scales of fish and fantastic animals (Oswald & Pryce, 1920, 84 and 85; Atkinson, 1914, pl., XVI).

Spanish *terra sigillata:* a reflection of approaching decline

In contrast to the rich decoration of Italian and Gaulish ware, decribed above, Spanish *sigillata* is far less elaborately decorated, with a preference for the geometrical simplicity of concentric circles. This is characteristic of the last phase of *sigillata* production, but initially the Spanish wares imitated of the Gaulish style of the late Claudian period.

At this point, fish representations are simply inserted where there is no place for human characters and simplicity is the main characteristic. Fish are depicted in isolation or sometimes with more fishes forming shoals, independently of the rest of the scene. The same observation applies to other marine animals like dolphins, seals or walrus, foreign to Hispanic geography. Their forms correspond to the exotic natural or intrinsic symbols that we have commented on above in the context of Gaulish ware and which would later be used in Christianity.

This type is represented by the Riojan products of *Tritium Magallum.* Drag. 37 forms from the later Flavian period (figure 7) show this type of decoration.

Southern workshops, on the other hand, never include these punches in their works, except during the last phase of engraving. First of all, we note how they are limited to the production of the Andújar workshop in Jaén during its last phase. These engravings are situated in the centre of the piece as the only motif, or repeated in the same piece (Sotomayor *et alii,* 1999, 24). This engraving is reminiscent of lozengic

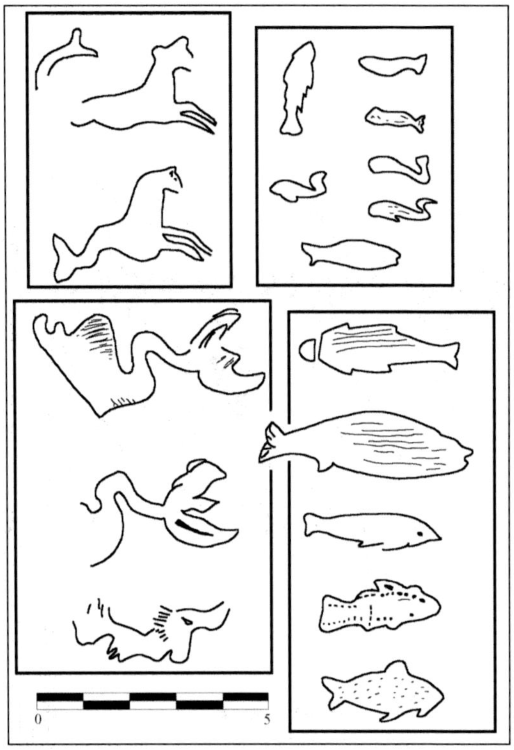

Figure 7. Fishes in Hispanic iconographic register (Mezquíriz, 1961, ill. 77).

symbols from the Campaniense B phase of Italian pottery from Campania. In the case of the Punic environment of Cádiz, this engraving reminds one of the forms of the *Kouass* type, having zoomorphic motifs (lobsters, squid, lions, cervids or chariots) or gods (Jupiter, Minerva or Mercury). This indicates a clear harking back to the past that is getting close to the indigenous stamped pottery productions from the south of the peninsula, where the workshops of Andújar played an important role in the production of painted Iberian pottery. These representations also reveal an iconographic syncretism with inclusion of pagan elements.

The iconographic register of *terra sigillata*

In general terms, we wanted to suggest a work line that has been in the background, because of the primacy of chronological and morpho-descriptive aspects of the pieces. Although zoomorphic representations have been sporadically studied, like the study of the camel by Martin their importance has not been realized; they could be a reflection of a typical landscape of the production or consumer region. Such studies could also be useful for other types of iconographic studies that could even be compared with studies of archaeozoology.

In the same way, we think that a dissociation of politics and productive activity in the first stages of *sigillata* production is inevitable (Sangriso, 1998; Bustamante, 2008). In this respect, we do not want to conclude that all these representation were linked to the Imperial powerful. From our point of view, a connection in productions from *Tiberius* to late Italic productions can be established in the militarized circuits towards the limes.

In general terms, the iconographic register of these pieces seem to reflect to a triple desire on the part of productive centres:

- Firstly, to distribute these pieces in markets linked to marine and fishing activities. In these pieces the fishing sector saw an acknowledgment of the increasing value of its labour. That is why these pieces are abundantly in coastal locations such as Ampurias (Vázquez de la Parga, 1934) or *Baelo Claudia* (Bourgeois & Mayet, 1991, pl. XXVII, no. 2655 and pl. XVIII, 2696).
- Second, the exotic nature of their iconographic repertoire may have increased demand for these wares in inland areas.
- Third, to accommodate an ideological discourse promoted by the elite. I have already commented the connection between the origins of this ceramic type and the political context of the epoch, mainly with the process of Romanization. I would also point to the star role that dolphins gained after the battle of *Naulochoi*, where the Roman fleet was commanded by *Marcus Agrippa* (Bustamante, 2008).

The case of the South Gaulish ware from the potters L. COSIVS and GERMANVS has to be studied as a paradigm, as well as a sign of the linking between these potters and the coastal environment. On the other hand, the Spanish potters used these representations an another resource due to the decorative monotony of the second century.

By contrast, I have not observed any traces of religious propaganda, such as can be observed in the proliferation of stamped production between the second and seventh centuries AD – Paleochristian productions such as orange or grey *paleochrétienne,* as well as African Red Slip Wares C-D that show scenes with fishermen or fishes, where the acrostic of Christ alludes to the Greek term for fishes *ichtys,* or to the image of Christ as "fisher of men" (Dölger, 1928).

On the whole, one may conclude that Roman society had assimilated marine motifs into the iconography of tableware in a world where the production of ceramic vessels for everyday use had reached formerly inconceivable figures by the middle of the first century of our era.

A wider study linking anthroponomy, iconography and the role of marine environmental resources in the Roman world could reveal new insights that would contribute to the understanding of ceramic decorative registers beyond mere decorative programmes. Hopefully such studies will be started soon.

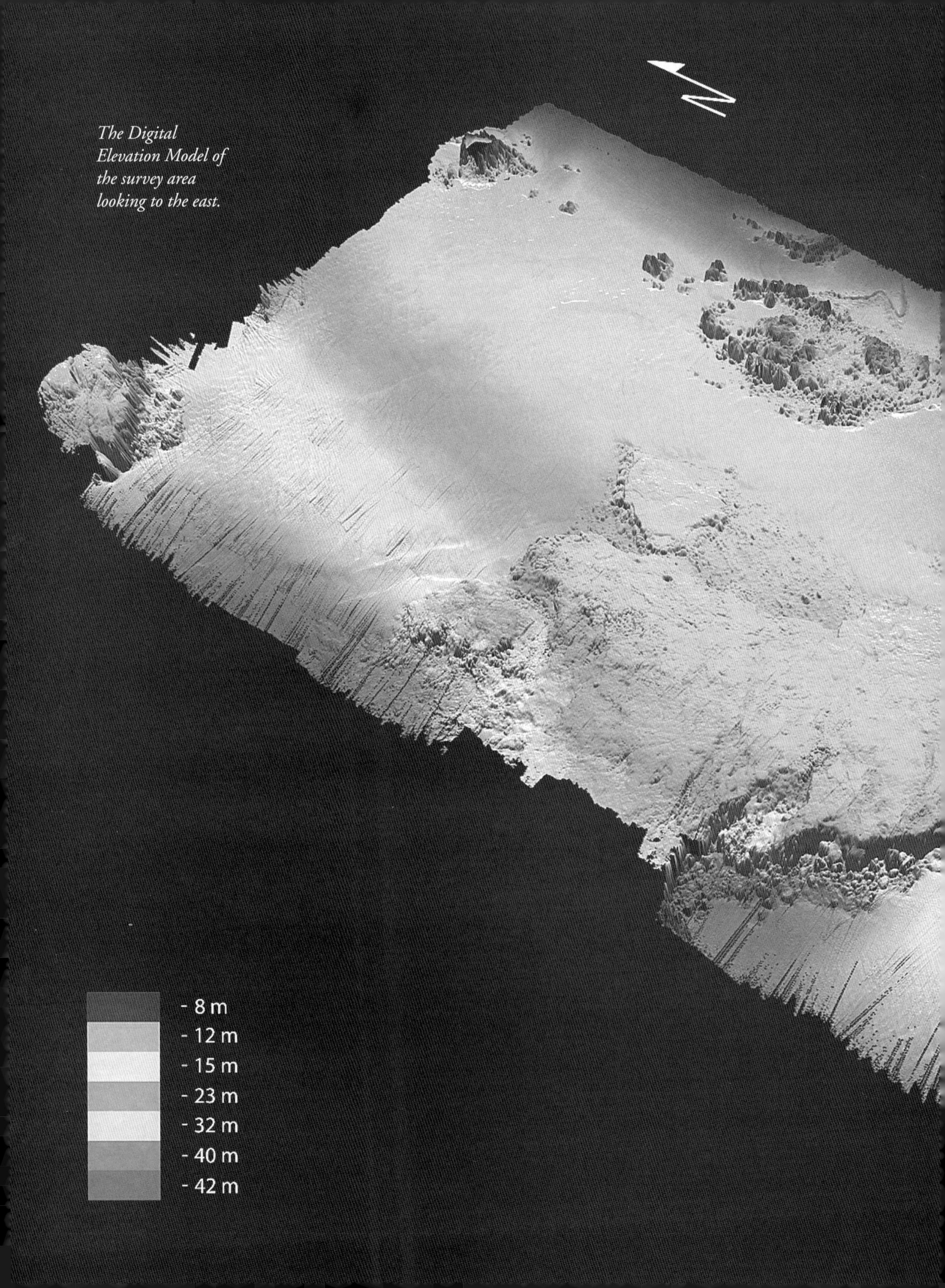

The Digital Elevation Model of the survey area looking to the east.

13. Archaeological Evidence for Ancient Fixed-Net Fishing in Northern Morocco

ATHENA TRAKADAS[1]

In 2002-2003, the Institute of Nautical Archaeology (USA) and Institut National des Sciences d'Archéologie et du Patrimoine (Morocco) conducted an underwater archaeological project, the Morocco Maritime Survey. The reconnaissance project surveyed the coastal waters of the Tanger Peninsula in order to investigate the ancient history and significance of the southern Straits of Gibraltar region, as reflected through the maritime archaeological record. The survey was realised through remote sensing (multibeam sonar), visual scuba diver survey (in waters shallower than 30m) and *in situ* documentation of artefacts and submerged cultural assemblages. Finds included ancient and historical shipwreck remains, ordnance, ceramics and numerous anchors (Trakadas, 2003; Trakadas, 2004; Erbati & Trakadas, 2008).

Cap Spartel

One of the areas surveyed during the project was the Atlantic face of Cap Spartel, the ancient *Ampelusia pr.* (Pliny, *N.H.* 5.2; possibly identical with *Soloeis*: Herodotus, 2.32.1; Hanno, *Periplus* 3) (figures 1-2). The headland is formed by 325m-high Jebel Kebir and physically defines the northwestern-most point of Africa and the southwestern boundary of the Straits of Gibraltar. To the south of Jebel Kebir extends the low sandstone bluff of Ras Achakar, where there is evidence of Neoli-

1 I would like to thank the organisers of the workshop for their kind invitation to and hospitality in Cádiz. I am grateful to my survey co-director, Dr. Elarbi Erbati, and the director of INSAP, Dr. Aomar Akerraz, for their assistance and support during the project, as well to the staff of the Tangier American Legation Museum for the kind use of their library's map collection.

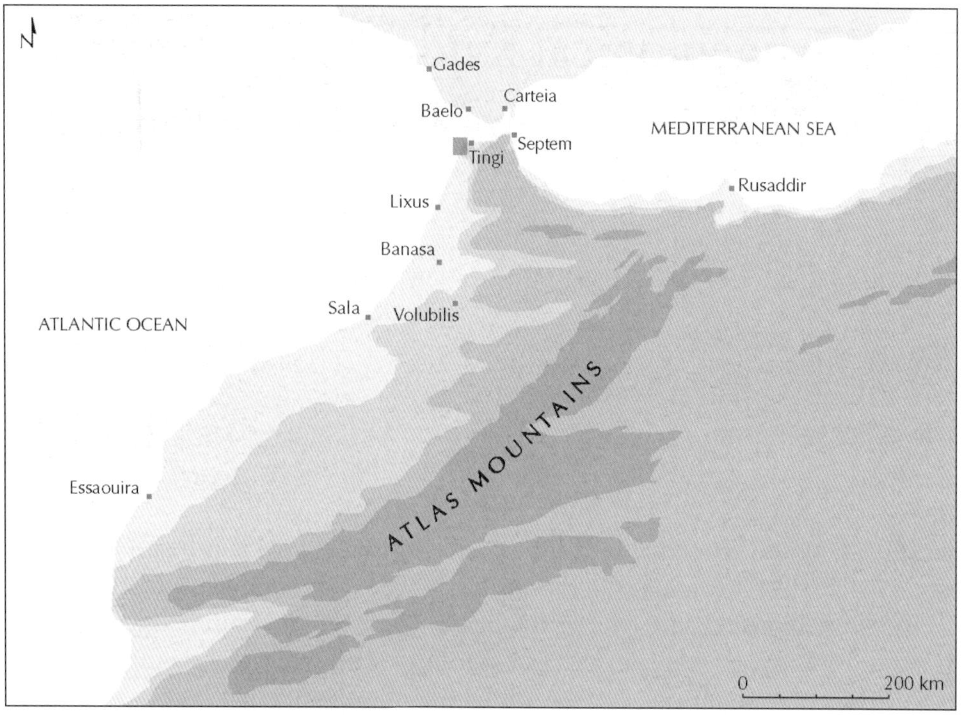

Figure 1. The Cap Spartel survey area (box), at the western Straits of Gibraltar, and some of the major Punico-Mauretanian and Roman settlements of the region (by A. Trakadas).

thic occupation but also a Phoenician and Punico-Mauretanian presence, the latter dating from the late eighth/seventh-first centuries BC (Gilman, 1975; Ponsich, 1970, 65-222; Ponsich, 1964b, 262-267). South of Ras Achakar are the remains of the beach fish-salting site of *Cotta*, which functioned from the first century BC to the third century AD (from the Punico-Mauretanian to the Roman periods; the region was annexed as the province *Mauretania Tingitana c.* 42-43 AD). Punico-Mauretanian necropoleis and Roman *villae rusticae* are located just further inland of *Cotta* (Ponsich & Tarradell, 1965, 55-68; Ponsich, 1970, 271-299; Ponsich, 1988, 150-159; Ponsich, 1964a; Ponsich, 1964b, 262-268).

The bathymetry offshore of Cap Spartel is comprised of a dramatic transition from shallower and gradually-sloping continental shelf of the Atlantic shoreline to the extreme depths of the Straits. The prevailing eastward surface current enters the Straits while an westward, deeper band of exiting Mediterranean outflow water occasionally creates pockets of upwelling along the headland. In this extremely active hydrographic zone, strong alternating north-south tidal currents also begin to run along the Atlantic coast (Hagen, 2001, S117; Snoussi, 2000, 20). In the 4km-long underwater survey area are a few submerged rock pinnacles to the north and a mixture of large sand fields and flat, eroded hardpan rock plateaus (lithified ancient beaches) to the south (Hinz *et alii,* 1982, 35).

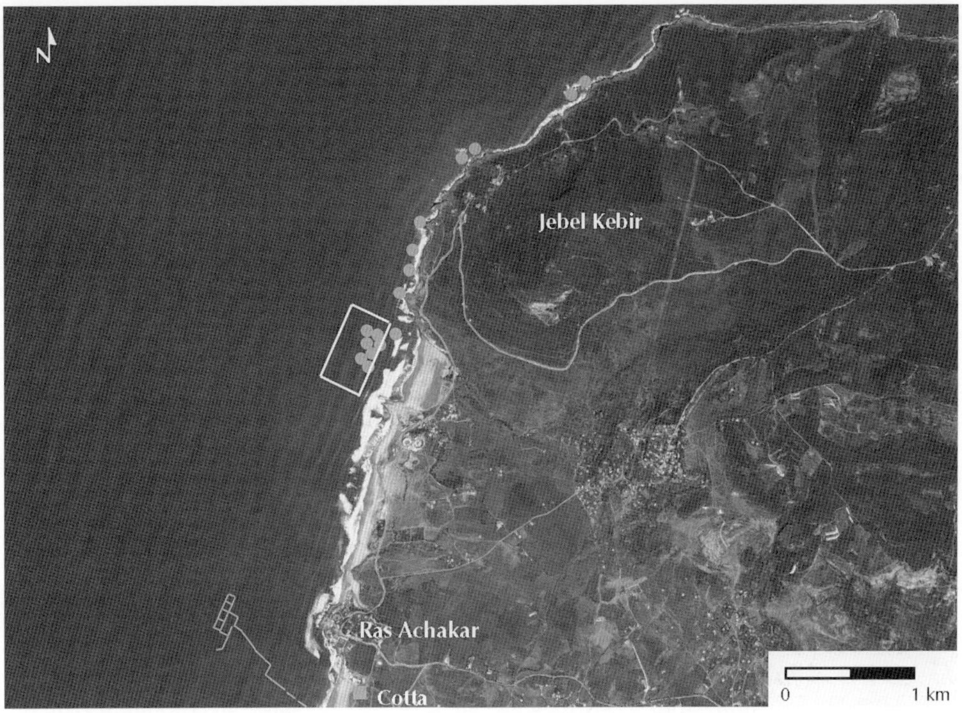

Figure 2. Detail of the Cap Spartel survey area. Find sites are indicated by dots, and the multibeam sonar survey area is indicated by the white box. The modern al-madraba (pound net) is highlighted (by A. Trakadas).

During the course of the Morocco Maritime Survey, 16 archaeological sites/assemblages were identified and recorded in the Atlantic waters off Cap Spartel and date from the fifth century BC to the 20th century AD. Unsurprisingly, an overwhelming majority of these finds were anchors. These included lead anchor elements (49 total) that were at one time fixed to wooden components, dating from the fifth century BC-first century AD (Kapitän, 1984; Haldane, 1985), and iron anchors (10 total), dating from the fourth to the fifteenth centuries AD (Joncheray, 1975, 117-138; Bass & Van Doorninck, 1982, 121-140) – figure 3. One of the main concentrated find sites is CSP062, where 50 ceramic and metal artefacts are distributed throughout an area of 120sq.m. Thirty-six of these artefacts are lead anchor elements whilst nine are iron anchors.

Navigation

The bathymetric Digital Elevation Model of the anchorage made from the survey's multibeam sonar data shows that site CSP062, in waters 16 to 30m deep, is the nor-

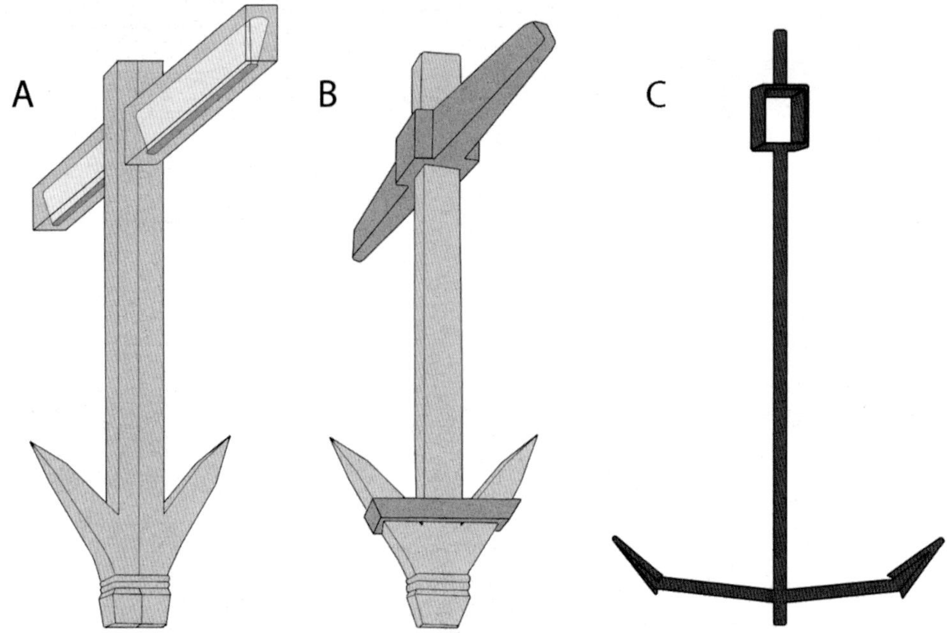

Fig. 3. The main anchor types found at Cap Spartel. A. Kapitän IIa-type lead anchor stock cores (fifth-second centuries BC) set into a hollowed-out wooden stock. B. A Kapitän IIIb-type lead anchor stock (second century BC to first century AD), shown attached to a wooden shaft. This example is depicted with a three-holed lead "collar" at its base. C. A Kapitän C/D Late Roman iron anchor (fourth-seventh centuries AD) with square hole for a removable wooden stock. It is possible that lead and wooden anchors, unlike the Mediterranean, were used in the region until the third century AD. Since the first century, however, iron anchors were more frequently and widely used, becoming the sole type throughout the Mediterranean by the fourth century AD (by A. Trakadas).

thernmost location where ships' anchors could find secure purchase on the hardpan rock outcrop exposed on the seafloor (figure 4). North and east of this hardpan rock is a large sand expanse, punctuated further to the north by rock pinnacles that are exposed at low tide. The site offers an ideal anchorage for sailing vessels awaiting favourable winds and changes in the tidal currents. This location also is directly in the western lee of Jebel Kebir, which provides shelter from the strong *chegui* winds that blow from the east in the summer (NIMA, 1994, 199; Ricard, 1927, 234) (see figure 2).

The numerous lead and iron finds indicate that vessels transiting this area, due to entanglement on the seabed or other recovery problems, were unable to retrieve their anchors and left them in order to continue their journeys. The majority of these anchors were lost during the period from the fifth century BC to the first century AD. From the Cap Spartel anchorage, vessels from the south could enter the Straits of Gibraltar, continue northward to the Atlantic coast of the Iberian Peninsula, or, if coming from the north or east, sail south along the Mauretanian coast to settle-

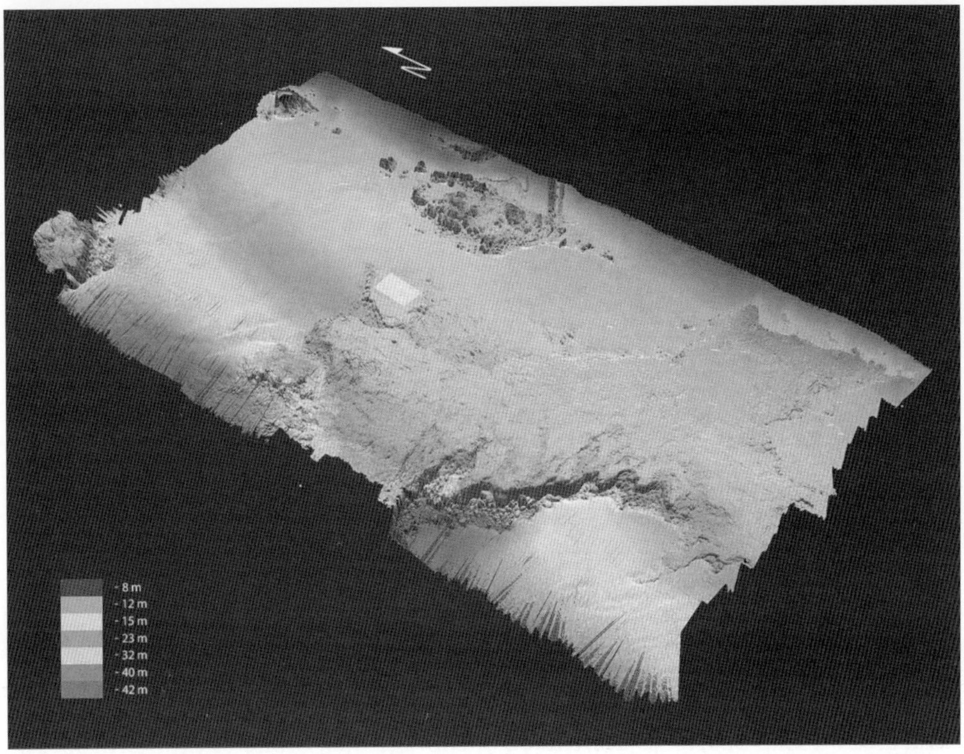

Figure 4. The Digital Elevation Model of the survey area looking to the east. Site CSP062 is the white box located near the centre of the image, at the edge of the hardpan rock plateau and sand field (by L. Huff).

ments. Voyages between the cities of *Gades, Baelo, Carteia, Septem, Tingi* and *Lixus* are mentioned by Strabo (3.1.8, 17.3.2-6) and Pliny (*N.H.* 2.168-169) (see figure 1), and the contact is well-documented archaeologically (i.e., Alonso Villalobos & Navarro Domínguez, 1998; Rouillard, 1992). In addition, the anchorage could have been used repeatedly by vessels visiting the Phoenician, Punico-Mauretanian sites and subsequent Roman settlements located south of Cap Spartel. Fresh water was also available here, from both small streams near Ras Achakar and springs on the southern face of Jebel Kebir; at the latter there is evidence of pre-Roman canalisation (Ponsich, 1964b, 267).

Vessels continued to use the Cap Spartel anchorage in the post-Roman period, as the iron anchors dating from the fourth to the fifteenth centuries AD attest, although there are far fewer finds of these. Perhaps iron anchors were easier to recover than lead and wooden anchors or, if lost, simply did not survive intact in the marine environment (Kapitän, 1984, 42-43; Haldane, 1990, 23-24). It is also quite likely that mariners who frequented the Atlantic coastline chose another area in which to anchor, such as to the south of CSP062 or to the west, where deeper waters offer more security to larger vessels (see figure 4). This scenario is most probable

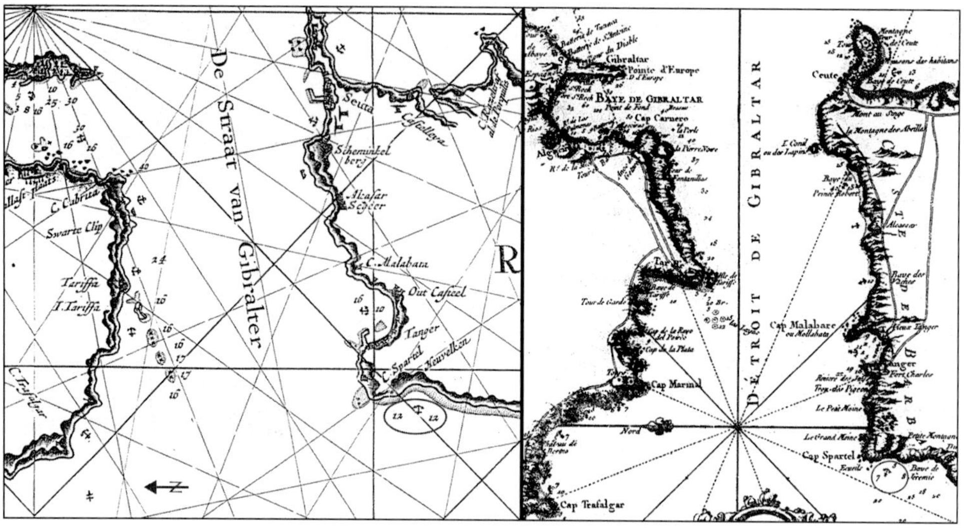

Figure 5. Historical charts and sailing instructions since the sixteenth century indicate this area as an anchorage with good purchase, sheltered from the east winds. On the left, a map by J. van Keulen from 1694 and on the right, a map by J. Bellin from 1764.

as many historical maps and navigational treatises do clearly indicate the same general area off Cap Spartel as a sheltered anchorage (i.e., Stone & Stone, 1968; Ricard, 1927, 234) – figure 5. Indeed, many more anchors were observed just west and south of CSP062 during the underwater surveys, but due to temporal constraints could not be recorded or mapped.

Fishing activities

Another activity that possibly took place offshore of Cap Spartel in antiquity was fixed-net fishing. In particular, some of the numerous lead and wooden anchors could have been used to fix so-called pound nets to the seabed. These chambered nets were primarily meant to trap members of the Scombridae family ("tunny", e.g., *Thunnus thynnus* (bluefin tuna), *Sarda sarda* (bonito) and *Thunnus alalunga* (albacore)), and set during the fishes' annual migrations (INRH, 2002, 130-135; Carrera Ruiz *et alii*, 2000, 44-47; Moreno Páramo & Abad Casal, 1971, 212; Timoule, 1985, 52). The southernmost find sites of anchor elements, especially those at and near site CSP062, are well situated for the establishment of a pound net. This area is approximately 300m offshore, and the relatively flat hardpan rock seafloor provided purchase for the nets' anchors in waters between 16m and 30m depth. Particularly in light of the strong alternating tidal currents at Cap Spartel, a fixed net would have been an ideal trapping method compared to mobile net fishing.

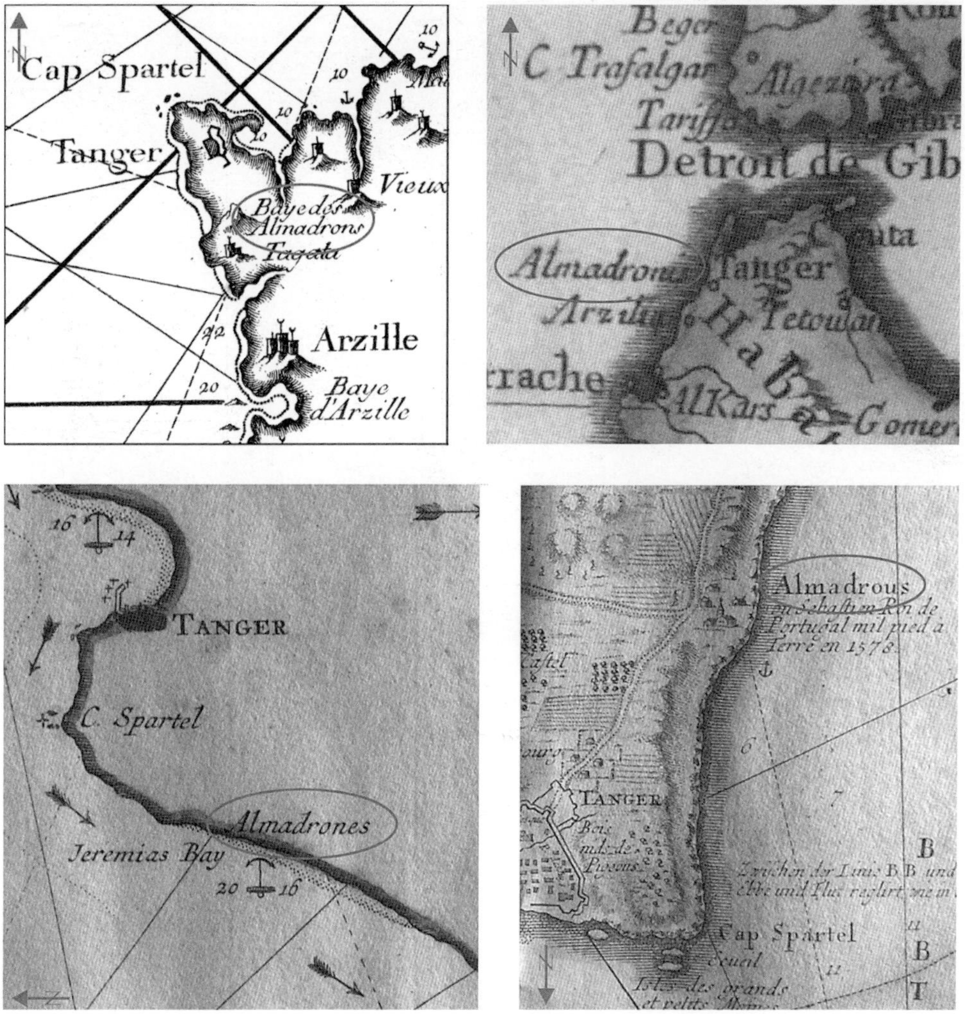

Figure 6. The toponyms "Almadrones" and "Almadrous" at Cap Spartel. Top left: a map by J. Roux from 1764; top right: a map by P. Santini from 1775; bottom left: a map by H. Lynslager from 1726; bottom right: a map by H. de Homan from 1756. The Lynslager and de Homan maps also show anchorages nearby the toponyms; perhaps this indicates that the dual use of the area for fixed nets and as an anchorage was quite possible, or that anchorage was available when the nets were not set.

Marine resource exploitation is known to have been practised in the Cap Spartel area since Prehistory, where *Scombridae* were most fished during their migrations from shore by hook and line or harpoon (Gilman, 1975, 85). Tunny bones have also been found at the fish-salting factory at *Cotta*, less than 3km from the anchor find sites (see figure 2). The factory, which operated from the mid-first century BC to the third century AD, is thought to have had a *thynnoskopeion* (a tunny watch-tower, described by Strabo in the Tyrrhenian Sea (5.2.6, 5.2.8) and on the

Figure 7. The al-madraba presently set at Cap Spartel, with the leading net from the shore oriented to the south for the late spring migration. Boats can be seen around the chambre de la mort in the upper left.

North African coast (17.3.16) on the southwest corner of the building) (Ponsich & Tarradell, 1965, 55-68; Ponsich, 1988, 150-159). The presence of such a feature suggests that the sea was watched for the distinctive surface shoals of migrating tunny or other Scombridae in order to determine when to set the leading nets from shore to the main pound net. Also, the tower could have served as a way to guard the net once it was set (Pliny, *N.H.* 2.181).

Fishing for tunny with this fixed-net method is known to have been practised in the area in the historical and modern eras. In the sixteenth century, Portuguese and Spanish fishermen frequented an area a little south of Cap Spartel, named "*Al-madrava*", where they would establish pound nets called *al-madrabas* (Ricard, 1955, 162; Bennis, 1996, 57-58). In the eighteenth century, this area was still noted on maps as "*Almadrones*" or "*Almadrons*", suggesting an occasional if not regular presence of pound nets (Lynslager, 1726; de Homan, 1756; Roux, 1764; Santini, 1775) – figure 6. Each year, between late April and August, an *al-madraba* is set to catch tunny less than 900m offshore of *Cotta*, in waters no deeper than 40m (INRH, 2002, 142-143; Timoule, 1985, 61; personal observations in 1999, 2002, 2003, 2005, 2007) – figure 7. The differing locations of these fixed nets in a general area over time in

Figure 8. An early third-century AD mosaic from Sousse, Tunisia, shows a fish (presumably tunny) captured inside a net depicted with small brown objects attached – possibly representing floats or weights. The fisherman appears to be about to hit the tunny with a club, echoing Manilius' statement that the fish were killed in the water (from Yacoub, 1995, figure 3a).

not unusual, as their placement follows the slightly changing paths of the tunny migrations – and is also documented to have occurred near Larache, Morocco, in western Sicily and southern Sardinia – (Ponsich, 1966, 394; Ponsich, 1981, 20, n. 13; Maggio, 2000; Graziani, 2002). It is likely that after lead-and-wood anchors began to go out of use in the first century AD, iron anchors were utilised to fix pound nets; perhaps if left on the seabed they did not survive in the marine environment, or, other areas closer to *Cotta* (and outside the survey area) were chosen to fix the nets.

Fixed-net fishing in Antiquity

Certain aspects of tunny behaviour were understood in Antiquity, particularly the oceanodromous species' annual migrations, when the large, normally pelagic schools pass close to shore. Ancient writers, such as Oppian (*Hal.* 3.620) and Athenaeus (7.315), describe the fishes' journey beginning in the Atlantic, passing through the Straits of Gibraltar into the Mediterranean in early summer and returning in the late summer. Strabo (3.2.7) also states that large numbers of "fat" (spawning?) tunny congregated outside the Pillars of Hercules.

One of the techniques to catch tunny during their migrations was to channel them into and trap them in a net consisting of several chambers and fixed to the seabed. Fixed tunny nets are mentioned by Alciphron (*Ep.* 1.17.1) as being heavy and held up by cork on the surface and weighted on the bottom by lead. More detail is given in the text of Manilius (*Astronomicon* 5.659-666), who describes a network of meshes of the net that trap the fish, which are then killed by a knife in the water (figure 8). Oppian (*Hal.* 3.637-642) describes a tunny net with a series of gates, gate wardens and inner courts.

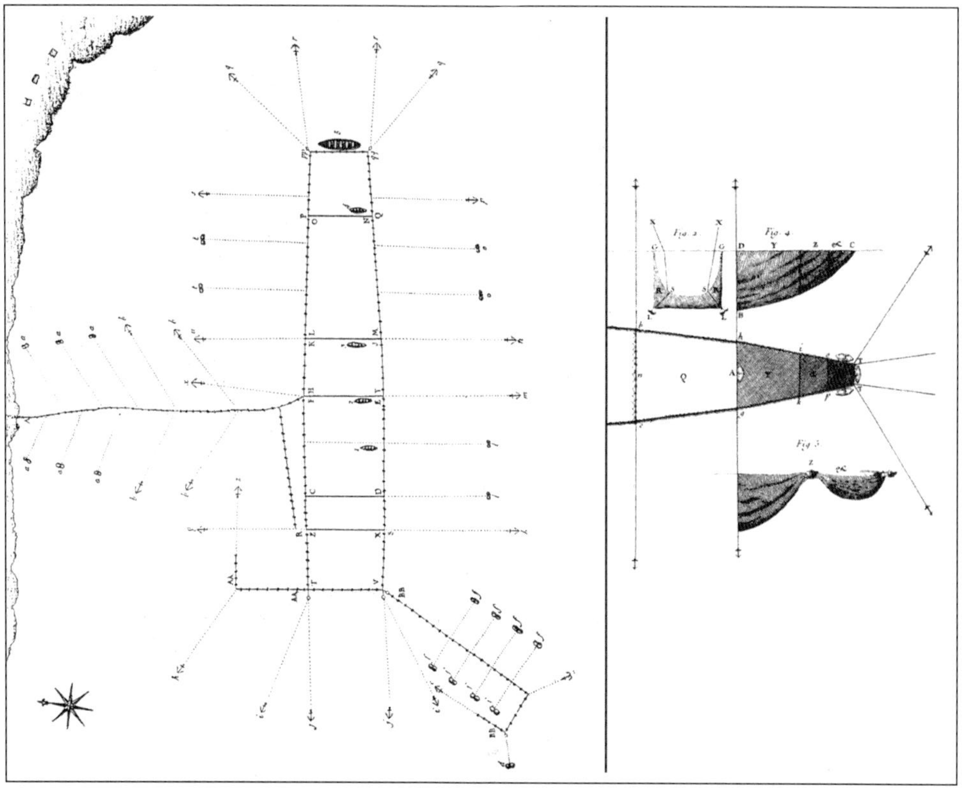

Figure 9. Left: Plan of an eighteenth-century al-madraba net with anchors, used off southern Spain (from Ponsich, 1988, figure 11). Right: Detail of the lift net contraption in the chambre de la mort from a 1777 plan (from Étienne & Mayet, 2002, figure 8).

These descriptions are similar to the type of fishing methods used with historical and modern fixed pound nets that are positioned perpendicular and near to shore in shallow waters. These are called variously *almadraba* (Spanish; different from an *almadraba de tiro,* see above, pp. 207-210), *madrague* (French), *tonnara* (Italian) and *armação* (Portuguese). In the western Mediterranean and eastern Atlantic, these nets are set in April/May and dismantled in September/October to coincide with the fishes' migrations. A net consists of a series of chambers and is held on the surface by floats and on the seabed by large anchors (Ponsich & Tarradell, 1965, 93-97; Moreno Páramo & Abad Casal, 1971, 212-213; Ponsich, 1988, 34-36; Carrera Ruiz *et alii,* 2000, 44-45; Étienne & Mayet, 2002, 31-34). A leading line of net, fixed from the shore to the main net, guides the fish into the first chamber. The fish are then moved or herded through a series of chambers by opening and closing gates in the net walls until they reach the final chamber. This last chamber, the *chambre de la mort,* has a floor that consists of a lift net (figure 9, right). This net is drawn up by fishermen in boats that encircle the last chamber; the fish are speared and clubbed within the net and brought aboard the waiting boats (figure 10).

Figure 10. The al-madraba nets off Cotta in the 1960s. Top: pulling up the lift net; middle: bringing the tunny to the surface, spearing and clubbing the fish; bottom: bringing the tunny aboard the waiting boats using gaff hooks (from Ponsich & Tarradell, 1965, pl. XXI).

Conclusion

Since Prehistory, the Cap Spartel area was undoubtedly an important region, both in its terrestrial and maritime aspects. The material assemblages of habitation sites, necropoli and *villae rusticae* indicate that the headland and the surrounding lands had been settled and the terrestrial resources had been exploited since the Neolithic and well into the Roman period. The maritime importance of the region did not go unnoticed, however. The anchors documented in this area reveal that the western lee of Jebel Kebir provided a sheltered anchorage. This site was a well-placed stopping point on navigational routes between numerous settlements on the African and Iberian coasts and close to fresh water. Additionally, the rich marine environment provided regular possibilities for sustenance, and some of the anchors from Cap Spartel could demonstrate that large-scale annual harvesting events which targeted migratory species took place. Certainly the placement of the Roman fish-salting factory of *Cotta* near the Cap Spartel anchorage was meant to take advantage of these resources, and the establishment or even continued use of a fixed net nearby would have provided abundant products to consume fresh or to process.

Hauling of nets. The technique used by the person hauling is called "to haul at the wave".

14. Fishing Gear, Open Boats and Preserving Skills

Atle Ove Martinussen

Introduction

Technology history within the field of fishing gear has to a great extent focused on technological changes and new inventions. From my point of view, it is especially interesting to study the transition from one raw material to another, i.e., from natural to synthetic fibres in fishing nets. One point which seems to have received less attention is the fact that the practical knowledge and experience of fishermen using the fishing gear really make a difference when it comes to studying the development and diffusion of new types of fishing gear.

Natural fibre nets have a long tradition all over the world. From prehistoric times onward we find natural fibres as the dominant material in fishing gear. This changed around 1950 with the introduction of synthetic fibres. In this paper, I shall focus on the huge technological time span in gears and open boats and show the connection between ancient and modern times in this research field. The fact that we still have reminiscences of ancient fishing gear and methods in our modern world gives us an advantage in our research when it comes to identifying and understanding the fishing gear of Classical Antiquity. Fishermen who have been working with traditional fishing gear made of natural fibres during the twentieth century have skills and know-how that will be interesting for researchers in this field. They possess a knowledge transmitted and accumulated through many generations.

Technological leaps

Over time, we have seen a lot of interesting technological innovations which have lasted for several hundred and even thousands of years before they disappeared or were replaced with new technology. Some of these technological leaps have led to revolutionary changes in society and everyday life in its economical, social and cultural respects. We could mention, for instance, the introduction of bronze, iron, the longbow – and synthetic fibres in fishing gears.

The fishing gear used in Classical Antiquity had many similarities with gears from modern times at least up to the 1960's, in method of use as well as in appearance. A recent archaeological find shows an impression in clay of an ordinary knot made of nettle fibre, used in natural fibre nets. The find is 25-27,000 years old (*Illustrert Vitenskap*, 1999, 7, 29). The same knot was commonly used in natural fibre nets until the 1950s and 1960s. With the introduction of synthetic fibres around 1950, the knot changed because the new material was more slippery. The knot was strengthened with one or more turns of the needle which changed the nature of the knot.

The huge technological time span shows how modern and ancient times are connected. This span of more than 25,000 years with accumulated knowledge of making this specific knot for nets equals over 1,000 generations of living transmission from person to person. In Scandinavia we have used oars and open boats for more than 2,000 years. Archaeological finds from Manger (Radøy, Hordaland) show several rowlocks from 85 to 245 AD; almost the same shape of rowlock is still used in traditional open rowboats in this area today (Christensen, 1995; Ekroll, 1989; Kolflåth, 1986, 1987 and 1988). If we make the same calculation as with the knots, we see that the knowledge of making and using open boats with rowlocks has a transmission history covering *c.* 80 generations.

If we take a look at different kinds of fishing gears in mosaics and finds from Classical Antiquity, we can see that items such as net needles, sinkers and hooks have changed very little up to modern times.

We, the people of the twenty-first century, are in fact in the position that we can decide to break this chain of knowledge, skills and live transmission. The preservation of skills is an important issue in all transmission of traditional craftsmanship and the Norwegian Crafts Development (NHU) has developed models for preserving skills within different traditional crafts through 20 years of fieldwork. UNESCO is also stating the importance of this in the *Convention for the Safeguarding of the Intangible Cultural Heritage* promulgated in 2003. The convention has to date been ratified by nearly 100 states (UNESCO, 2007, 5).

Historical background

The use of nets made of natural fibre has a long tradition all over the world. In our part of the world it is especially well known from the salmon and herring fisheries of the Viking Age and the Middle Ages, where seines were used both along the

Figure 1. Copy of the knot (mesh knot) made of natural fibre dated to 25-27,000 years ago (copy and photo Atle Ove Martinussen, NHU).

Figure 2. Copy of the mesh knot made of artificial fibre (nylon) which changed the knot in the 1950s (copy and photo Atle Ove Martinussen, NHU).

Norwegian coast and the coast of Bohuslän in present-day Sweden (Iversen, 1937, 4, 60; Nordgaard, 1928, 6-7). Archaeological findings from the Czech Republic shows a print in clay of an ordinary knot used in nets made of natural fibre (nettle). The anthropologist Olga Soffer claims that these nets were used by women for small game hunting in the Ice Age (25-27,000 years ago) (*Science-Week*, August 21, 1998). The nets were made of hemp and nettle. We also know prehistoric finds of parts of nets from Finland from the Stone Age, most probably made of nettle (Clark, 1948, 56). The knot that Olga Soffer found was commonly used in nets made of natural fibre in the 1950's and the 1960's, and it was also used in nets made of synthetic fibre but then it was strengethened with several turns. It is called by several names, for instance *fisherman's knot, standard knot* or *mesh knot.* The knot had been unchanged for almost 27,000 years but was altered because of the introduction of new fibres, an artificial fibre. I will give a brief description of the innovation processes to show how quickly the conversion from natural to synthetic fibres actually took place after World War Two (figures 1 and 2).

In the traditional fisheries for cod in Norway we know nets made of natural fibre to have been used from the seventeenth century onwards. Before this time, the cod was caught with hand fishing lines (KLNM 4.56). This method was still used in the 1950's. The hooks used at this time had much the same design as in ancient times (Olsen, 1998). Long lines and cod nets were most probably introduced into the cod fisheries in Norway in the sixteenth and seventeenth century and used alongside hand lines (Nordgaard, 1928, 20). There was great opposition against this new fishing gear both in Lofoten and other fishing district along the coast of Norway (Iversen, 1937, 44). The nets were expensive and required bigger boats, and only a few fishermen could afford to invest on that scale (Iversen, 1937, 43).

Hemp was almost the only material for fishing gear till 1900 when it encountered competition from cotton. By the 1950's, cotton was the all-dominant material for herring nets and seines in Norway, while hemp and cotton were the preferred material in cod nets. It took some time until cotton made a breakthrough in the cod fisheries. In the seasonal fisheries for cod off Lofoten some fishermen had actually started to use nets made of cotton as early as 1935. Still we find that cod nets of hemp outnumber nets of cotton until 1950 (Strømsheim, 1970, chapter 14, 4).

This gives us reason to believe that among natural fibres, a transition from hemp to cotton was taking place at the same time that the synthetic fibre was introduced into the arena around 1950. We can look at cotton as a technical material innovation, an improvement in comparison with hemp, which amongst other things is heavier than cotton when wet. The diffusion of cotton as a new material for cod nets was going really slowly and it took almost fifty years from 1900 to 1950 before it made any real impact in the cod fishery. During the 50's this changes, new synthetic fibres are introduced and become dominant in net fishing, especially for cod. We have focused on cod nets but there was an ongoing conversion from natural to synthetic fibres in other types of fishing net as saithe nets and also long lines, ropes and so on. The switch from natural to synthetic material was rapid compared with that from hemp to cotton; it started with governmental tests of synthetic fibres in fishing gear shortly before 1950. Similar tests took place shortly afterwards on a private basis (Martinussen, 2006b, 33-36) – figure 3.

By 1960, 95% of all fishing nets used to catch cod in Norway were made of synthetic fibre (mainly nylon/polyamide). As I see it, the conversion from natural to synthetic fibre in net fishing for cod was almost complete at this time. The transition was going on at an accelerating pace around 1960 and onwards for nets, trawls and lines used for targeting other species than cod. If we look at all types of fishing gear we must conclude that the conversion is going slower after 1962. In 1957 4% of all fishing gear in Norway was made of synthetic fibre, in 1962, 62%, but not until 1975 did we reach 95% for all kinds of fishing gear (Nordstrand, 1980, 24).

Polyamide (nylon) is the most popular fibre in fishing gear, and if we look at 1971, polyamide accounts for 76% of all synthetic fibre in fishing gears in Norway (SF, 1972, 7). The table shows the different synthetic materials used for fishing gear.

Gear \ Fibre	Polyamide	Polyester	Polyethylene	Polyprophylene	Polyvinylalcohol
Net	×				
Seine	×				
Line	×	×			×
Ropes	×	×	×	×	
Trawl	×	×	×	×	

Table 1. Synthetic fibres used for fishing gear (Source: "Syntetfibre i fiskeredskap", SF, 1972).

We have seen that the introduction of synthetic fibres in fishing gear from around 1950 transformed a way of net-making that are going back thousands of years as far as materials and knots are concerned. The use of synthetic fibre in the fisheries must be characterized as revolutionary when it comes to the amount of catches, which increased almost 400% in net fishing for cod compared with hemp and cotton (Martinussen, 2006b, 29).

Today, research and development in the chemical field are partly concerned with the environmental consequences of using artificial fibres in fishing nets and the researchers are striving to make synthetic fibres with the properties of natural fibres. Lost fishing nets in thousands are standing on the bottom all over the world, fishing for eternity as the nets are continuously catching fish, and unable to rot because they are made of artificial fibres. As it is filled with fish, the net sinks to the bottom, the fish rot and the nets rise again and start fishing till they are filled once more, and so it starts all over again. This is very appropriately known as *ghost fishing*.

In my opinion the motives behind the chemical research mentioned above are not only environmental: the most important issue is the fact that all the fish caught in "ghost fishing nets" are lost potential catches that provide no profit to the fishing fleet. We are actually speaking of thousands and thousands of tons of fish caught by ghost fishing (Martinussen, 2006a, 11). Maybe it is time to start all over again, downscale the fisheries and restore ancient fishing techniques and gear in order to make a sustainable development for the future. But this is another story.

This overview of the development from natural to artificial fibres in fishing gear also highlights the fact that we have to take action now if we want access to the fishermen with reminiscences of the knowledge and skills inherited and transmitted from ancient times. In my opinion it will increase our understanding of how the fishing gear of Classical Antiquity were used and give us some interesting and new approaches to this research field. Is it possible that the skills and knowledge of fishermen who have experienced the last period of manual fishing in the twentieth century have anything to contribute to the research in ancient fishing gear and technology? Can we create an arena where skills and knowledge can be taught, transferred and preserved?

Figure 3. Hauling nets made of artificial fibres outside the Western Coast of Norway in 2003 in the NHU project "Use of open boats". The triangle model was used to transmit action-borne knowledge in this project (photo Atle Ove Martinussen, NHU).

Preserving skills

Knowledge and skills are passed on from one generation to another and the practical knowledge and experience of fishermen is accumulated and transferred through many generations. We are the first generation actually having a choice, either to stop this accumulation of skills and knowledge, or to pass some of it on to new generations as living tradition and to record it in archives. We could even demonstrate it in museums and technological "know-how" centres if we were able to pass on the skills and knowledge to new persons (Martinussen, 1993). If we look to UNESCO and the "Convention for the Safeguarding of the Intangible Cultural Heritage" (2003), the following quotation underlines the importance of this work on a worldwide basis:

> Traditional craftsmanship seems in many ways to be the most tangible of domains in which intangible heritage is expressed, but the focus of the Convention is not on craft products as such, but rather on the skills and knowledge crucial for their ongoing production. Any efforts to safeguard traditional craftsmanship must focus not on preserving craft objects – no matter how beautiful, precious, rare or important they might be – but on creating conditions that will encourage artisans to continue to produce crafts of all kinds, and to transmit their

Figure 4. Audunarstova. A Norwegian Medieval wooden building from 1315 reconstructed at Holar, in the northern part of Iceland. This project would not have been possible to do without almost 10 years of training and passing on action-borne knowledge to the craftsmen (both carpenters and blacksmiths) who made this reconstruction. The project was done in cooperation between Norway and Iceland from 1997-2001 (photo Atle Ove Martinussen, NHU).

> skills and knowledge to others, especially younger members of their own communities (Ségur/ZED, 2007)

By 23 February 2007, 75 states had ratified the convention (UNESCO, 2007). Norway ratified the convention in December 2006.

The skills of craftsmen need to be passed on through interaction in a practical learning situation. Norwegian Crafts Development (NHU) has developed models of competence building for craftsmen in traditional crafts through massive fieldwork, mostly in Scandinavia, over the last 20 years. We have also for several years had projects concerning fishing gear and open boats, reconstructing fishing gears and boats to use them together with old fishermen to record and preserve their skills and knowledge and document them for posterity (figure 3 & 4). NHU projects generate large quantities of documentation material in the form of photographs, video and text. The entire craft process is documented from raw material to the finished product (Martinussen, 2005).

Developing new methods for passing on action-borne knowledge

NHU have developed, through the last 20 years, new methods for passing action-borne knowledge on to younger generations. We have placed particular stress on finding people who are part of a living tradition and whose skills and knowledge have been passed on to them directly from a senior master. Unless we are able to pass them on, these living and action-borne skills are bound to disappear as the old practitioners of these crafts pass away. This model or method has been tested through years of fieldwork and different projects relating to traditional crafts. We think it has proved that it is efficient and well suited for preserving and passing on action-borne knowledge in different traditional crafts (Martinussen, 2005). Our experience from projects related to fishing gears and boats have given the same results.

In the NHU project "Use of open boats" which took place during the years 2003-2006 the aim was to transfer action-borne knowledge regarding the use of open rowboats for fishing, both in rowing, sailing and fishing with both crab pots (creels), different kind of fishing nets, long lines and hand lines. We used both four-oared boats and a six-oared boat of the type "Oselvar", which was the most common open fishing boat for inshore fishery off the coast of Bergen (western coast of Norway) up till the 1960s. These four-oared boats had a crew of two or three persons. An important aim was to provide the young boatbuilders, who today build the traditional Oselvar boat, with a knowledge of how the boats function through living transmission and in interaction with old fishermen who had used this type of boat together with fishermen born in the nineteenth century (their fathers and grandfathers).

Later in the project, the boatbuilder reconstructed one of the boats and the task was to make it with the same properties as the old one. In figure 5 we can see a young person who has just started to haul a net while the old tradition-bearer (the fisherman) rowing the boat is instructing him how to do it. An interesting point here is that the boat is actually constructed to assist the fisherman in the task of hauling the nets or long lines. The shape of the bottom and the front of the boat is made deliberately so that the waves of the sea will lift the gear with the boat; at the crest of the wave, the fisherman waits for a second and when the boat goes down he hauls in the slack of the net or line. To be able to do this the lifting capacity must be built into the boat during the construction and the fisherman must have learnt to use the boat in this way from tradition bearers by living transmission through action-borne knowledge.

The boats were made especially for the different kinds of fishery according to what characteristics the fishermen had asked for when they ordered the boat. To do that, the boatbuilder had to know a lot about what kind of fishery the boat was meant for and what requirements the fisherman had for its use. It was quite common for a fisherman to spend a week with the boatbuilder when he started to build the

Figure 5. Hauling of nets. The old boat has been built with specific lifting capacities. The technique used by the person hauling is called "to haul at the wave" (photo Atle Ove Martinussen, NHU).

boat, to make sure they had the same view of how the boat should function and what properties should be built into it (Martinussen, 2004).

This long process starts in the mind of the boatbuilder when he gets the order for a boat. The boatbuilder actually built the boat in his mind to try to find optimal solutions and make the boat in a way that would meet the needs of the fisherman who had ordered it. If the boatbuilder was unable to do this the fisherman could cancel the deal and send the boat back to the boatbuilder (Martinussen, 2004). In the project mentioned above, NHU was able to pass on this vital action-borne knowledge to the young boatbuilders by using the "triangle model" (figure 6) and they were actually able to make a copy of the old four-oared Oselvar with the same properties as the old one. We established that they had been successful by testing the two boats against each other as part of the project.

The triangle model works especially well within planned and organised teaching arenas which NHU design for each project. The model is based on the old relation between the senior master of a craft and a new craftsman, where the senior master is responsible for the training of the craftsman in a 1:1 relation. This is actually an old model going back to the medieval training system of craftsmen and NHU has developed a third leg to this model to make it function in a new manner for our purpose of preserving and passing on action-borne knowledge. The model can be described as a triangle (figure 6) consisting of three relations; first we have the senior master called the *tradition bearer*, second a new *craftsman* and then we have the third person the *documentator*.

The tradition bearer is a person carrying the knowledge and skills of several generations within him- or herself, and he or she is part of a living tradition. These are craftsmen forming part of a cultural tradition where knowledge has been passed on from father to son and mother to daughter through generation after generation in a cultural and knowledge fellowship.

The new craftsman is willing to take part in the project to learn what the old person is able to learn him. He is then obliged to bring the skills further and practice his skills and teach other person what he has learned.

The documentator has the task of documenting and recording – in videos, photos and writing – all that takes place between the tradition bearer and the craftsman. He/she is the proverbial "a fly on the wall" and not allowed to interrupt the processes going on between the tradition bearer and the craftsman. The documentator must know the work processes and take part in the work before (s)he is allowed to document what is going on. NHU is very strict on this point because it is very important that the documentation is done by a person well aware of and familiar with the different stages in the processes from the selection of timber in the forest to the finished boat. In that way the documentation is more accurate and more suitable for keeping the skills of the new craftsman alive when the old master has passed away. The new craftsman will have access to the documentation material once the project is over (Martinussen, 2005). It becomes his or her "memory book" and the objects he or she has made together with the master becomes, so to say, their "textbook".

The field work and projects carried out by NHU for the last 20 years have given us the clear impression that it is by no means sufficient to preserve the knowledge and skills only in writing and photos; yes, we have experienced that even hours of videotaping is not enough to learn and preserve the action-borne knowledge and skills, they *must* be passed on through the work processes and by interaction and cooperation between living persons. You will not be able to learn a craft fully by watching videos and reading books or articles about the processes. To NHU the most important documentation medium is the new craftsman as a living person able to pass on the skills to new generations (Martinussen, 2007, 56-57).

By now, you will surely already have asked yourself the question: "What is action-borne knowledge?" I will try to define it as we use it in the context of NHU. Action-borne knowledge is the sum of experience and skills that have been passed on from one generation to another in a knowledgeable fellowship of action and

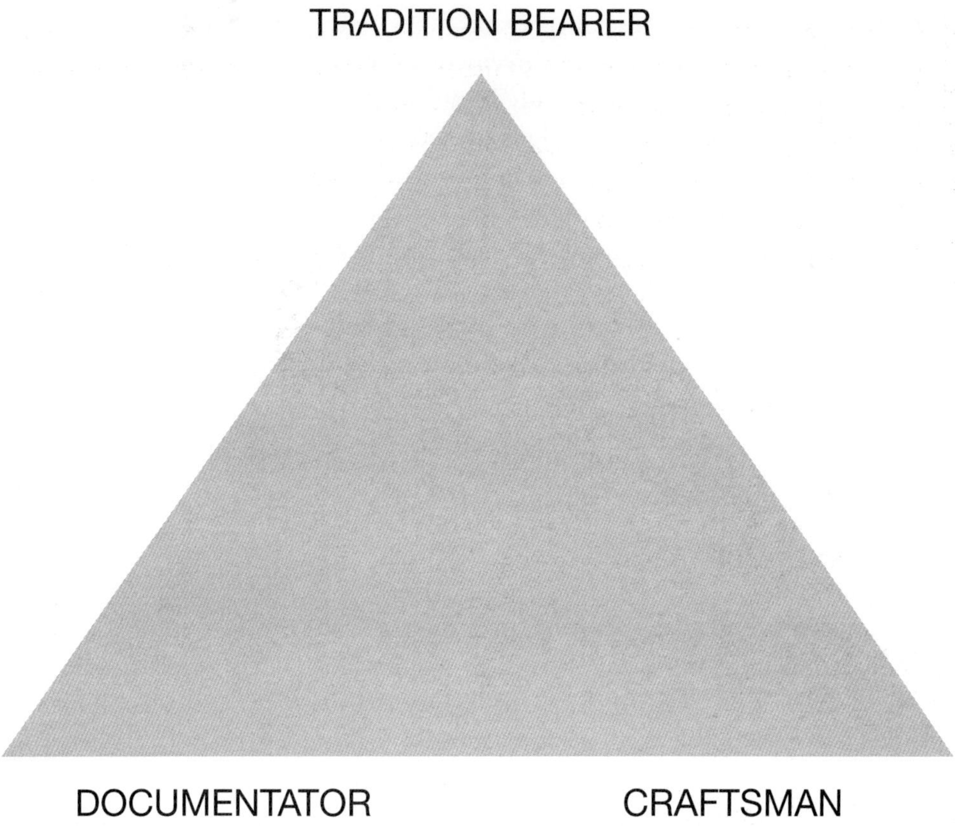

Figure 6. Triangle model for preserving and passing on action-borne knowledge developed by NHU.

culture. Both craftsmanship, patterns of acting and perceptions are inherited from generation to generation. In the transmission of action-borne knowledge the basic learning process is copying, combined with experimenting and personal experience. It is the same way that children learn to speak and walk (Martinussen, 2007, 55-56). This is the short version of what action-borne knowledge is and there will be no space in this paper to develop this further. The same phenomenon is described in other connections as; situated learning, silent knowledge, non-verbal knowledge, tacit knowledge, learning by doing, traditional knowledge and experienced knowledge, just to mention some of the expressions used in different research fields.

NHU has with good results used this model for passing on skills at different levels. The model has to be adapted to each project if it is to function, because each project has its own specific requirements when it comes to people involved and the type of craft. The documentation material such as items, tools, material samples, photos, video, drawings and written material are being archived at the national museum Maihaugen at Lillehammer where NHU is situated. Bigger items such as buildings and boats we prefer to leave in the region where they are built, in cooperation with local museums, organisations, enterprises or private owners. In such cases the point is to reach an agreement which gives us access to the item if we or the new craftsman need to study it in the future.

By using this triangle model and giving active, skilled craftsmen access over to different projects and contact with tradition bearers over several years, we are actually able to preserve and transmit action-borne knowledge which is of great importance for preserving our cultural heritage. Craftsmen taking part in our projects on a regular basis reach a level of craftsmanship that goes far beyond the level of a normal Craft Certificate. Many of the projects result in the establishing of new business for the craftsmen involved. A spin-off of the projects is that they often become motors in rural development.

Concluding remarks

There are clearly links between modern and ancient times when come to the use and design of fishing gear. I have shown how this is closely connected to the transferring of skills and knowledge. I have called it *action-borne knowledge* because the knowledge is only active through the action of making the product. When a fisherman made a net he had to have a lot of knowledge and experience of what kind of fish he was going to catch, how the wind and streams and bottom conditions of the sea were at the fishing locations if he was going to make a net of good condition and function. He put all his own experience as fisherman and all he had learnt from his father and grandfathers into the process of making the net.

In Lofoten during the 1950s the cod fishing nets used early in the season for cod fishery were mounted differently than later in the season (Martinussen, 2006b, 29-44). And why, you would ask? The fishermen knew from experience and from the knowledge passed on through generations from their ancestors that the cod moved much faster and was more aggressive early in the season when it reached the coast; later, just before the spawning period it was more lazy and relaxed. To be able to catch the cod the fishermen had to mount the net differently during the season. This shows that there is more to making nets than just a needle and a thread. This is the same also for other crafts such as boatbuilding, blacksmithing and so on. As an academic, you can know everything about net making such as the size of all kinds of meshes, what mesh size goes with what kind of fish, when and where the different nets are used, what kind of material is used, names of all the parts of the nets, what sinkers and floaters are used, how the nets are shot and hauled, what kind of waters they fish in, what boats are used plus anything you theoretically could think of, but could you *make* a net or use it to fish? No, you would not be able to do so without experience and action-borne knowledge.

Figure 7. The traditional way of hauling of nets by hand reconstructed in the project "Use of open boats" in 2003. Compared with the same scene in mosaics from Classical Antiquity (p. 346) it is hard to believe the span of time which is over 2,000 years.

Figure 8. Interaction and live transmission of action-borne knowledge connected to the use of open boats.

This is the reason why I like to speak of the practical world, where you really have the skills to do or make the objects, as being *in* crafts and the theoretical world as speaking *about* crafts. In fact, these are two different worlds of knowledge: a practical and a theoretical world. What is really interesting for me in my research is that the level of abstraction connected to practical knowledge (action-borne knowledge) is not lower than in the theoretical world. This is also the reason why you, as a researcher, have to take part in the practical work to get hold of, and understand, this partly unknown world of practical knowledge called action-borne knowledge. I think that the time is ripe for some interdisciplinary research projects which involve craftsmen in the research process not only as someone who demonstrate crafts but as one who contributes in the process to get the most interesting research results as a partner (figure 8).

Earlier I asked the question: is it possible that the skills and knowledge of fishermen who have experienced the last period of manual fishing with gear made of natural fibres in the twentieth century have anything to contribute with in the research of fishing gear and technology in ancient times? I think the answer is *yes* and that we as researchers should be more aware of this and use the competence of fishermen to understand more about fishing gears and methods of fishing. I also find reconstruction of ancient fishing gears interesting as a source for knowledge and through practical trials we may find answers to some of our questions about fishing gear in ancient times. The practical skills and knowledge have been passed on and accumulated through thousands of years through interaction between generations. This approach may detect some of the true stories of the use and production of fishing gear in ancient times. We just have to open our eyes and discover it as a piece of evidence or a piece in the puzzle of nets and fishing gear in Classical Antiquity (figure 7).

Bronze coin minted in Ilipa in the mid-second century BC. Reverse. Sábalo towards the right, underneath ILIPENSE *legend between two horizontal stripes.*

15. Corrals, *Sabaleras* and *Pulperas*: Three Types of Fishing in the Bay of Cádiz

J.J. López Amador & J.A. Ruiz Gil

From the methodological perspective of Cultural History (Kristiansen & Larsson, 2006, 409-413), three types of fishing gear can be distinguished. This study will focus on static types, those located on the coast (corrals), the surface (the *sabaleras*), and at heart (the *pulperas*). The three types have pre-Roman roots and can also be documented in subsequent periods.

Corrals

Corrals can be described as large traps with straight or curved side walls in drystone construction (figure 1). The existence of a similar structure, known as a *farush* in the Persian Gulf has been published recently, dated to the Late Islamic period (Beech, 2004, 47, figure 15). According to Lagóstena (2005) the technique goes back to the Roman period, as seen from an *armadilha*-type trap found on the shore at Silvade (Espinho, Portugal).

In the Gulf of Cádiz, similar traps, here called *corrales*, are found, known to have been used as early as the sixteenth century in Sanlúcar de Barrameda and up to the present day in Chipiona and Rota. The *corrales* of Sanlúcar have now been declared a Natural Monument of Andalucía.

Until recently, the corral of Merlin, in Sanlúcar, was held to be even older. In fact, it was assumed that the *corrales* located towards the estuary of the Guadalquivir were older than those of the Bay itself, those at Santa Catalina in El Puerto de Santa María and those of Vives in Cádiz. Recent data indicate that this is not the case. In El Puerto de Santa María, a corral known as "de las Almejas" existed from the beginning of the sixteenth century. It looked like a sandbox – coarse grain sand for works, as its name suggests. From its mention in a permit granted by the Duke of

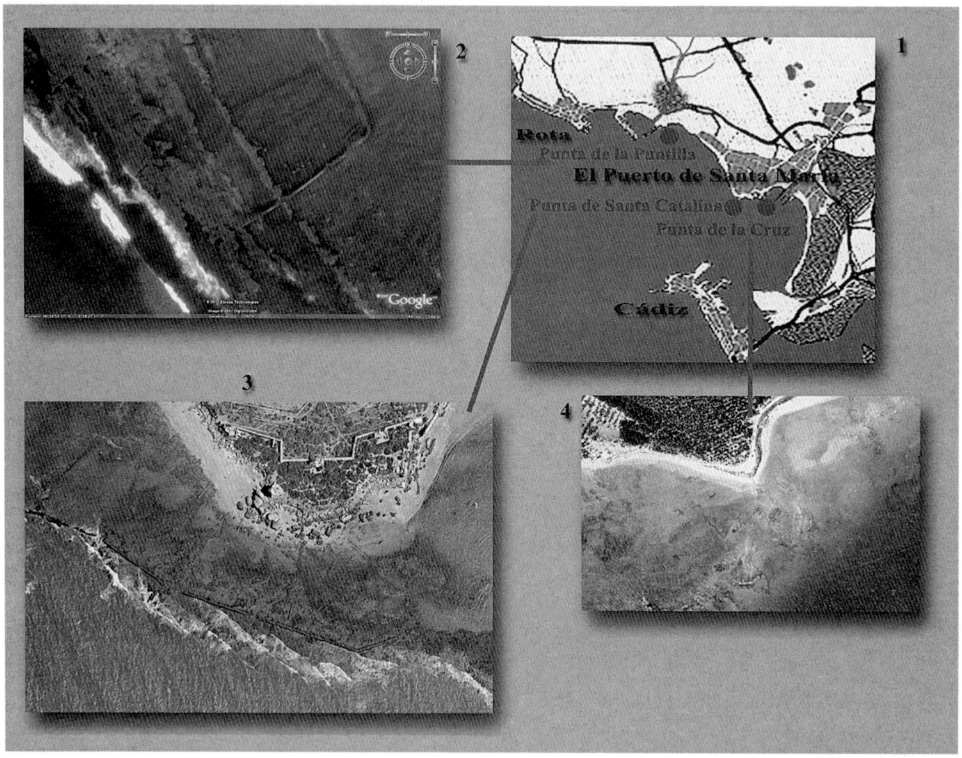

Figure 1. 1. Corrales in the Bay of Cádiz. 2. Catalan's corral; 3. Santa Catalina's corral; 4. Punta de la Cruz's corral.

Medinaceli, we know that was located in the Punta de la Cruz, under the asphalt at the beach of Puerto Jerez.

There is no certain knowledge of its use in Turdetanian times (Iron Age II). Nevertheless, Moreno and Abad (1971) have considered those of Barbate to be as old. One is connected to the coast by corridors excavated in the rock. Farther away are circular constructions of masonry, 20 to 30 metres in diameter (whereas the corrals of the Bay of Cádiz that are polygonal). In the example that we have exposed at the Puntilla del Salado (within the present Naval Base of Rota), the main structure abuts on other, smaller adjacent basins that may have served for keeping live fish (Pliny, *N.H.* 9.49).

***Sábalo* fishing**

We do not know what impact the migrations of *sábalos* (*Alosa alosa*) had on the fisheries of the Bay. The certain thing is that for its capture in the river, the normal gear

Figure 2. As from Caura. First half of second century BC. Obverse: helmeted male head. Reverse: sábalo towards the right. Underneath, CAVRA legend between two horizontal stripes. Under line of exergue crescent moon to left, A to the right.

was the *sabalera,* similar to the *jábega* (a traditional large net for fishing from shore), the only difference being that one is used in the river, the other on the coast. Fisheries dedicated to this fish may have existed, but they would have to be located on the middle and lower Guadalete, and only operated on a seasonal basis.

In the description of the Guadalete river given by Pascual Madoz in his dictionary of 1849, he mentions that in the spring, *sábalos* were abundant at the tide-line near the Cartuja bridge. The seasonal fishing activity employed many people and provided Jerez and the markets and towns of the immediate surroundings with fish.

The importance of this fish in the middle of the twentieth century is evident from the work of Ferrer (1995). For example, in year 1952 the city council of El Puerto de Santa María allotted 35 fishing areas for *sábalo.* The fishermen, who had to bring their own nets and tools, came in great numbers from the neighbouring city of Jerez. As many fishermen were concentrated in a small area, the amounts of *sábalo* must have been even more abundant than in the Guadalete, described above.

In Antiquity, *sábalo* must have been as important for the towns of the shore of the Guadalete as the salmon was for Nordic peoples. Their annual presence in the river from spring to summer allowed the exploitation of this economic resource to continue until the 1970's, when pollution and the barriers constructed for flood control prevented the *sábalo* from entering the rivers to spawn.

In fact, in antiquity several cities represented sábalo on their coinage. Mértola (in Portugal), sited on the Guadiana river produced coins showing two *sábalos* (http://www.tesorillo.com/hispania/2latinas3.htm#murtili). The sábalo also figures on coins of *Corumbaria* on the Guadalquivir river, the same *Caura* (Coria del Río), that in the second century BC produced asses and semises with a male head and a *sábalo.* In *Ilse* and in *Ilipa* city (Alcalá del Río), again on the Guadalquivir, second-century BC coins with ears of corn and *sábalo* were produced.

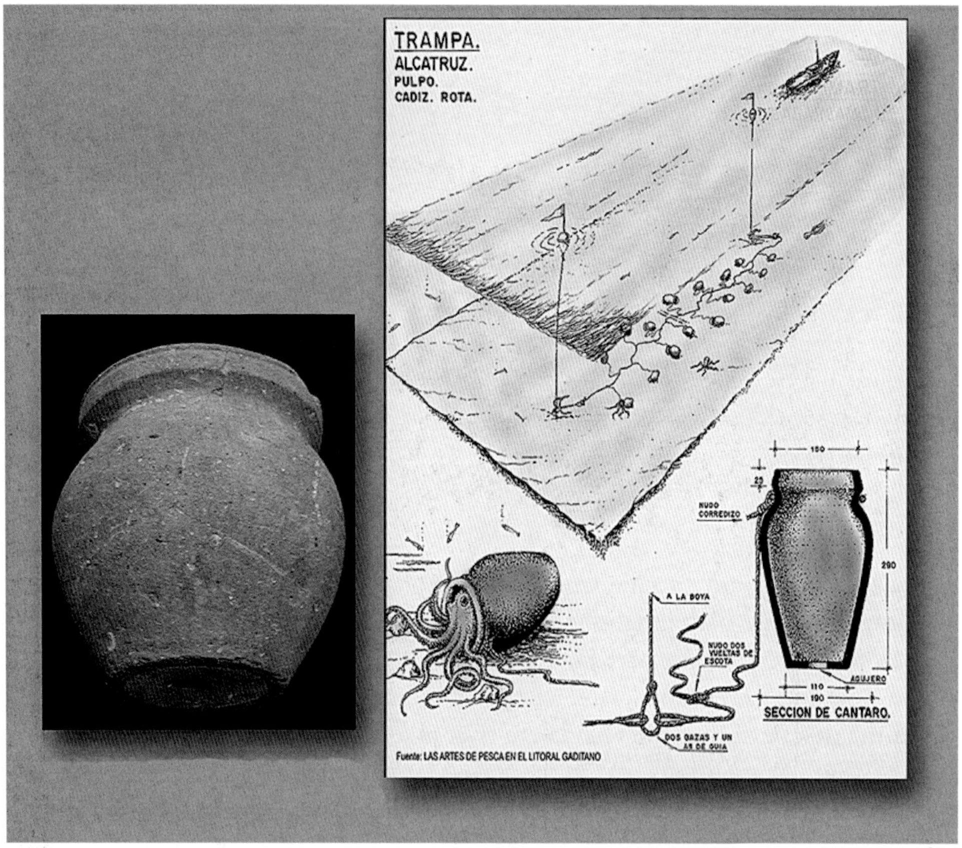

Figure 3. Pulperas in the Bay of Cádiz, with sketch of use (from Las Artes de Pesca en el litoral gaditano) and pottery vessel from El Puerto de Santa María Museum.

From the 1940's, in El Puerto de Santa María, the *sabalera* replaced a local type of fishing gear called *zarampaña* that we can see in the image by J. Ferrer (1995). At the moment, not a lot of data are available about the place of this fish in the economy of the villages along the Guadalete river. This rich resource must have been obvious to the inhabitants of the zone. Future excavations in towns and ancient cities on the Guadalete will surely provide more data.

Octopodes

Octopodes are typically caught in nets, traps or between underwater rocks (as in Galicia); the use of *"pulperas"* or pottery vases is well known (figure 3). In Valencia they are known as *"cadufos"* (Gallart, Escariche & Fito, 2004, 151), a name derived by the similarity to an *alcadafe*, a pottery scoop for a water-wheel. We are convinced that the same applies to the fishing gear known as *alcatruz* in Rota and Cádiz, where a specific type of vessel is used to construct underwater labyrinths and passages, some times formed by as many as 250 vessels. By offering a shelter, these attract species like the octopodes. In an unpublished paper (*Ostricultura Romana*), L. Lagóstena attributes to Aristotle the first mention of the use of these pottery traps.

Phoenician gaff (fifth-fourth centuries BC) from factory P-19 in El Puerto de Santa María (Cádiz).

16. The SAGENA project. Fishing equipment in *Baetica* in Classical Antiquity

D. Bernal, M. Bustamante, J.J. Díaz, E. García Vargas, J. Hernando, J. Lagóstena, J. Ramos, A.M. Sáez, M. Soriguer & C. Zabala

Introduction[1]

This paper presents the SAGENA Archaeological Research Project, supported by the Regional Ministry of Innovation, Science and Business of Andalusia, through its Excellence programme funds (2008-2011). The main aim of the project is to implement a research agenda on fishing equipment in the ancient world, a subject which has been much neglected in Atlantic-Mediterranean contexts. A second goal is a comprehensive, diachronic compilation of all the available evidence, between the Phoenician period and Late Antiquity, as the study of technological change in this, very functional, kind of equipment is otherwise impossible. Thirdly, we aim to follow an interdisciplinary approach, archaeological and zoo-archaeological, ensuring a full understanding of each archaeological site, not only in relation to fishing techniques and but also to industrial aspects of the exploitation of marine products. Hence the interdisciplinary research team, comprising archaeologists and biologists. Finally, we shall present a preliminary prospective assessment of this original research agenda, focusing on the ancient Roman province of *Baetica*.

The SAGENA Research Project, "Fishing instruments in Andalusia in Classical Antiquity. Archaeological research method modelling and initial compilation of evidence" aims at the archaeological study of the origins of many of the traditional fishing techniques carried out in Andalusia, in the south of the Iberian Peninsula. It has been approved and funded, through a competitive process, by the programme entitled *Proyectos de Investigación de Excelencia en Equipos de Investigación*

1 This work was developed within the framework of the SAGENA Project (HUM-03015), supported by the Regional Ministry of Innovation, Science and Business of Andalusia and Research Group HUM-440 of the 4th PAIDI.

de las Universidades Públicas y Organismos de Investigación de Andalucía (*Boletín Oficial de la Junta de Andalucía* 63, March 29, 2007, 11), and is to be carried out between 2008 and 2011. An interdisciplinary team has been designed in order to articulate a comprehensive compilation of all the available evidence and its interpretation according to current fishing criteria. The research approach followed, searching and archaeologically analyzing all the evidence on ancient fishing equipment available from Andalusian archaeological sites, pioneers not only in Andalusia and Spain, but in the whole of the Euromediterranean region, as recently shown in the international workshop *Nets and Fishing Gear in Classical Antiquity. A first Approach* (University of Cádiz, November 2007). At the same time, a comprehensive database of all faunal remains which have been subject to study – which does not mean all the available ones, for they are present almost on every coastal site – is needed, it could shed light on the fishing gear used. Thirdly, with the design of an interdisciplinary team, we aim to combine the historical perspective and the biological approach, testing the results against current fishing criteria. Finally, we aim at drawing historical inferences about the origins of the different fishing techniques, their transformation over time, and the external influences, for which a comparative study with the evidence available for other areas of the Mediterranean and the Atlantic seems essential.

The study will span from Prehistory to Late Antiquity, enabling a diachronic assessment of the processes through several millennia divided into four major periods: Prehistory (from the human origins to the Late Bronze Age), Protohistory (eighth to third centuries BC, focusing on the Phoenician world), Roman (second century BC-second century AD) and Late Antiquity (third to seventh centuries AD). A wide historical perspective is paramount for an effective evaluation of technological change through population contact in the remote past. The extension of the period under study to the middle and modern ages will be left for future stages of research, for the abundance of written records exponentially increases the possible approaches.

This approach will allow for an objective assessment of issues of enormous historical significance, such as the origins of the well-known Andalusian *almadrabas*, or the intensely discussed topics of overexploitation of our fishing grounds and the problems of seasonal fishing. It will also provide a solid historical and archaeological knowledge of ancient fishing techniques, enabling an objective evaluation of the "traditional" nature of many current fishing techniques – we just have vague references to their origins currently en vogue –, of the specificity of Andalusia in relation to other Mediterranean regions during Antiquity and of the reasons for the disuse of certain techniques over time.

In addition, the project will recover part of the Andalusian heritage which is currently undervalued, allowing dedicated museums to open monographic halls for the display of archaeological and ethnographical collections, and thus overcoming the traditional divorce between these two disciplines in the understanding of our pre-Islamic past.

Background

Although the topic of preserves production in Antiquity has often been archaeologically studied (Curtis, 1991 and 2001; Etiénne & Mayet 2002, for a synthesis), this line of research lacks a specific scientific agenda. The same applies to *Hispania*, where only a few pioneer studies have been produced, not followed by others (Moreno & Abad, 1971; Gracia, 1981-82). Normally, the evidence for fishing activities is published along with the bulk of material culture, its study being limited to archaeological contextualisations, in accordance with the implicit belief that fishing instruments are handicraft products subject to no typological variation over time and, therefore, allowing for no chronologic conclusions. Sometimes, however, the evidence is indeed subject to systematic study, especially in submarine contexts, as was the case with the well-known Byzantine shipwreck of Yassi Ada, dated to the seventh century (Kuniholm, 1982), although normally the evidence is merely presented without further historical inferences. This situation has not changed, notwithstanding the remarkable development of the so-called Archaeology of Production (Mannoni & Gianniccheda, 1997), as shown by our research group's recent assessment of the state of the art around the strait of Gibraltar: to date, research goes no further than mentioning the availability of new evidence for fishing equipment, normally hooks, weights and needles for mending the nets (Arévalo, Bernal & Torremocha, 2004, 23-59). For the prehistoric period, some suggestions have been made about bone-made tools, apart from some isolated cases, as the Cave of Nerja, in Málaga.

Only in areas very distant from the Atlantic-Mediterranean region approaches have been adopted, including the archaeological record – fishing implements, zooarchaeology, faunal remains – and palaeoenvironmental approaches; the results, on species selection and mortality profiles in North America in the nineteenth century, have been very interesting (Greenspan, 1998).

This deficiency is not only to be found in the Atlantic areas and the *Mare Nostrum*, it has also hindered research in one of the most renowned fishing areas during Antiquity, the Black Sea (Bekker-Nielsen, ed., 2005; 2009).

This absence of a specific research agenda, and the consequent lack of specialized studies, has led us to develop a specific research group made up of archaeologists and biologists. The former have been selected with full chronologic coverage in mind, including prehistorians (Dr. J. Ramos), and specialists in the Phoenician period (A. Sáez), the early Roman Empire (Dr. E. García Vargas, J.J. Díaz and M. Bustamante) and Late Antiquity (Dr. D. Bernal and J. Lagóstena). All three biologists (Dr. M. Soriguer, Dr. J. Hernando and Dr. C. Zabala) have extensive experience in zoo-archaeology, having worked with archaeologists in the identification of faunal remains from prehistoric, protohistoric and Roman sites from the Gibraltar area for more than five years. In addition, their current field of specialisation, in fish population dynamics in intertidal areas and, particularly, their latest work on fishing equipment selectivity, made them ideal candidates for this research group.

The project team is clearly interdisciplinary in nature, as 70% of its members belong to two Research Groups ascribed to the P.A.I. of the Regional Government of Andalusia, in the area of Humanities (HUM-142 and HUM-440), and the remaining 30% belongs to the Natural Resources and Environment group RNM-243, "Fish Population Dynamics", representing three areas of knowledge (Prehistory, Archaeology and Zoology); furthermore, it is likely that some archaeometric analysis will be needed, such as metallographic analysis, ethnoarchaeological studies or the restoration of archaeological remains, incorporating specialists from other areas of knowledge.

The project guidelines are historical-archaeological, including the cited biologists, who will carry out the interpretation of biofacts. The history of past cooperation between members of the team is enough illustration for an interdisciplinary project, both consolidated and viable. This will result in a transversal approach to a single topic – fishing gear in Antiquity – through the study of all the available evidence, basically, the literary sources, the archaeological record (seen diachronically, encompassing Prehistory, Protohistory, the Roman period and Late Antiquity), zoo-archaeological remains (ichthyofauna and malacofauna), the ethnoarchaeological evidence (traditional fishing gear) and historical retrospective (current day fisheries and fishing arts).

Aims of the Project

The general aim of the SAGENA Project is to increase the historical knowledge of fishing technology in Andalusia during Antiquity, encompassing the period between human origin and the Late Antiquity, prior to the Islamic conquest of the Iberian peninsula in the early eighth century. The research agenda includes the compilation of all available evidence on fishing gear and its interpretation with modern fishing criteria. In the following paragraphs we break down our five specific aims.

Compilation of the archaeological record for fishing techniques in Andalusia

A previous comprehensive knowledge of the archaeological evidence for fishing instruments is paramount for their systematic study and interpretation. As mentioned in above, the study of this topic has followed no consistent agenda, in Andalusia or elsewhere in Spain. Only a few preliminary papers can be found, such as the aforementioned case of Gracia (1981-82) – as shown in the latest synthesis of the state of the art in Spain (Fernández Pérez, 2002) – forcing recurrence to *topoi*, data extrapolation from other regions, or implementation of generalist information obtained from the literary accounts, such as Oppian's *Halieutika* (second century AD) or Aelian *On Animals* (mid-third century AD). The team members' previous work has made us fully aware of this situation for Andalusia, but also of the remarkable wealth of unpublished and unstudied material and, therefore, its research potential (Arévalo, Bernal & Torremocha, 2004).

The slow typological evolution of fishing implements – basically hooks, weights and needles for mending nets – during Antiquity has made the setting of a wide time span advisable, in our opinion the only way to detect technological change and innovations in this sort of equipment. That is the reason behind the diachronic approach, including Prehistory, Protohistory and the Roman and Late Roman periods; the middle and modern ages, however, will be left for future projects, because there is less archaeological evidence (especially for the middle ages) and also because they involve a heavy philological task due to the abundance of written records.

The archaeological evidence is quite scattered, for almost every site from late prehistory onwards contains fishing-related material. The lack of interest in this sort of evidence to date has resulted in the absence of specific inventories in Museums, which store these finds within the metallic (hooks) and ceramic (weights) collections. Only needles are occasionally individually catalogued along more selected items, due to their size and appearance.

An intensive survey of museum collections is therefore required, especially focusing on the coastal provinces of Almeria, Cádiz, Huelva, Granada and Málaga. Both provincial and local museums, all linked to the Andalusian museum network, will be surveyed, up to a total of twenty. The aim is to obtain a comprehensive insight into the most significant sites for preserves production during the period under scrutiny; due permission for the archaeological study of the collections will be requested from the Regional Government of Andalusia. The hundreds of items to be studied make this task a potentially profitable one. For obvious reasons, Cádiz, where the team members have carried out most of their fieldwork, is the better known and most accessible area. Fieldwork is indeed one of the project's cornerstones, and the explanatory aims of the project rely heavily on its success and thoroughness, the aim of the initial stage being merely the first compilation of the evidence and the design of the strategy to be followed in later phases.

Compilation of the zoo-archaeological record in Andalusia in the periods under scrutiny

Archaeofaunal remains have been extensively studied in Andalusia (Morales & Roselló, 1988), including the well-known publications by A. von den Driesch, A.

Morales, R. Moreno, J.A. Riquelme or E. Roselló, and several comprehensive syntheses for ichthyofauna (Roselló, 1989) and malacofauna (Moreno, 1994). However, this evidence has not been considered so far in relation with ancient fishing gear, being limited to taxonomic and taphonomic considerations.

The Project does not aim at detailed archaeofaunal studies, but at the compilation of all the evidence available in publications and reports and its general interpretation. The determination of the species captured in each area and their origin (benthic, demersal) will be the key to the interpretation of the nets and other fishing equipment used in their capture. Osteal evidence from fish and shellfish need to be collected and combined with the archaeological evidence in order to track specialized practices – as the capture of *muricidae* or shellfish gathering for obtaining *patellae* – particularly the difficult capture of deep sea species. Recent works have shown the potential of this approach, as in the Roman city of *Carteia*, where evidence for selective shellfish gathering (sea snails and other marine shells) has been found in combination with the exploitation of *muricidae* (basically *Hexaplex trunculus*) in the mid-fourth century (Bernal *et alii*, 2008).

Archaeological study of the origin, development and typological evolution of fishing equipment

Fishing instruments are poorly defined from an archaeological perspective, as their classification relies on isolated finds and not on systematic archaeological collections. Therefore, not even an initial classification can be made prior to a significant compilation of evidence from archaeological sites. Secondly, typological and chronological assessments of the different instruments will be produced, taking into account the chronology of their associated contexts, the typological evolution during their period of use, and the chronology of their abandonment/replacement with different types of tackle. That will allow for relevant historical inferences, such as the geographic origin of innovations – particularly significant during the Phoenician-archaic and Roman republican periods – and other phenomena of cultural interaction.

The items are divided into three groups – hooks, weights and other tackles – plus other equipment (foremost, needles for repairing the nets); many of the items found are inadequately identified. We present them briefly below.

Hooks

These are the most common type of find on Phoenician and Roman sites. To date, no chronological inference has been drawn from their typological evolution, and one of our aims is to change that. As shown in figure 1, their shape changes little

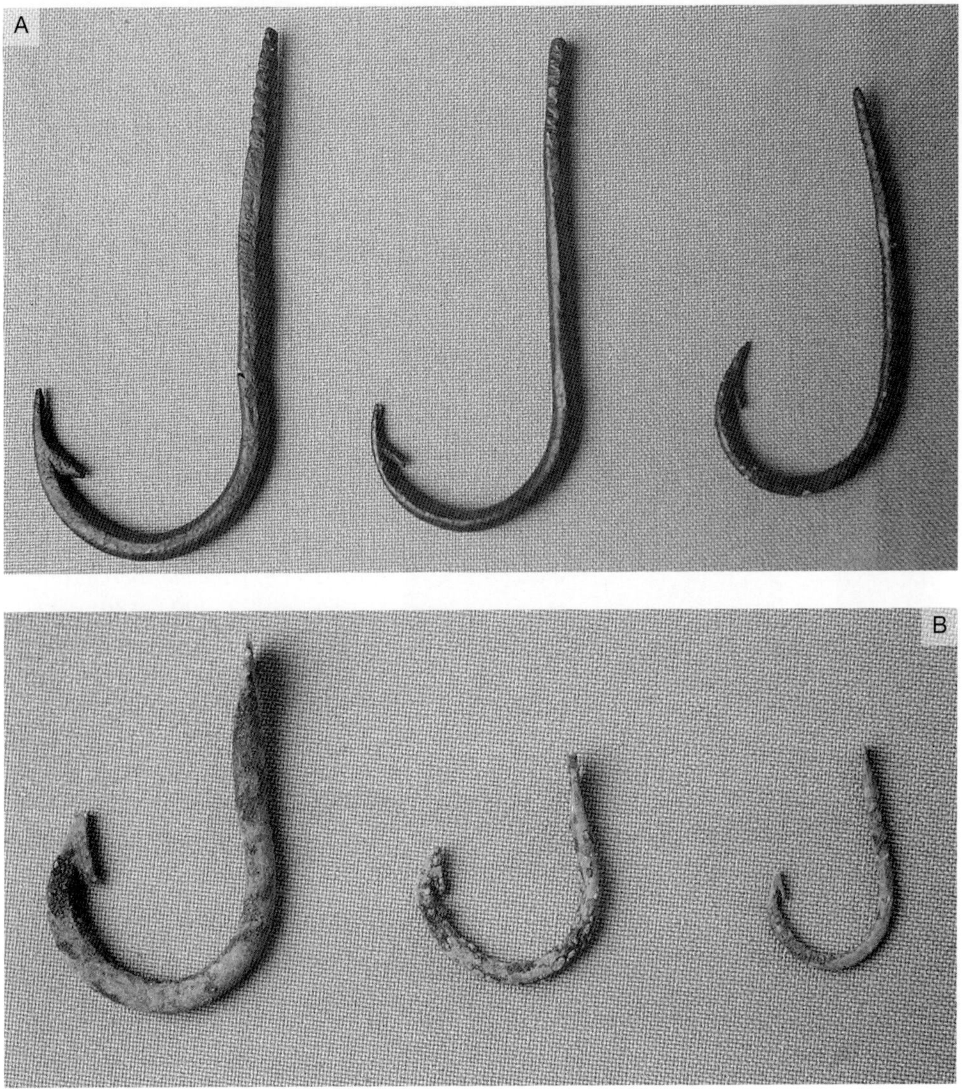

Figure 1. Hooks, dated to the Phoenician period (fourth century BC) from the factory P-19 (A) and from the early Roman imperial period from Baelo Claudia (B) (Arévalo, Bernal & Torremocha, 2004, 141 and 161).

over time, although some differences in size and typology may be detected, as well as in the way they are attached to the fishing line.

It is especially important to develop a research protocol following the guidelines of current fishing gear selectivity. The presence of specialised biologists in the team is thus justified. The generation of such a protocol will allow for the development of a new tool for fieldwork archaeologists, who will be able to apply the new methodology to new finds. Another of the study's basic aims is the determination of catch size according to the size and shape of hooks.

Figure 2. Phoenician weights (fourth century BC) from factory P-19 (A) (Arévalo, Bernal & Torremocha, 2004, 139), and unpublished stone weights from the Roman period, from Baelo Claudia (B).

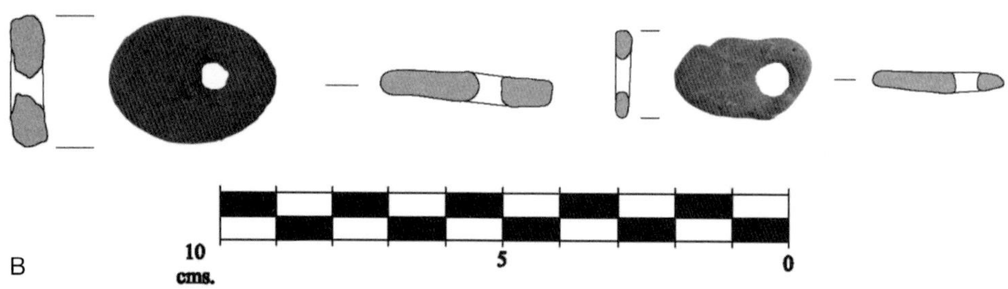

Weights

Also an abundant find in archaeological sites, their study so far has been limited to a morphologic description. They can be of different types. Raw material is also important, with clay, stone and lead, used for different purposes, as shown in the examples presented in figure 2.

In this case, the research approach will try to go beyond typological studies, trying to develop a characterisation of net types used throughout history, through the interdisciplinary work of specialized archaeologists and biologists. Secondly, we aim at producing the first chronologic assessment of these artefacts in Antiquity, based on typology, for future use in dating sites and, potentially, in drawing inferences of exogenous population inputs.

Other fishing tackle

Here, we include materials seldom considered in the Spanish bibliography and badly studied due to their rarity, such as bites, handled nets, or multiple hooks. Figure 5 shows a gaff, known at least from the Phoenician period (fifth to fourth centuries BC), as shown by the finds in the so-called P-19 factory in El Puerto de Santa María (Cádiz). The study parameters will be those traditionally followed in archaeological research. Fishing-related items hitherto incorrectly characterised archaeologically will be subject to specific insights. This includes items of diverse nature (made of stone, for prehistoric contexts, and of metal for Protohistoric and Roman contexts) traditionally interpreted as not related to fishing, because of the contexts in which they were found. On the one side, microliths and other archaeological finds from some of the sites mentioned, recently linked to fishing activities, as fishing rather than hunting projectiles, following their contextual relations and the functionality studies practiced (Clemente *et alii,* this volume). On the other hand, we will consider the well known "arrow-heads" of the Macalón type – seventh to fifth centuries BC – traditionally interpreted as military gear, which find in fishing-preserves production contexts, as in P-19 factory in El Puerto de Santa María has led some to suggest their use as harpoons in the rich ancient fisheries. The number of available artefacts will increase exponentially thanks to the evidence compilation we aim at, derived from the team members' own excavation experience.

Fishermen's equipment (needles, shuttles)

Very common in the archaeological record. They come in two main categories, as shown in figures 3 and 4. On the one side, the so-called shuttles or net needles (see p. 64 and 356 for examples), with notches at both ends, and bronze needles, of different shapes and sizes, on the other. The task in this case is also to develop an intensive typological and chronological assessment, aiming at a deeper insight into net-repairing activities and, indirectly, into the sort of fishing arts used on each period (we must not forget that the presence of needles, and especially shuttles, stands as clear evidence for the use of nets, and may also be used to infer mesh size). Other elements, such as gaffs (figure 5), rods or floats/buoys, the later very well attested iconographically but poorly represented in the archaeological record, will also be considered.

The validity of archaeometric analysis of the above mentioned material will be tested. The focus will be on four main issues:

- Archaoemetallographic analysis. Instruments made of metal – specifically bronze or iron – will be subject to metallographic characterisation studies in order to

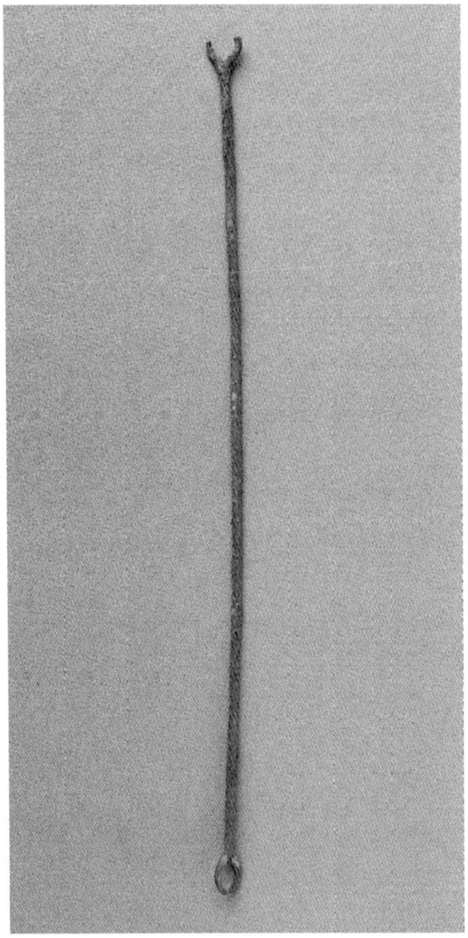

Figure 3. Bronze net needle from Calle San Nicolás, in Algeciras, dated to the early sixth century AD (Arévalo, Bernal & Torremocha, 2004, 115).

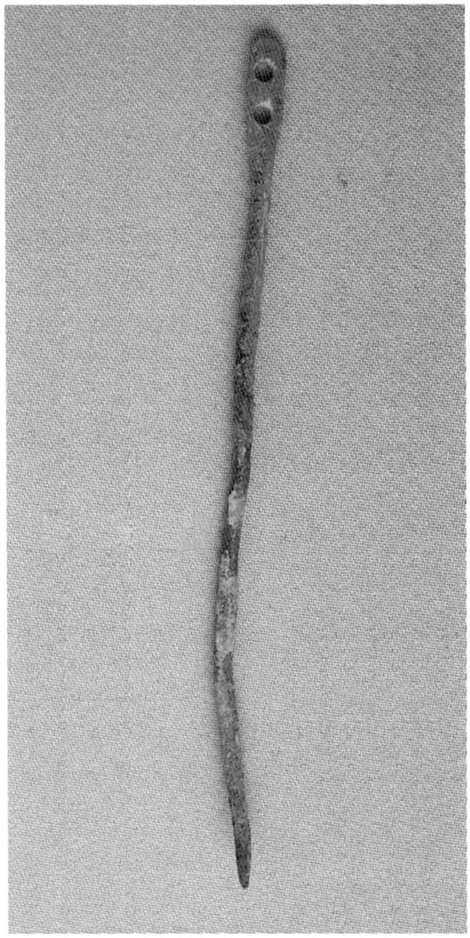

Figure 4. Bronze needle from Calle San Nicolás Fish-Plant, in Algeciras, dated to the early sixth century AD (Arévalo, Bernal & Torremocha, 2004, 115).

determine their provenance, and at the same time, to obtain a deeper insight into their function and possible improvements in their manufacture over time.

- Functional analysis. The analysis of traces of use preserved in the active parts of these tools is particularly important, especially in Palaeolithic – and less so in Protohistoric – stone industries, for the determination of their functionality. Also, the eventual find of traces of certain matters (such as cereal remains or phytoliths), will help us to rule out those items not used in fishing activities. The validity of these analyses for other, such as metallic, finds, will also be tested.
- Archaeobotanical analysis: Weights will be analysed in search of traces of friction or of actual remains in their inner surfaces, in order to infer the sort of fibres used for nets (esparto, flax, hemp).

Figure 5. Phoenician gaffs (fourth century BC) from factory P-19 (Arévalo, Bernal & Torremocha, 2004, 141).

- Ethnographic comparative studies. To study the archaeological evidence in combination with traditional fishing instruments in the region is particularly important in order to draw analogies. Specifically, we will try to track possible Roman origins for some Andalusian traditional fishing arts, such as the so-called "Corrales" in Rota/Chipiona or the almadrabas, to verify whether the typically simplistic statements about their remote origins can be tested empirically.

These additional questions will increase the interdisciplinary nature of the project even further, for they will bring physicists, chemists (for metallographic and other archaeometric studies), biologists, archaeobotanists, chemists (for functionality studies) and medieval, modern and contemporary historians (for the comparative ethnographic studies), into the team.

Mediterranean contextualisation of the results. The geographic key

Integrating the results into an Atlantic-Mediterranean context is the key to a full understanding of the importance of cultural exchange – with the consequent technological transfer – among ancient communities. The introduction of a geographic key to the project is as important as the chronologic dimension aimed at with the evaluation of technologic change over time. In order to accurately evaluate the importance of Andalusia in Antiquity, is indeed key to establish the degree of technologic evolution in other Mediterranean contexts. In this sense, we focus on four regions:

- Baltic countries. They are particularly relevant for prehistoric evidence, as countries such as Sweden and Denmark enjoy unparalleled conditions for preservation. That is the reason behind the institutional relationship maintained between the University of Cádiz and the University of Southern Denmark.
- Black Sea. Future cooperation, along the lines already followed with the University of Southern Denmark, is planned.
- Atlantic. The importance of Atlantic fisheries, from the British Isles to the Canary Islands, is also vital for the understanding of fishing arts in Andalusia during Antiquity. It is, therefore, important to consider data coming from the important fisheries located in the Lusitanian coast; or from Morocco, where, apart from the known *cetariae* of *Lixus*, *Cotta* or Tahadart, new Roman preserves factories have been found around the mouth of the river Martíl – Tetuán area; the fishing equipment recovered will be studied along the Baetican examples and the data recovered in Ceuta, on the African shore of the strait.
- Mediterranean: Obviously, this is the most important context for us, especially during Protohistory and the Roman period. But research on ancient fisheries is still in its early stages in northern Africa and the eastern Mediterranean, whereas in Italy and southern France it is somewhat more advanced. The project will be linked with a project recently started in the Campania by Cádiz University and the Università Ca Foscari of Venecia, called "Fishing and *Garum* in Pompeii and *Herculanum*. The exploitation of sea resources around the Vesuvius" (2008-2012), directed by D. Bernal, D. Cottica and A. Zaccaria.

Expected Results, Diffusion And Exploitation

The preceding paragraphs summarize the scientific dimension of this project; for details, see the projects website (www.sagena.es). The expected results are as follows:

1. Compilation of all archaeological evidence related to fishing gear in Andalusia, between Prehistory and the Late Antiquity (seventh century).
2. Compilation of all marine faunal remains found in Andalusian archaeological sites in order to draw diachronic inferences about exploited species.

3. Typological and chronological assessment for fishing-related archaeological finds in Spain, and its application to Euromediterranean contexts (especially for the Phoenician and Late Roman periods), enabling the production of a catalogue of traditional fishing arts in Andalusia during Prehistory and Antiquity.
4. Approach to the evolution of fishing technologies in Andalusia in Prehistory and Antiquity.
5. Recovery of our region's fishing heritage, through the archaeological characterisation of traditional fishing instruments in Andalusia during Antiquity.

The methodological insights obtained from the implementation of the model of archaeological characterisation of Andalusian fishing gear in Antiquity will lead to the important task of comparison with data from other Atlantic and Mediterranean contexts. This characterisation should similarly be compared with materials from other contexts and new archaeological excavations, for some conclusions will be based in arguments *ex silentio* (absence of certain arts in specific periods). Getting a deeper insight into the effects of seasonality in fisheries will also be important, and into how can technological conditions influence societies, especially in early stages of human history. Future developments include testing suggested technological transference between different cultural contexts (Phoenicians, Turdetanians, Iberians, Romans, etc.) and relationships with other areas of economic history, such as agriculture and mining.

Within the examination of fishing techniques in ancient Andalusia, evidence for fish-farming, so far only attested in the bay of Gibraltar in the sixth century AD, will also be considered (Bernal, 2007). Finally, we will develop a first proposal for the origins of traditional Andalusian fishing gear, and will make some suggestions about the reasons for their adoption or eventual abandonment. After the project's four-year duration, the aim is to widen the geographical scope, in relation with other Spanish and European projects.

Finally, the project has developed a communication strategy on several levels (researchers, academics, the public and schools), including the internet, conferences, an international workshop at the end of the project, a programme of local/regional publications, journals and, finally, an exhibition in which didactic brochures will be provided.

Regarding the exploitation of the results, we believe that they are of great interest for the Regional Ministries of Culture and Agriculture and Fishing. The project can provide a historical and archaeological justification for the attention paid to traditional fishing and related gastronomy. The continuation of the project is foreseeable, in cooperation with the above public agencies. In addition, the project will reinforce Andalusia's position as one of the most important fishing regions in the world, by recovering its fishing practices, which are without parallel in the Atlantic-Mediterranean world.

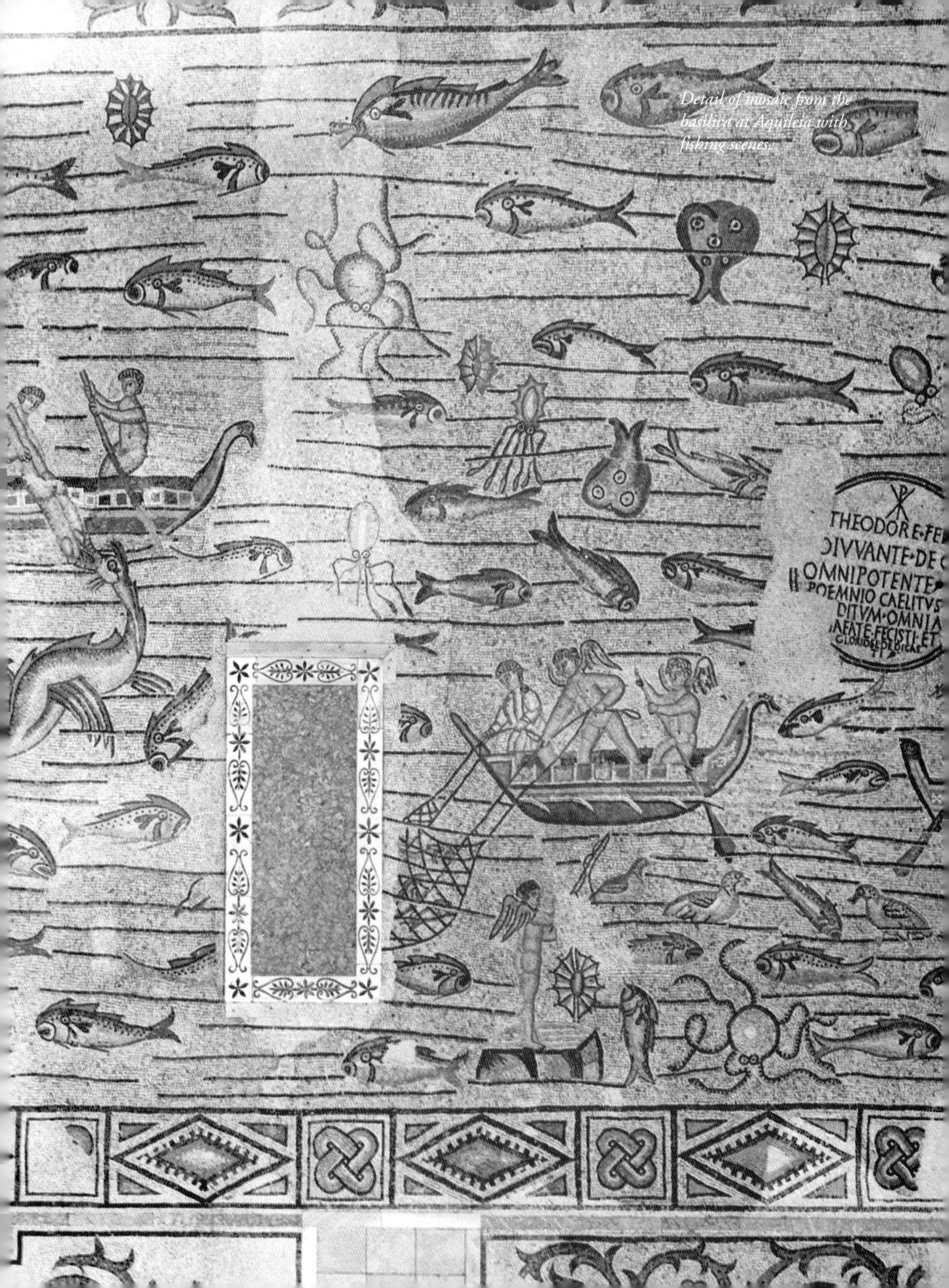

Detail of mosaic from the basilica at Aquileia with fishing scenes.

17. Spheroid clay weights from the Venetian Lagoon

DANIELA COTTICA[1] & LUIGI DIVARI

The issue

This contribution aims to draw attention to a group of clay weights of still disputed function: roughly spheroid objects, undecorated, with vertical perforation (figure 1). Comparable clay weights dating from the pre-Roman to the medieval period are known from a number of published archaeological contexts (e.g., Bianchini Citton, Gambacurta & Ruta Serafini, 1998, figs. 201-202; Bietti Sestieri, 1992, tav. 26, forma 33; Alarcão, Étienne, Moutinho Alarcão & da Ponte, 1979, pl. X n. 143; Davidson, 1952, pl. 78; Fozzati & Gobbo, 2007, figs. 26 and 67). Scholars interpret these weights either as spindle-whorls (e.g,. Bortoletto, Spagnol & Toniolo, 2000, nos. 13-14, 29-30) or, alternatively, as fishing net weights (*cf.* Celuzza, 1985). The fact that published reports often lack quantitative data referring to weight, occurrence and distribution of spheroid clay weights makes it more difficult to assess their technical features and use in Antiquity.

As far as spindle-whorls are concerned, numerous typologies are known: the main published shapes include ovoid, spheroid, hemispheroid, conoid, bi-conoid and "irregular squat" types (for a recent overview with archaeological evidence typical of Anatolian tradition, *cf.* Symington & Collon, 2007 with figures). Regardless of their dating, recent studies (*cf., inter alios* Barber, 1991; Symington & Collon, 2007) have shown that spindle-whorls are often decorated and vertically perforated; furthermore

1 Sections 1-3 of this paper are by Daniela Cottica, section 4 jointly by the authors. The authors wish to thank Ernesto Canal, who found the clay weights illustrated in figures 2 and 7 as well as Luana Toniolo, who drew the archaeological finds accompanying this contribution. Finds in figures 1-2, 4a-b, 6-8 are published with the kind permission of the Soprintendenza per i Beni Archeologici del Veneto (prot. N. 6516).

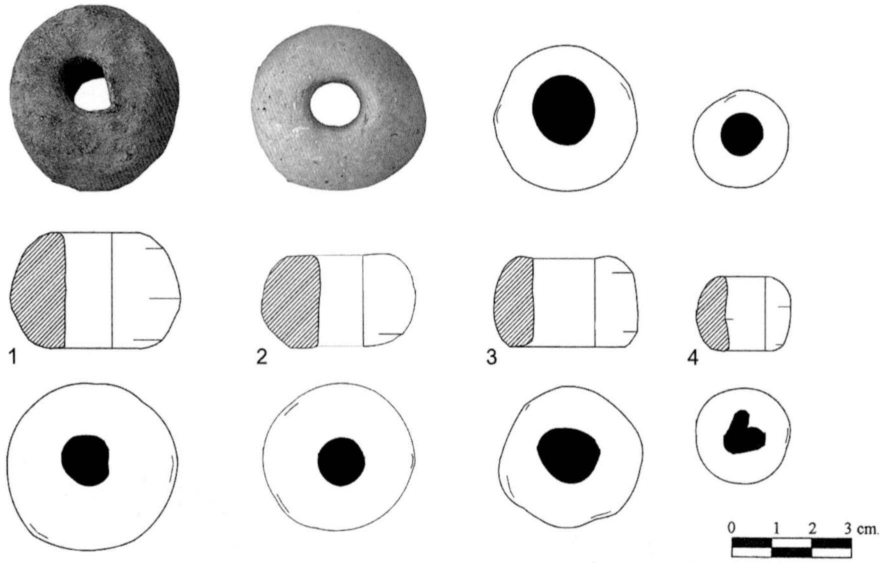

Figure 1. Typical spheroid clay weights from the Venetian Lagoon.

Figure 2. Detailed views of a typical spheroid clay weight from the Venetian Lagoon.

they usually present a flattish upper end provided with a concave depression (*cf.*, e.g., Symington & Collon, 2007, fig. 287, no. 1839 and figs. 292-293).

On closer examination, the clay weights discussed here differ from typical spindle-whorls in many aspects: they are spheroid or doughnut-shaped (i.e., resembling flattened spheres), they lack symmetry, are undecorated and characterised by the irregular diameter of their central perforation. Moreover, they appear crudely hand-made and fingerprints are visible on the less worn examples (figure 2). Similar objects are well attested at various locations along the western shores of the northern Adriatic sea. Some published examples from Caorle (in the Gulf of Venice) have been interpreted as net-weights (Fozzati & Gobbo, 2007, fig. 26: here a terracotta spindle whorl is also depicted, allowing to clearly distinguish between these two different classes of finds). The same function has been assigned to a couple of similar objects from the Comacchio wreck *Fortuna Maris* (figure 3), datable to the end of the last century BC or the beginning of the first century AD (Berti, 1990, 269 and tav. LXXVI, nos. 257-258).

The presence of *clay weights* in a wreck is not confined to the case of Comacchio nor to the Roman period. Roman terracotta *loom weights*, interpreted as fishing gear, have been found, *inter alia*, at Pisa (Bruni, 2000) while two re-used Roman pyramidal and lentoid terracotta loom weights found in the Serçe Limani wreck (eleventh century AD) have been interpreted as spinning implements used "...for some kind of fishing twine..." (Bass, Matthews, Steffy & van Doorninck Jr., 2004, 429 and figs. 2-17). In this latter context, the presence of net-mending equipment, including bone spindle-whorls and bronze net-needles (Bass, Matthews, Steffy & van Doorninck Jr., 2004, 418-423) adds new, puzzling evidence to the discussion on the function of implements, normally classified as spinning or weaving tools, found on board ancient sunken ships.

Spheroid clay weights in the Venetian Lagoon

Interestingly, hundreds of crudely made spheroid terracotta weights, comparable to the Comacchio and Caorle finds mentioned above, have been retrieved at various locations in the Venetian Lagoon. These objects (figure 4 a-b) are roughly manufactured from local/regional clays, they are normally well worn with an asymmetrical profile (figure 2): the vertical hole is not centrally placed and appears to be irregular in diameter. The external surfaces, together with the internal margins of the hole, usually appear well worn and smooth as if repeatedly exposed to water action (figure 4 a-b). The weight can vary from *c.* 6g to 50g, depending on the object's size. The majority weigh between 20 and 35g and the average size is 4.5cm across and 2.5-3.0cm in height. The diameter of the hole, often well worn, usual-

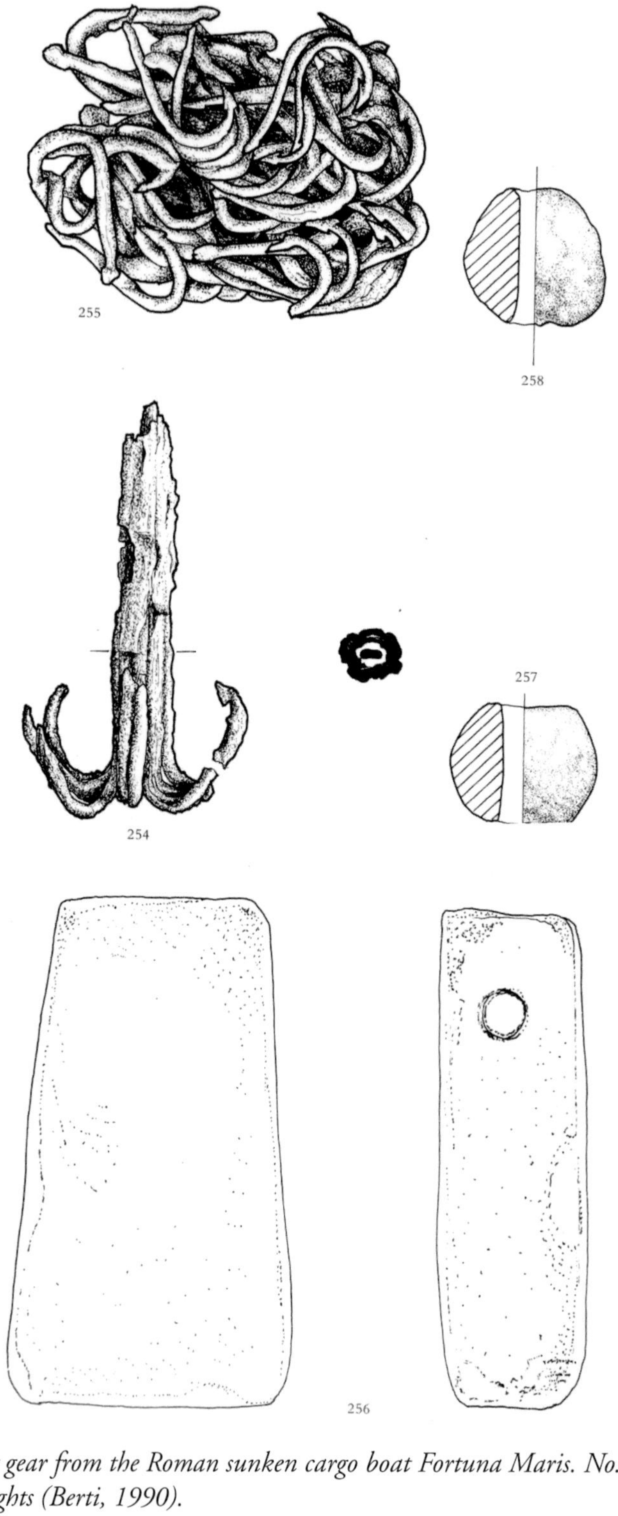

Figure 3. Fishing gear from the Roman sunken cargo boat Fortuna Maris. No. 257 and 258: spheroid clay weights (Berti, 1990).

Figure 4. Typical spheroid clay weights from the Venetian Lagoon; a): front view, b): top view.

ly varies from 1.2 to 1.6cm; all weights are undecorated. These finds are mostly unpublished (for some published examples, *cf.* Cottica, 2003, fig. 3 nos. 12-15; Bortolin, 2005, 146, tav. XIX, F/P 1-3) and rarely mentioned in excavation reports and papers (for a notable exception, see Buchi, 1987, 128). Indeed, spheroid clay weights can still be found as stray finds along the shore of the islands in the lagoon.

Of particular interest are two groups of spheroid clay weights respectively from the Roman town and port of *Altinum* and the offshore site of Sette Soleri in the la-

Figure 5. Nautical Map of the Lagoon of Venice. Detail of the Northern Lagoon, indicating the sites of Altinum and Sette Soleri (Laguna Veneta, Carta Nautico Turistica, scala 1: 50.000, Rimini).

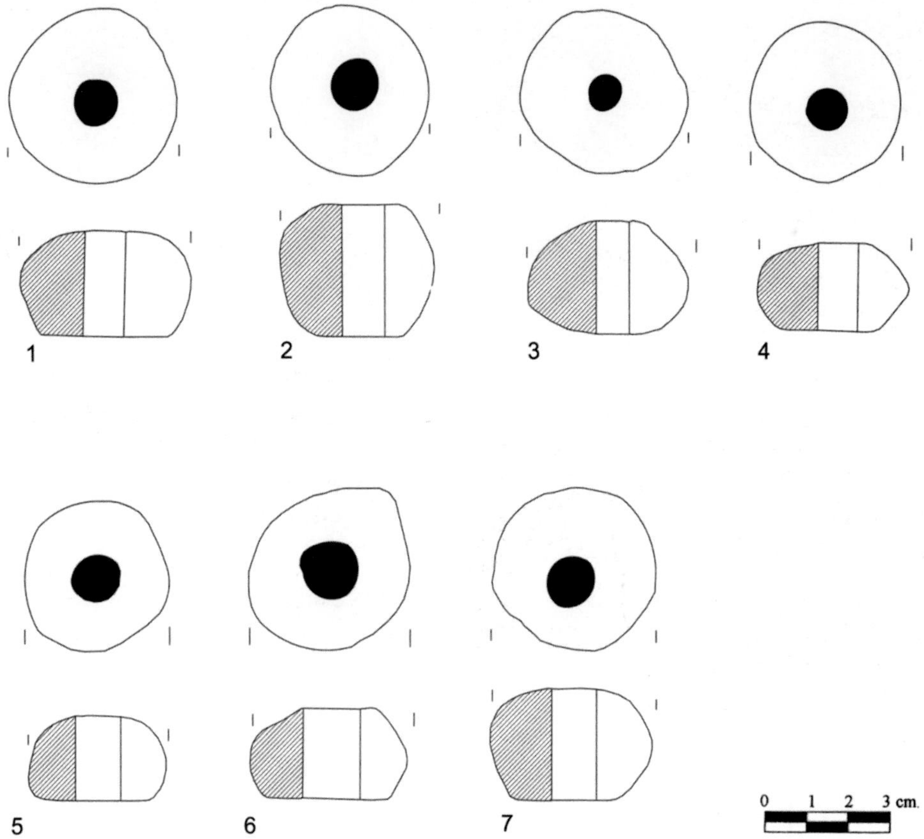

Figure 6. Selection of spheroid clay weights from Altinum (inv. nos. AL 29040 and AL 4152).

goon of Venice (figure 5). Hundreds of spheroid terracotta weights from *Altinum* were examined by D. Cottica in 2001. The largest group of finds (inv. nos. Al 20426-Al 20445) was found in 1972 in "area Ziliotto" and excavation records in the archives date them to the first century AD (figure 6). Interestingly, some of these weights were partly fused with each other, as if they were kiln wasters. The second group of globular clay weights was examined by D. Cottica in 2008 (figure 7) as part of the research activities related to "Project Costanziaco", an archaeological project focused on the area of the northern lagoon of Venice, located just off the coast of *Altinum* and named in some medieval documents as *Constanciacus* (*cf.* Cottica, Fozzati, Traviglia & Goti Vola, 2008; Cottica, Traviglia & Busato, 2009; Cottica, Fozzati & Tirelli, 2010). At Sette Soleri, mentioned as *"Septe(m) Salaria"* in an eleventh-century document (see below), an area named after the salt-pans once present there, several spheroid clay weights were retrieved during field investigations carried out by Ernesto Canal, a pioneer of archaeology in the Venetian Lagoon (Canal, 1995 and 1998). These clay weights were found together with abundant Roman pottery.

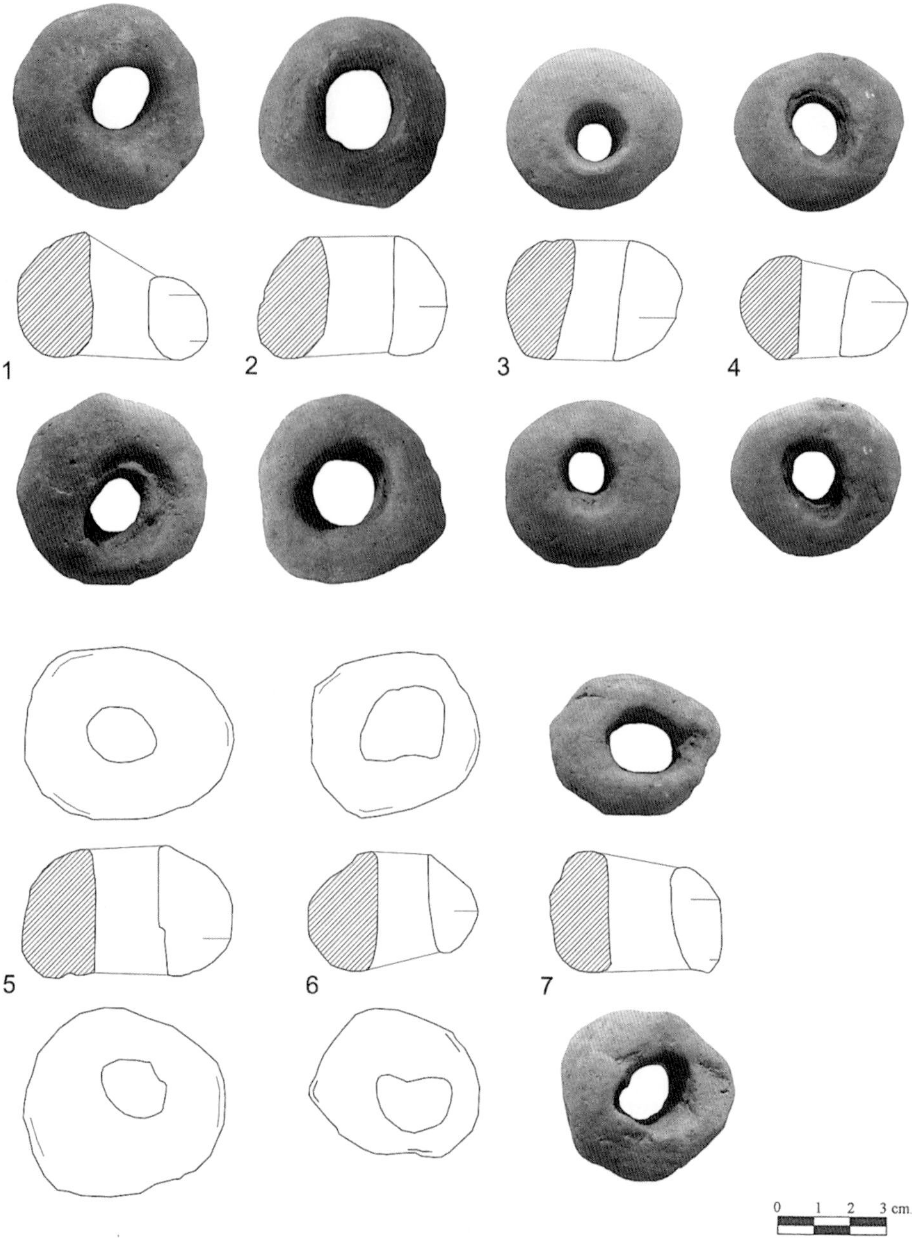

Figure 7. Spheroid clay weighs from Sette Soleri in the northern Lagoon of Venice.

Today, Sette Soleri and Costanziaco are part of a natural reserve, characterized by a natural landscape dominated by the shallows that emerge from waters during low tides, covered by vegetation typical of brackish environments (*"barene"*). Geomorphological studies have demonstrated that the sea level of the northern lagoon has changed considerably through time: significant environmental changes occurred in the Late Roman period, in the early middle ages, in the twelfth and again in the sixteenth century AD (Cavazzoni, 1995), with obvious consequences for the patterns of human settlement and occupation. In the Roman period, several sites of the lagoon, now submerged or almost submerged (including Sette Soleri), were cultivated and inhabited land, as revealed by archaeological evidence (Canal, 1995 and 1998; Dorigo, 1995; Cottica, Fozzati, Traviglia & Goti Vola, 2008) and literary sources (see below).

Altogether, in the area stretching from Comacchio to Caorle, globular clay weights are attested at least from the Roman period to the middle ages, apparently without significant changes in their morphology. A few specimens dating to the medieval and post-medieval periods have been published from recent excavations at Caorle and in the Venetian Lagoon (*cf.* Fozzati & Gobbo, 2007, fig. 26; Bortoletto, Spagnol & Toniolo, 2000, 30-31, nos. 13-14 from an excavated context at San Francesco del Deserto island, dating to the fifth or sixth centuries AD and classified as spindle whorls, and 34-35, nos. 29-30 from a stratified context on Torcello island, dating to the first half of the seventh century AD, classified as spindle whorls). Although ancient sources do not provide us with any clues to the interpretation of these finds, details of their morphology such as their lack of symmetry, and therefore balance, the irregularity of the pierced hole (the diameter of which usually varies within a same specimen) and their weight, make these objects unsuitable for spinning purposes. On the other hand, their distribution and frequent occurrence in regions characterized by abundant shallow waters and lagoons, in addition to their presence on board of cargo boats, as in the case of the Comacchio wreck, reinforce their interpretation as fishing gear rather than spinning implements.

As far as raw material is concerned, we may note that clay, contrary to lead, was an abundant and cheap raw material. This circumstance might have played a significant role in choice of raw material for communities depending primarily on fishing for their daily subsistence, rather than as a large scale business (*cf.* Cassiodorus on the case of settlers living in the Venetian lagoon, see below).

Concerning marine activities in the Venetian Lagoon, recent studies have already focused on the role of the integrated system of communication active in the stretch of land connecting *Ravenna* with *Altinum* and *Aquileia*. Along the Northern Adriatic littoral, Roman land roads were complemented by a network of navigable rivers, artificial and natural navigable channels (Uggeri, 1992; Rosada, 1992; Dorigo, 1994 and 1995) connecting the sea, with its lagoons, to mainland sites and fluvial ports. In the Northern Lagoon of Venice, together with abundant ceramics from all over the Mediterranean (*cf.* Toniolo, 2007 and 2008), archaeological investigations have brought to light a range of Roman and Late Antique structures and infrastructures aimed at controlling and exploiting the natural landscape with its resources (Canal, 1995; Fozzati & Toniolo, 1998; D'Agostino & Medas, 2005). Nowadays, Roman structural remains normally lie about 2m below sea level, un-

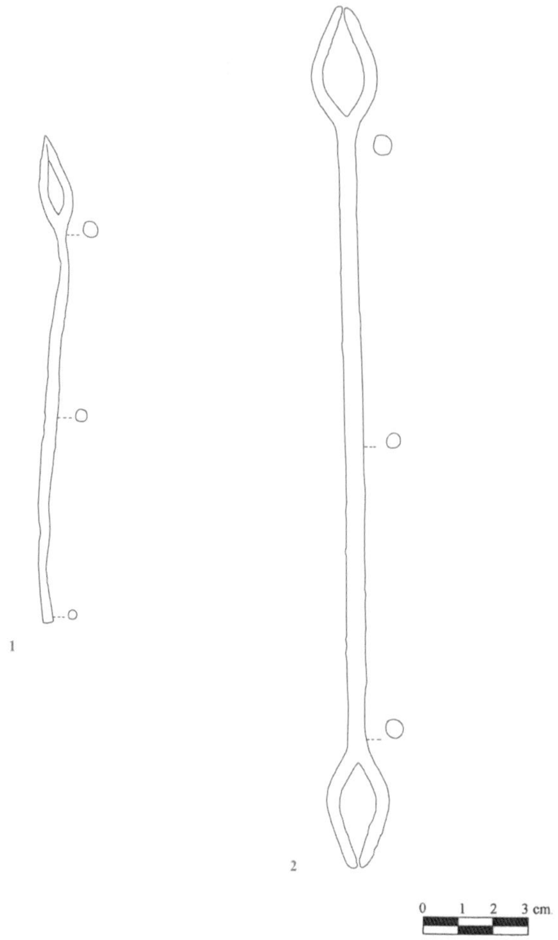

Figure 8. Net needles from Altinum (inv. nos. AL 34214 and AL 14133).

less obliterated by medieval and post-medieval occupation or destroyed as a result of recent activities. In this area, further indicators of exploitation of marine resources in Antiquity are the numerous archaeological finds (mostly unpublished) relating to fishing discovered at different locations in and around the lagoon, especially in the Roman town of *Altinum*, including net-needles (see figure 8 for examples) and fishing hooks.

Fishing in the Northern Lagoon of Venice in Antiquity and Beyond

Roman and Late Antique literary sources, together with medieval and post-medieval documents and legal sources, confirm not only that the lagoon of Venice was settled but also that fishing was practiced. In the first century AD period the poet Martial compares the littoral of *Altinum* (in *Venetia et Histria*) to the villas of *Baiae* in *Campania* (*...Aemula Baianis Altini litora villas...* Martial, *Epigrams* 4.25).

Centuries later, a letter of 537-538 sent by the praetorian prefect *Magnus Aurelius Cassiodorus* to the tribunes of the coasts provides the following description of the Venetian lagoon in the sixth century AD:

> (...) The Venetian districts ... enjoy the pleasures of the Adriatic coasts, where alternate tides in their movement now cover, now expose the face of the land ... here you have your home like sea-birds. For a man is seen now as a mainlander, now as an islander ... Now the inhabitants have one source of supply: they cram themselves with fish alone. There rich and poor feed together on equal terms. One food keeps all alive; a similar dwelling houses everyone ... All your rivalry, though, is in the salt-works ... A man may have small interest in seeking gold, but there is no one who does not wish to acquire salt – rightly so, since all kind of food owe to it the pleasure they give... (Cassiodorus, *Variae* 12.24).

Latin sources confirm that *Venetia et Histria*, and the Northern Adriatic region, were rich in different species of table fish, including the following:

- *gobi ("In Venetis sint lauta licet convivia terris/ principium canae gobius solet",* Martial, *Epigrams* 13.88),
- *lupi ("Lanueus Euganei lupus excipit ora Timavi/ aequoreo dulces cum sale pastus aquas",* Martial, *Epigrams* 13.89),
- *trichiae ("... trichiae .... in Hadriaticum mare defluunt..."* Pliny, *N.H.* 9.53),
- *rhombi ("... eadem aquatilium genera aliubi atque aliubi meliora... rhombus Ravennae..."* Pliny, *N.H.* 9.168-169),
- *pectines ("... pectines maximi et in his nigerrimi aestate laudatissimi, hi autem Mytilenis, Tyndaride, Salonis, Altini...",* Pliny, *N.H.* 32.150).

In addition, archaeological investigations have revealed the existence at *Altinum* of a possible *ostriarium*: indeed an ancient artificial channel full of oysters on both its bottom and edges was interpreted by the excavators as an artificial pond where oysters were kept for breeding and for the local market (Tirelli, Balista, Gambacurta & Ravagnan, 1988, 358; Balista & Sainati, 2003; Cao, 2003).

Medieval and post-medieval documents attest the widespread importance of fishing and fish-breading in the economy of the Venetian northern lagoon. A lease dating to the year 1001 concerning the area of Sette Soleri includes the following: *"...et offerimus similiter ipsas terrenas de Septe Salaria qui fuit // Vitalis Senatori cum aucellatione et piscatione sua omnio ex omnibus nostro possessum fuit..."*. The contract, therefore, testifies to the existence of a fishing and bird-hunting preserve in the area concerned (Lanfranchi Strina, 2006, sent. 43).

Spherical terracotta weights as fishing implements: some working hypotheses

Unfortunately, in the present state of knowledge, there is no archaeological evidence for net-fishing techniques used in the Venetian lagoon in Antiquity. Indeed, the only pictorial representation of a fishing scene is the well- known mosaic at Aquileia depicting Jonah being swallowed by a marine monster (figure 9). Here a seine (the Latin *sagena, cf.* Bekker-Nielsen, 2002a, 221; 2005, 92) appears, together with other catching devices. The mosaic is located in the *aula Theodoriana* of the basilica at Aquileia and dates to the first half of the fourth century AD. Beside its symbolic meaning, the mosaic is an interesting later example of an iconographic theme well known in the Roman world. Indeed, the Jonah episode is part of a more complex scene taking place in a sea populated by abundant fishes and molluscs of different species.

On the other hand, in the medieval and post medieval periods there is abundant evidence for traditional fishing techniques in both written and iconographic sources. A significant example is given by an illustrated codex from the *Biblioteca Marciana* in Venice (BNMVE, *Cod. lat.* XIV.77 (= 2291)), depicting a primitive Venice with wooden houses and churches clustering together on small islands surrounded by shallow canals: boats and fishing nets are integrated within this landscape (*cf.* Turri 1995, 6). A similar, but later, view can be appreciated in the famous painting by Carpaccio *"Caccia in Laguna"* ("Hunting in the Lagoon"), depicting the typical economic and natural landscape of the fifteenth-century Lagoon. The scene includes wooden houses, stationary fish nets, standing in shallow waters, and bird hunters on boats.

Modern studies in maritime anthropology confirm that up till around 1960, fishing was still a common economic activity in the lagoon islands and fish was the main source of sustenance of the local population (Vianello, 2004). All of this resulted in a long-standing tradition in fishing techniques and gears: according to a recent survey, more than 80 different types of fishing gears/techniques are known in the Venetian Lagoon, ranging from fishing with bare hands to permanent and semi-permanent nets and traps, from traditional to modern gear and practices from shore/water and boat (Pellizzato & Giorgiutti, 1997).

Therefore, in order to link the clay weights – the subject of this contribution – with a specific type of fishing gear/technique, results of recent studies in maritime anthropology and experimental archaeology have been combined under the supervision of the native fisherman Luigi Divari, who at present is one of a few living depositaries of a long-lasting local tradition in fishing practice and gear. With his advice, and in view of the peculiarities of the local environment and fauna, an attempt has been made to understand the possible usefulness of the clay spheroid weighs as fishing gear. The data collected allowed us to elaborate the working hypotheses presented below.

In principle, clay weights could have been less practical, but cheaper, substitutes for lead weights on the sinker line. Their spheroid shape and limited average weight

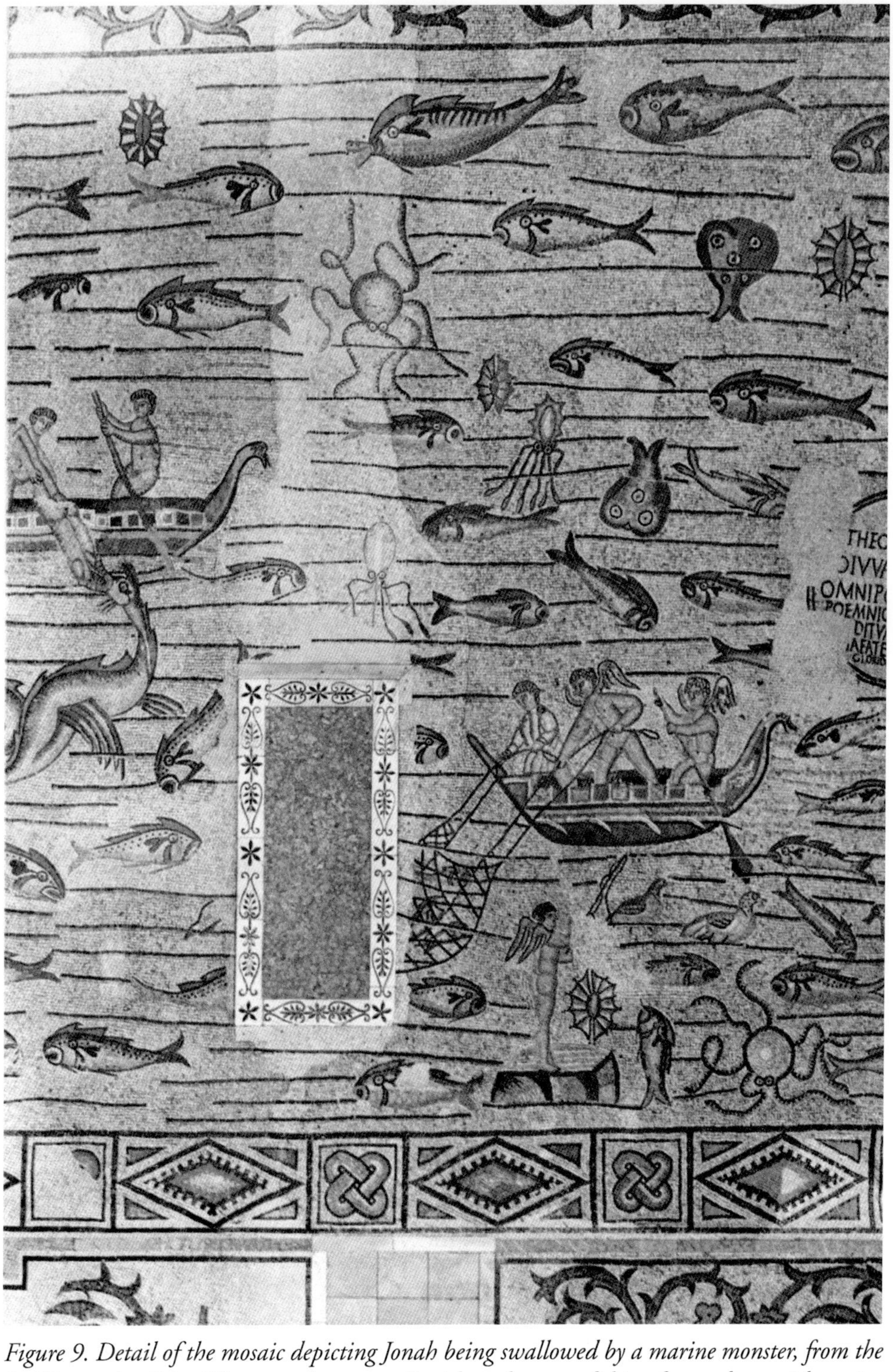

Figure 9. Detail of the mosaic depicting Jonah being swallowed by a marine monster, from the basilica at Aquileia. The scene shows two Cupids and a man fishing from a boat with a seine (Cuscito, 1974).

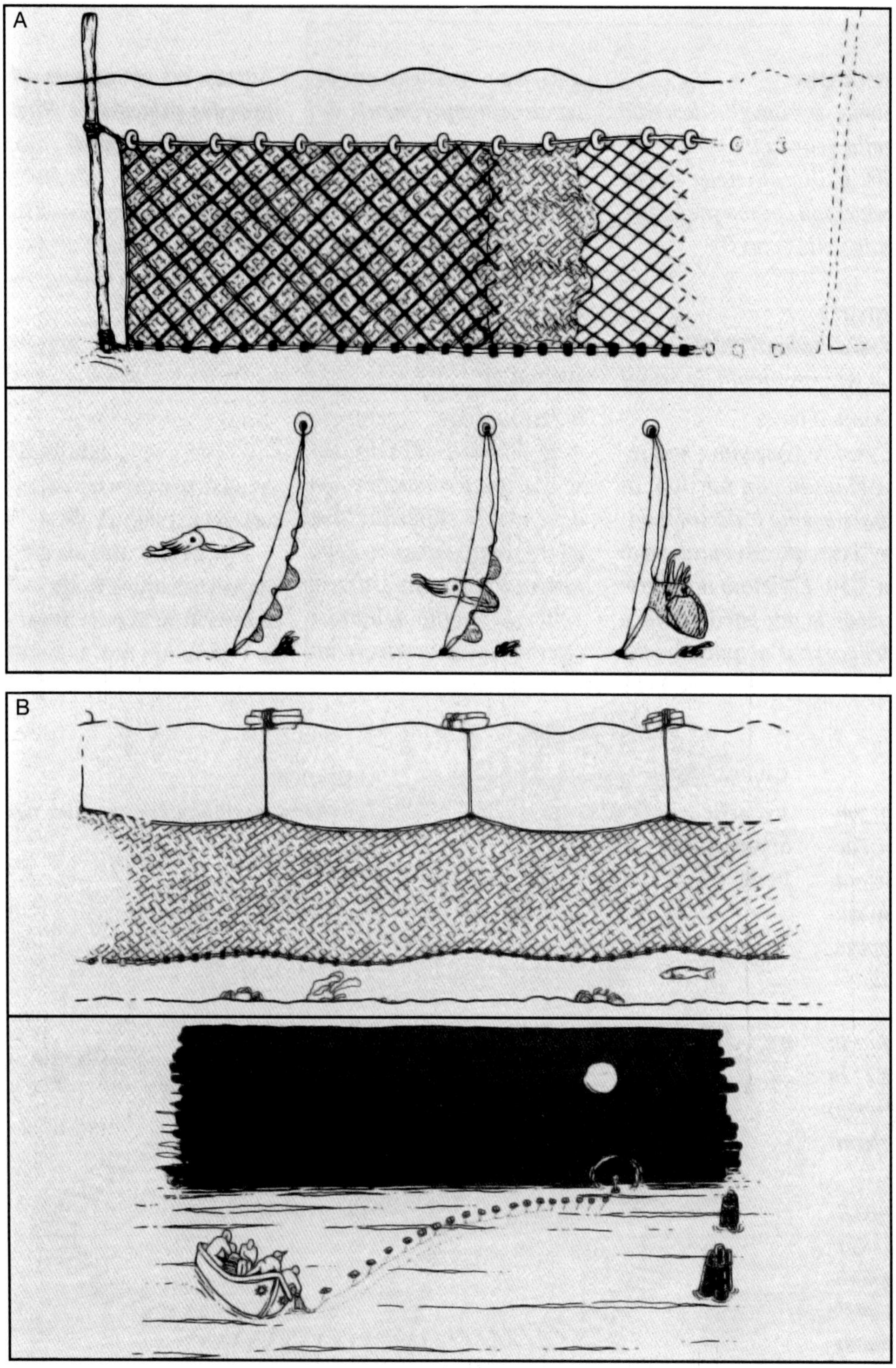

Figure 10. Drawings of stationary nets typical of the Venetian Lagoon: a) so-called "tramaglio/trimaglio", b) so-called "ré da imbrocco" (Pellizzato & Giorgiutti, 1997).

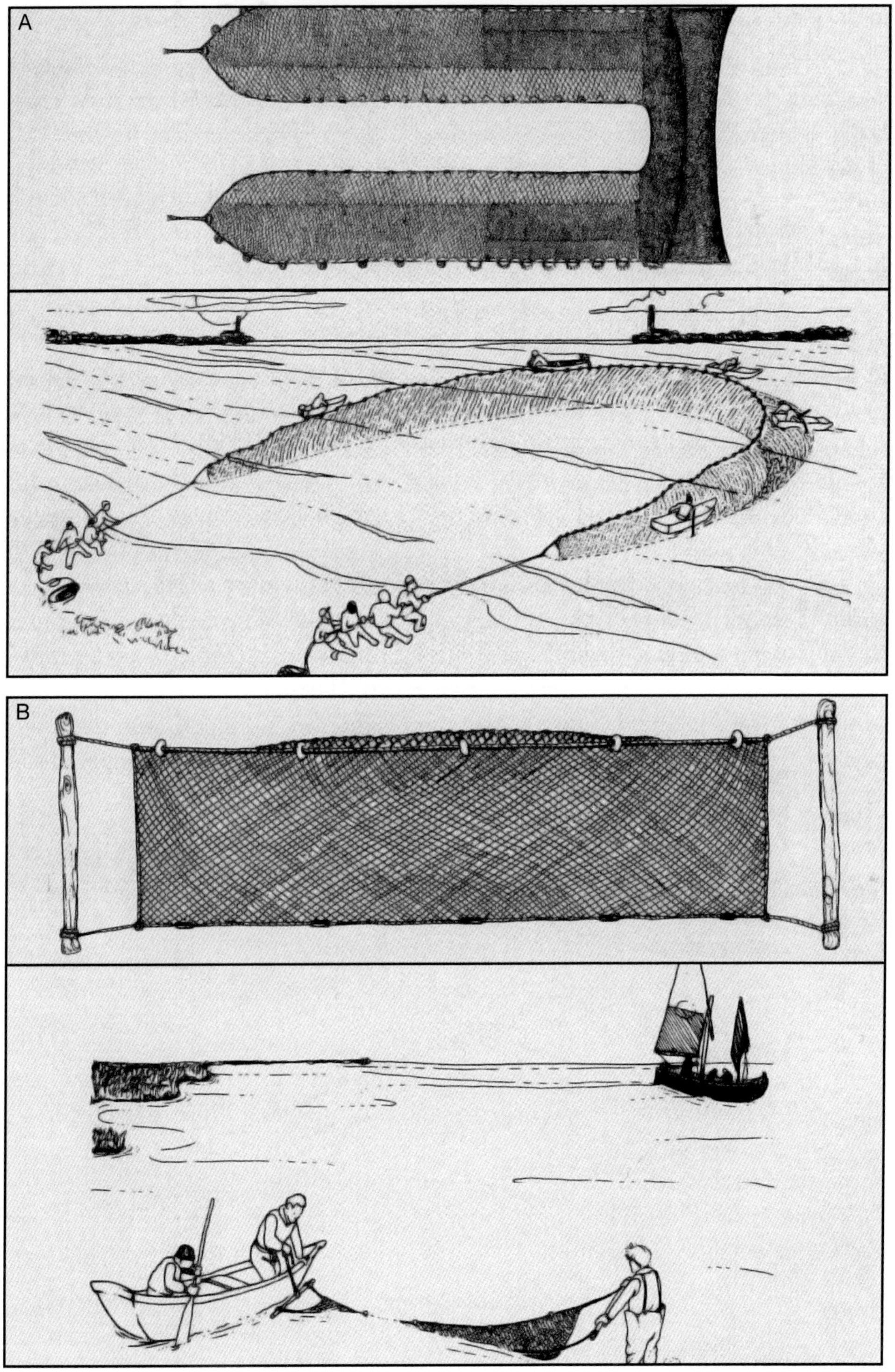

Figure 11. Drawings of dragging nets typical of the Venetian Lagoon: a) so-called "tràtada canal"/"tràta", b) so-called "bragotìn" (Pellizzato & Giorgiutti, 1997).

allow us to suggest that they could have been efficiently employed as weights for three different types of nets used in shallow waters.

Terracotta weights could be employed on stationary catching nets of different typologies: sinking to the bottom (figure 10a) or floating in shallow waters (figure 10b). Similar types of nets today are known as "*ré serberài*" and "*ré da imbrocco*" (figure 10b). These nets require a very fine, but strong, vegetal thread, to create a thick trammel. Nets with a 30-40mm mesh can be used to catch cuttlefish, sole and flounder, which normally do not have violent reactions. On the other hand, a close-mesh net (with a 18 to 26mm mesh), being more solid, is well suited for catching young mullet, red mullet and the like. Floating net types ("*ré da imbrocco*" *cf.* figure 10b) can be used for sardine and mackerel. These nets break easily and require constant repair: a limited use of similar types is attested in the Venetian Lagoon over the last three centuries, however, it is likely that they were employed well before. Documented specimens present two lines *c.* 6-8mm in diameter: the sinker line normally presents lead weights *c.* 35g each, achieving an average weight of *c.* 80-100g per metre.

Terracotta weights could also have been used on dragnets and beach seines of various sizes, depending on the fishing technique employed (from boat, from shore, in shallow water or in a canal). Local Venetian fishermen call these "*tràte*"/"*tratùri*" (figure 11a), "*strassìni*" and "*bragòti*" (figure 11b). These can reach one, or several, hundred meters in length, and a maximum height of *c.* 5-6 metres. On the sinker line the fishermen normally mount lead weights, not exceeding 30-40g per metre. Similar nets are used to catch a wide range of fishes: from small species to anchovies and bass-fishes of large size.

It has been calculated that a traditional "*trata da canal*" (i.e., dragnet, *cf.* figure 11a), a long net measuring up to 680 meters in length and 7 metres in height at the centre, requires 30 grams of weight per metre and presents a sinker line *c.* 14mm in diameter. Obviously smaller versions require less weight and a sinker line of smaller diameter. In principle, in the lagoon these nets can be used about eight months a year, allowing to catch, with little effort, the quantity of fish necessary for daily subsistence and even a surplus for the market. Moreover, we may suppose that in Antiquity, as well as in medieval and post-medieval times, during periods when migrating fishes came close to the shore, dragging nets could take particularly important catches, which in turn promoted the production of salted fish and similar products, as testified by Cassiodorus (*cf.* above). The diameter of the spheroid terracotta weights, varying from 12 to 16mm, fits well with the average diameter of the sinker lines employed.

Finally, clay spheroid weights could also be used as weights on the sinking-line forming the edge of a casting-net (*amphiblêstron*, *cf.* Bekker-Nielsen, 2002a, 216; 2005), a type of throwing-net with a long tradition, used in Antiquity from boat or from shore and still in use (though to a very limited extent) for shore fishing in

Figure 12. Drawing of a typical casting-net still in use in the Northern Adriatic, the so-called "rezzaglio" (Pellizzato & Giorgiutti, 1997).

the northern Adriatic. In the Venetian dialect, this net is called a "*rezzaglio*" (figure 12); it has been calculated that a casting-net with a circumference of 17m requires about 2kg of weight on the sinking-line, approximately 20g for every 18cm. This net is used to catch mullets.

In conclusion, we may suggest that clay weights could have been a viable substitute for more expensive lead weights in those regions characterized by a natural landscape dominated by shallow waters, lagoons, canals and fishing valleys, such as the northern Adriatic coast from Comacchio to Venice and Caorle. In this type of marine environment, in Antiquity (as testified to by *Cassiodorus*) as well as in modern times, a large proportion of the local population based their living on the integrated exploitation of the natural resources available: fishing, hunting and bird catching. Within this context, local production and use of clay spheroid net-weights seem to agree with the economic logic of a "family scale" business, where fishing was almost an everyday routine, all the year round, and a traditional practice for island settlers aiming to fulfil the needs of daily subsistence.

Mosaic of the Fishing Cupids. Villa del Casale (Sicily). Late third-early fourth century AD (detail).

Perspectives

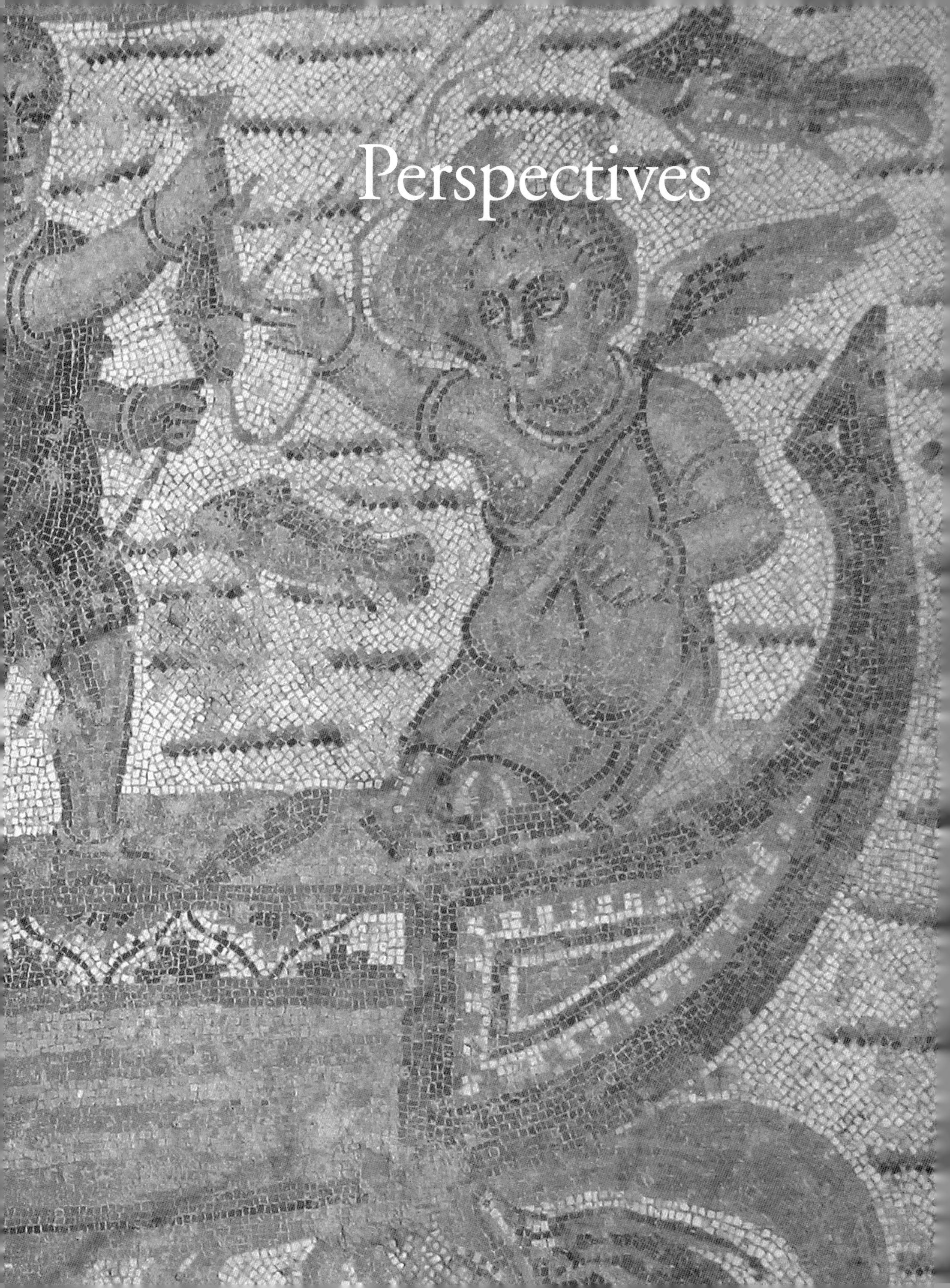

Detail of the Mosaic of the Sea Creatures. first century AD. National Archaeological Museum, Naples (originally in the House of the Faun, Pompeii).

18. Nets and Fishing Gear in Classical Antiquity: Past, Present and Future Scholarship

Athena Trakadas

Marine resources – primarily fish but also shellfish and marine mammals – in the Graeco-Roman world are more frequently becoming the subject of historical scholarship. As early as 1832 classical philologists and ancient historians began to collect and debate textual references or inscriptions related to these resources – predominantly those pertaining to the manufacture of *salsamenta* and salted-fish sauces such as *garum* within the Black Sea and Mediterranean (e.g., Köhler, 1832; Cuvier & Valenciennes, 1832; Smidth, 1876; Besnier, 1877; Eberl, 1892; Zahn, 1910). Almost concurrently, research began to be undertaken into the identification of marine species mentioned in classical sources, such as Aelian's *De Natura Animalium*, Oppian's *Halieutika* and Book 9 of Pliny's *Historia Naturalis* (e.g., Clarke, 1888; Barbier, 1925-26; Barbier, 1927-28; Barbier, 1933-36; Wood, 1927; Wood, 1928a and 1928b; Höppener, 1931; Strömberg, 1943; Cotte, 1944; Thompson, 1947; de Saint-Denis, 1947; Andrews, 1949; Gow, 1968).

After the Second World War, archaeological studies have contributed greatly to these initial strands of philological and historical research. On the Strait of Kerch in the northern Black Sea, the excavations of the remains of a large number of fish-salting facilities within the settlements of Chersonesos, Tyritake and Myrmekion were first published (sites and publications reviewed in Højte, 2005). The first major work in this area in the Mediterranean is Ponsich and Tarradell's 1965 publication, *Garum et industries antiques de salaison dans la Méditerranée occidentale*, in which fish-salting sites in southern Iberia and northwest Africa are discussed (see also Ponsich, 1988 and comments in Étienne & Mayet, 1998a; Wilson, 2006). Following these precedents, in the last four decades other excavations of individual sites such as *Baelo Claudia* and *Tróia* or surveys of entire salting areas have also been undertaken, primarily in the Iberian Peninsula but also in regions such as southern France, eastern Tunisia and more sites in the northern Black Sea (e.g., Arévalo & Bernal, 2007; Étienne *et alii*, 1994; Trousset, 1992 and 1998; Étienne & Mayet, 2002; Højte,

2005; Sternberg, 1998; Trakadas, 2005; Curtis, 1991). In several cases, the processing at these salting sites of purple dye from specific shellfish has received scholarly attention (e.g., Fernández Uriel, 1995; Alfaro Giner, 2002b; Wilson, 1999).

Fortuitously, icthyo-archaeological studies have also been gradually incorporated within some analyses of the fish-salting industry (see Desse-Berset & Desse, 2000; Larje, 1995; Wendrich & Van Neer, 1994). These have helped not only to identify scientifically the types of fish found at the processing sites but also those found in *salazón* transport amphorae, mainly preserved from shipwrecks particularly from southern Spain, southern France and the Straits of Bonifacio (see Delussu & Wilkens, 2000; de Grossi Mazzorin, 2000; Sternberg, 1998).

Analyses of production events and the economics of the industry are also well established. Topics include how processing was conducted, what resources were needed, where and how the products were shipped, who was involved in their transportation and the different scales of the trade. Source material includes texts, epigraphy (stamps and *tituli picti* in particular, but also some inscriptions) and archaeological data ranging from the processing sites themselves to fish-salting amphorae distribution and kilns that manufactured these types (e.g., Étienne, 1970; Ben Lazreg *et alii,* 1995; Étienne & Mayet, 1998b; di Stefano, 2002; Sternberg, 2000a; Edmondson, 1987; Lowe, 1997; Jardin, 1961; Curtis, 1984 and 1984-86; Arévalo González *et alii,* 2004; Hesnard, 1998; Martínez Maganto, 2000; Ørsted, 1998).

The studies discussed above pertain almost solely to the state of knowledge of the fish-salting industry, with archaeology providing a solid evidential basis for critical analysis of literary and iconographic sources. But the art of fishing in Antiquity was of interest to a few modern scholars albeit with the initial discussions, like those of fish-salting, based on textual analysis. The largest compilation of authors on the subject, both ancient and modern, first appeared in *Bibliotheca Piscatoria* in 1883 (Westwood & Satchell, 1883; additions in Maston, 1901). Within this early period of philological treatment, references to ancient exploitation methods were collected and sometimes more often discussed as leisure or sporting activities undertaken for recreation and not for providing sustenance. Some more recent histories include contemporary iconographic representations and extend beyond the Graeco-Roman world to include Mesopotamia and ancient Egypt (e.g., Blümner, 1869; Bunamann, 1910; Radcliffe, 1921; Horn, 1929; Butler, 1931; Bohlen, 1937; Corcoran, 1957; Corcoran, 1964; Dumont, 1977 and 1981; Donati & Pasini, 1997; Bekker-Nielsen, 2002a; Brewer & Friedman, 1990; Darby *et alii,* 1977).

Finds of fishing equipment have long been recovered from excavations of settlements and fish-salting sites, although generally they have received limited attention. Fortuitously, several archaeological excavations and surveys from the past few decades have focused specifically upon the physical evidence of Graeco-Roman fishing technology and brought to light new examples and revisited these earlier finds. Some recent studies have collected evidence for fishing equipment such as hooks, spears, net weights and nets from specific regions or broadly throughout the Roman Mediterranean. Research has also examined a few fixed fishing installations; these means of passive fishing, such as weirs or catch basins, trapped and kept fish and shellfish in lagoons, rivers or the tidal zone (Kuniholm, 1982; Wendrich & Van Neer, 1994; Ayodeji, 2004; Alves *et alii,* 1988-89; Moreno Páramo & Abad Casal, 1971; Rustico, 1999; Gazda & McCann, 1987; Trakadas, 2006; Galili & Rosen, 2008). It is now also more common in many regions to include ichthyo-faunal analyses in settlement excavations, not only to examine sustenance patterns and determine procurement methods but to assess environmental change as reflected through marine faunal remains (Roselló & Morales, 1992; Grau Almero *et alii,* 2001, 204-223; Rodríguez Santana & Rodrigo García, 2005; Van Neer *et alii,* 2004; Prowse *et alii,* 2004).

Ancient fishing vessels still remain better understood from texts and iconography, as the only archaeological finds in the Mediterranean region are presently known from Marseille, France (sixth century BC), Lake Kinneret, Israel (first centuries BC/AD) and Fiumicino, Italy (second century AD) (e.g., Donati & Pasini, 1997; Carlson, 1999 and 2002; Pomey, 1995 and 2000; Testaguzza, 1970, 132; Boetto, 2006b; Wachsmann, 1990a).

Undoubtedly the earlier scholarship regarding marine resources, cursorily reviewed above, is more focused on the fish-salting industry, yet this corpus offers enormous contributions to our knowledge of how past populations interacted with the marine environment. The direct relationship between fishing and the salting industry is obvious, but it is often not adequately considered. Timely studies on ancient fishing methods and technologies are gaining momentum, and this emergence clearly demonstrates the potential the subject offers at re-assessing past sustenance patterns, past economies and the state of the marine environment.

Against this background, the workshop held in Cádiz in November 2007, *Artes de pesca en la antigüedad clásica: un primer balance,* was to provide, as its name suggests, a "first look" at fishing technology in antiquity. As stated in the opening remarks by Darío Bernal and Tønnes Bekker-Nielsen, the primary purpose of the workshop was to create an overview of the present state of scholarship on the subject. The presentations that form the contributions in this volume certainly demonstrate a rather rich and diverse body of evidence.

In addition to establishing a "first look", the Cádiz workshop organisers outlined a future cooperative project: the systematic documentation of fishing equipment from the Phoenician-Punic periods to the Islamic conquest (800 BC-AD 700). This project is to have a broad, intra-regional and synchronic scope, and is intended to include contributions from different specialists of archaeo-historical sub-disciplines. Ultimately, the collected material will contribute to a database centred at the University of Cádiz and digital resources to be compiled on the HMAP-Mediterranean (History of Marine Animal Populations) website (http://hmapcoml.org/projects/m&b/) – figure 1. A second workshop synthesising some of this material is also planned for the future.

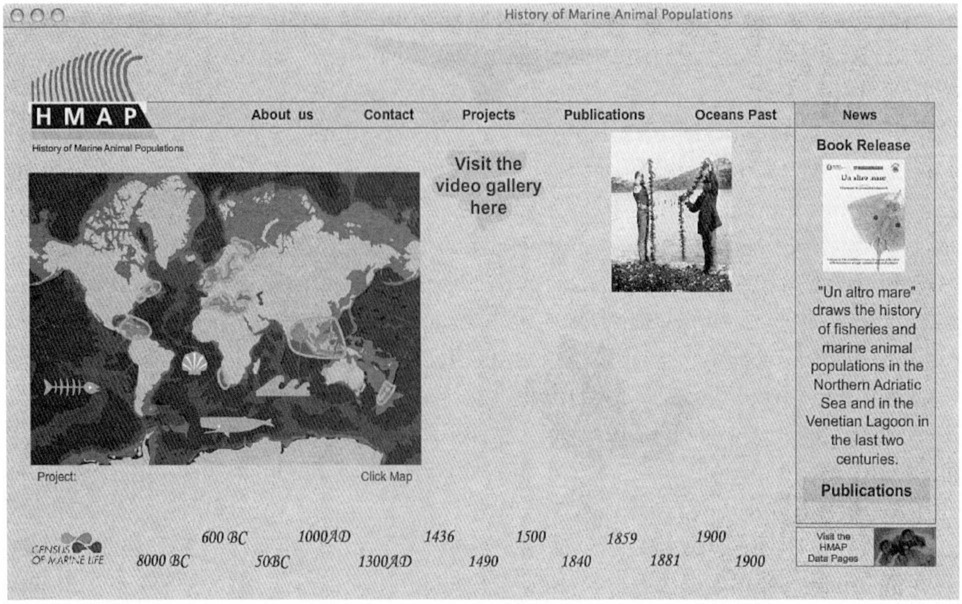

Figure 1. Main page of the website http://hmapcoml.org

The proposal for such a project is timely and well-founded. This present volume shows that there is a solid basis on which to build a more nuanced and broad-reaching corpus of information. The workshop in Cádiz also successfully created an environment in which specialists from various disciplines – ancient historians, zoologists, art historians, ethnographers, and archaeologists specialising in ships, trade and textiles – were able to have focussed discussions about the variable evidence of fishing technology. This situation opened new paths of thinking for many scholars, and underscores the need for an inter-disciplinary approach.

Of particular significance is the chronological extension of this proposed documentation project – almost a millennium and a half – which will provide a broader resolution to determine change, continuity, re-emergence or transference of fishing technologies. Extending also this examination and documentation throughout the Mediterranean, and even beyond to the Black Sea, eastern Atlantic and Red Sea, will, I believe, illustrate the potential of synchronic comparison. With a more complete record of fishing gear, it will hopefully allow for the determination of similarities and differences in methods that might exist, and perhaps even demonstrate technology transfer or isolation. This approach will also be exceedingly helpful for scholars to examine data similar to their own in a slightly different context, perhaps even allowing for gaps in their own record to be filled through cautious comparison. Systematic documentation might also allow for empirical studies in which productivity and the anthropogenic impacts of fishing on past marine ecosystems can be assessed.

Also discussed in Cádiz was the fact that there will be limitations to how thorough a record this proposed project will provide. Disparate geographical distribution of archaeological finds, differences in archaeological sampling, the lack of correspondence in many cases between the archaeological finds and the textual sources, and general preservation biases ensure that lacunae in the record will certainly exist. I believe that these are metadata issues that can only be addressed in the future, once the project's compilation is well underway and the nature of the information is better understood. We cannot underestimate that the proposed project, however, will answer numerous outstanding questions and raise new paths of inquiry. Such a study will have enormous impact upon how we perceive and approach not only the foodways and economy of the ancients, but also anthropogenic impacts on marine life.

DICCIONARIO HISTÓRICO
DE LOS ARTES
DE
LA PESCA NACIONAL

POR EL COMISARIO REAL DE GUERRA DE MARINA

DON ANTONIO SAÑEZ REGUART,

Socio de Mérito de la Real Sociedad de Amigos del Pais de Madrid, y de San Lúcar de Barrameda.

TOMO PRIMERO.

Cover of Sáñez Reguart 1791.

MADRID MDCCLXXXXI.

EN LA IMPRENTA DE LA VIUDA DE DON JOAQUIN IBARRA.

CON SUPERIOR PERMISO.

Bibliography

Abbreviations

AA	Antiquités Africaines.
AAE	Arabian Archaeology and Epigraphy.
ABeja	Arquivo de Beja: Boletim da Câmara Municipal.
AEspA	Archivo Español de Arqueología.
AJ	The Antiquaries Journal.
AJA	American Journal of Archaeology.
AJPh	American Journal of Philology.
AMM	Archaeologia Maritima Mediterranea, An International Journal on Underwater Archaeology.
ANSER	Ancient Sea Routes (Anciennes Routes Maritimes Méditerrannéenes), Programme Interreg IIIB Medocc.
AntO	Antiguo Oriente.
AR	L'Africa romana.
ASubacq	Archeologia subacquea. Documenti, studi e ricerche.
BAM	Bulletin d'Archéologie Marocaine.
BAR (IS)	British Archaeological Reports, International Series.
BIFAO	Bulletin de l'Institut Français d'Archéologie Orientale du Caire.
BPH	Bibliotheca Praehistorica Hispana.
BSR	British School at Rome.
C&M	Classica et Mediaevalia.
CAS	Cahiers d'Arqueologie Subaquatique.
CASC	Centre d'Arqueologia Subaquàtica de Catalunya.
CASCV	Centro de Arqueología Subacuática de la Comunidad Valenciana.

CC	Cahiers Corsica, publiés par la Federation d'Associations et Groupements pour les etudes corses.
CIETA	Centre International d'Études des Textiles Anciennes.
CIL	Corpus Inscriptionum Latinarum.
CJ	The Classical Journal.
CJFAR	Canadian Journal of Fisheries and Aquatic Research.
CMGR	Colloque sur la Mosaïque Gréco-Romaine.
CMRE	Corpus de Mosaicos Romanos de España.
CMRP	Corpus dos Mosaicos Romanos de Portugal.
CMT	Corpus de Mosaïques de Tunisie.
CoML	Census of Marine Life.
CPh	Classical Philology.
CQ	The Classical Quarterly.
CRAI	Comptes-Rendues de l'Academie des Inscriptions et Belles Lettres.
CSIC	Consejo Superior de Investigaciones Científicas.
DRASSM	Département des Recherches Archéologiques Subaquatiques et Sous-marines.
DS	C. Daremberg and E. Saglio, Dictionnaire des Antiquités Grecques et Romaines, Paris 1873-1919.
GGM	Carolus Mullerus (ed.), Geographi Graeci Minores, Paris 1855-1861.
HMAP	History of Marine Animal Populations.
ICAZ	International Council for Archaeozoology.
ICCAT	International Commission for the Conservation of Atlantic Tuna.
ICES	International Council for the Exploration of the Sea.
IEJ	Israel Exploration Journal.
IGR	Inscriptiones Graecae ad Res Romanas Pertinentes.
IGSK	Inschriften Griechischer Städte aus Kleinasien (= IK).
IJNA	International Journal of Nautical Archaeology.
ILS	Inscriptiones Latinae Selectae.
INA	Institute of Nautical Archaeology.
INRH	Institut National de Recherche Halieutique.
JAS	Journal of Archaeological Science.
JEA	Journal of Egyptian Archaeology.
JRA	Journal of Roman Archaeology.
JRS	Journal of Roman Studies.
KLNM	Kulturhistorisk Leksikon for Nordisk Middelalder.
Lattara	Lattara: Mélanges d'Histoire et d'Archéologie de Lattes.
LIMC	Lexicon Iconographicum Mythologiae Classicae.
MARQ	Museo Arqueológico de Alicante.
MEFRA	Melanges de l'École Française à Rome. Antiquité.

MEFRM	Melanges de l'École Française à Rome. Moyen-Age.
MM	Madrider Mitteilungen.
NESAT	North European Symposium for Archaeological Textiles.
NIMA	National Imagery and Mapping Agency, Bethesda, MD.
NJZ	Netherlands Journal of Zoology.
Paléorient	Paléorient: Revue Interdisciplinaire de Préhistoire et Protohistoire de l'Asie du Sud-Ouest.
P. Oxy	The Oxyrhynchus Papyri.
Praktika	Praktika tis en Athenais Archaiologikis Etaireias.
RE	Pauly-Wissowa-Kroll, Real-Encyclopaedie der klassischen Altertumswissenschaft.
REA	Revue des Études Anciennes.
RLR	Revue des Langues Romanes.
Saguntum	Saguntum: Papeles del Laboratorio de Arqueología de Valencia.
SF	Statens Fiskeredskapsimport.
SFECAG	Société Française d'Étude de la Céramique Antique en Gaule.
SURVAS	Synthesis and Upscaling of sea-level Rise Vulnerability Assessment Studies.
TopOO	Topoi Orient-Occident.
ZSAK	Zeitschrift für Schweizerische Archäologie und Kunstgeschichte (Revue Suisse d'Art et d'Archéologie).

AGATHARCHIDES OF CNIDUS: *Peri ten Erythras Thalasses*, translated by S.M. Burstein, London, Hakluyt Society, 1989.

AIMARD, J. (1951): *Les chasses romaines des origines à la fin du siècle des Antonins*, Paris.

ALARCÃO, J., ETIENNE, R., MOUTINHO ALARCÃO, A. & DA PONTE, S. (eds.) (1979): *Fouilles de Conimbriga*, Paris.

ALEXANDER, A. *et alii* (1973): *Utique*, CMT 2/1, Tunis.

ALEXANDER, A. *et alii* (1980): *Thuburbo Maius*, CMT 2/1, Tunis.

ALEXANDER, A. *et alii* (1987): *Thuburbo Maius*, CMT 2/3, Tunis.

ALFARO, C. (1983-84): "Notas sobre una redecilla romana de Medina Sidonia (Cádiz)", *Boletín del Museo de Cádiz* 4, pp. 77-81.

ALFARO, C. (1984): *Tejido y cestería. Historia de su técnica e industrias desde la Prehistoria hasta la romanización*, BPH 21, Madrid.

ALFARO, C. (2001): "Recent discoveries of gold textiles from Augustan Age (*Gadir*, Cádiz)", P. Walton, L. Bender & A. Rast-Elcher (eds.): *The Roman textile industry and its influence. A birthday tribute to John Peter Wild*, Oxford, pp. 76-83.

ALFARO, C. (2002a): "Étoffes cordées du site néolithique de Tell-Halula (Syrie-VIIIe millénaire avant J.-C.)", *CIETA Bulletin* 79, pp. 17-25.

ALFARO, C. (2002b): "*Ebusus* y la producción de púrpura en el Imperio romano", *AR* 14, pp. 681-696.

ALFARO, C. & COSTA, B. (2008): "Methodological aspects of purple dye production on Ibiza: the new site of Cala Olivera", C. Alfaro & L. Karali (eds.): *Purpureae vestes: 2nd International Symposium on Textiles and Dyes in the Ancient Mediterranean World, Athens, 24-26 November 2005,* Valencia, pp. 195-208.

ALFARO, C. & L. KARALI (eds.) (2008): *Purpureae vestes: 2nd International Symposium on Textiles and Dyes in the Ancient Mediterranean World, Athens, 24-26 November 2005,* Valencia.

ALONSO VILLALOBOS, C. & NAVARRO DOMÍNGUEZ, M. (1998): "*Baelo Claudia:* sus posibilidades portuarias y la navegación por el Estrecho de Gibraltar", G. Pascual (ed.): *III Jornadas de Arqueología Subacuática. Puertos antiguos y comercio marítimo, Valencia, 13, 14, y 15 de noviembre de 1997,* Valencia, pp. 131-140.

ALTORFER, K. (1999): "Neue Erkenntnisse zum Neolithischen Türflügel von Wetzikon ZH-Robenhausen", *ZSAK* 56, pp. 217-230.

ALTORFER, K. & F. MÉDARD (2000): "Nouvelles découvertes textiles sur le site de Wenzikon-Robenhausen (Zürich, Suisse). Sondages 1999", D. Cardon & M. Feugère (eds.): *Archéologie des textiles. Des origines au Ve siècle, Actes du Colloque de Lattes, oct. 1999*, Montagnac, pp. 35-75.

ÁLVAREZ, B.T. (1999): *Plantas de acción ictiotóxica usadas en España.* Unpublished thesis, Universidad Autónoma de Madrid.

ÁLVAREZ DE TOLEDO, M.L. (2007): *Las almadrabas de los Guzmanes*, Sanlúcar de Barrameda.

ALVES, F.J.S., ALVEIRINHO DIAS, J.M., ROCHA DE ALMEIDA, M.J., FERREIRA, O. & TABORDA, R. (1988-89): "A armadilha de pesca da época romana descoberta na praia de Silvalde (Espinho)", *O Arqueólogo Português* 4, pp. 187-226.

AMELLI, A.M. (1896): *Miniature della Enciclopedia Medioevale di Rabano Mauro (Codice di Montecasino No 132 dell'anno 1023)*, Montecassino.

AMORES CARREDANO, E., GARCÍA VARGAS, D., GONZÁLEZ ACUÑA, D. & LOZANO FRANCISCO, M.C. (2007): "Una factoría altoimperial de salazones en *Hispalis* (Sevilla, España)", L. Lagóstena Barrios, D. Bernal Casasola & A. Arévalo González (eds.): *Cetariae 2005. Salsas y salazones de pescado en Occidente durante la Antigüedad, Actas del Congreso Internacional, Cádiz 7-9 de noviembre de 2005,* (*BAR* (IS) 1686), Oxford. pp. 335-339.

ANDERSEN, S.H. (1987): "Tybrind Vig: A Submerged Ertebølle Settlement in Denmark", J.M. Coles & A.J. Lawson (eds.): *European Wetlands in Prehistory*, Oxford, pp. 253-280.

ANDERSON, R.D, PARSONS, P.J. & NISBET, R.G.M. (1979): "Elegiacs by Gallus from Qasr Ibrim", *JRS* 69, pp. 125-155.

ANDREAE, B. (2003): *Antike Bildmosaiken,* Mainz.

ANDREWS, A.C. (1949): "The 'Sardinian Fish' of the Greeks and Romans," *AJPH* 70, pp. 171-185.

ANTÓN SOLÉ, P. (1965): *Los pícaros de Conil y Zahara,* Cádiz.

APARICIO PÉREZ, J. & CLIMENT MANO, S. (1985): "Sobre la pesca en la Edad del Bronce", *Arse* 20, pp. 11-15.

ARANEGUI, C. (1982): *Excavaciones en el Grau Vell (Sagunto, Valencia). Campañas de 1974 y 1976 (Trabajos Varios del SIP* 72), Valencia.

ARBEX, J.C. (1990): *Pescadores Españoles,* Madrid.

ARÉVALO, A. & BERNAL CASASOLA, D. (eds.) (2007): *Las cetariae de Baelo Claudia. Avance de las investigaciones arqueológicas en el barrio meridional (2000-2004),* Cádiz.

ARÉVALO, A., BERNAL CASASOLA, D. & TORREMOCHA SILVA. A. (eds.) (2004): *Garum y Salazones en el Círculo del Estrecho,* Cádiz.

ARIAS GARCÍA, A.M. (2005): *Corrales de Rota. El monumento natural de Andalucía,* Chiclana.

ARNOLD, B. (1995): *Pirogues monoxyles d'Europe centrale. Construction, typologie, évolution,* (*Archéologie Neuchâteloise* 20), Neuchâtel.

ARTEAGA, O. (2002): "Las teorías explicativas de los 'cambios culturales' durante la Prehistoria en Andalucía: Nuevas alternativas de investigación", *Actas del III Congreso de Historia de Andalucía,* Córdoba, pp. 247-311.

ATKINSON, R. (1914): "A hoard of Samian Ware from Pompeii", *JRS* 4, pp. 27-64.

AURA, J.E. & PÉREZ, C. (1998): "¿Micropuntas dobles o anzuelos? Una propuesta de estudio a partir de los materiales de la Cueva de Nerja", J.L. Sanchidrián & M.D. Simón (eds.): *Las culturas del Pleistoceno Superior en Andalucía,* Málaga, pp. 339-348.

AURIGEMMA, S. (1960): *L'Italia in Africa. Tripolitania I: I Mosaici,* Roma.

AYALA, G., HORRY, A. & LAURENT, F. (2005): "Au cœur de Lyon, mille ans de navigation fluviale", *Archéologia* 419, pp. 40-48.

AYODEJI, K. (2004): *Fishing Equipment and Methods in the Roman World,* PhD Thesis, University of London.

BAIRRÃO OLEIRO, J.M. (1992): *Conimbriga: Casa dos Repuxos,* (*CMRP* I), Conímbriga.

BALISTA, C. & SAINATI, C. (2003): "*Ostrea* non *pectines* ad Altino: le evidenze archeologiche", G. Cresci Marrone & M. Tirelli (eds.): *Produzioni, merci e commerci in Altino preromana e romana,* Roma, pp. 331-346.

BALL, W. (2000): *Rome and the East: the transformation of an empire,* London.

BALLARD, R.D. (ed.) (2008): *Archaeological Oceanography,* Princeton.

BALME, J. (1983): "Prehistoric fishing in the lower Darling, western New South Wales", C.J. Grigson, & J. Clutton-Brock (eds.): *Animals and Archaeology 2: Shell middens, Fish and Birds,* (*BAR* (IS) 183), pp. 19-33.

BARBAGLI, D. (2005): "The Wreck C", Camilli, A. & Setari, E. (eds.): *Ancient shipwrecks of Pisa: A guide,* Venice, pp. 46-47.

BARBER, E.J.W. (1991): *Prehistoric Textiles: The development of cloth in the Neolithic and Bronze Ages,* Princeton.

BARBIER, P. (1925-26): "Noms de poissons", *RLR* 63, pp. 1-69.

BARBIER, P. (1927-28): "Noms de poissons", *RLR* 65, pp. 1-53.

BARBIER, P. (1933-36): "Noms de poissons", *RLR* 67, pp. 275-373.

BARNARD, H. (2005): "Sire, il n'y a pas de Blemmyes: A re-evaluation of historical and archaeological data", J.C.M. Starkey (ed.): *People of the Red Sea: Proceedings of Red Sea Project II Held in the British Museum, October 2004,* (*BAR* (IS) 1395), Oxford, pp. 23-40.

BARRAL I ALTET, X. (1978): *Les Mosaïques romaines et médiévales de la Regio Laietana,* Barcelona.

BARRETT, J.H., LOCKER, A.M. & ROBERTS, C.H. (2004): "The origins of intensive marine fishing in medieval Europe. The English evidence", *Proceedings of the Royal Society of London, Series B* 271, pp. 2417-2421.

BASCH, L. (1987): *Le musée imaginaire de la marine antique,* Athens.

BASS, G.F. (1975): *Archaeology Beneath the Sea,* New York.

BASS, G.F. (2004): "Fishing Spear", G.F. Bass, S.D. Matthews, J.R. Steffy & F.H. van Doorninck Jr. (eds.): *Serçe Limani: An Eleventh-Century Shipwreck, 1: The Ship and its Anchorage, Crew, and Passengers,* College Station, Texas, pp. 429-430.

BASS, G.F., MATTHEWS, S.D., STEFFY, J.R. & VAN DOORNINCK JR., F.H. (2004): *Serçe Limani: An Eleventh-Century Shipwreck, 1: The Ship and its Anchorage, Crew, and Passengers,* College Station, Texas.

BASS, G.F. & VAN DOORNINCK JR., F.H. (eds.) (1982): *Yassi Ada: A Seventh-Century Byzantine Shipwreck,* College Station, Texas.

BAZZANELLA, M. & A. MAYR (2003): "Scheda n. 13", M. Bazzanella, A. Mayor, L. Moser & A. Rast-Eicher (eds.): *Textiles: Intrecci e tessuti della preistoria europea. Museo Civico di Riva del Garda-La Roca, 24 maggio-19 ottobre 2003,* Trento, p. 172.

BAZZANELLA, M. *et alii* (2003): "Storia delle ricerche", M. Bazzanella, A. Mayor, L. Moser & A. Rast-Eicher (eds.): *Textiles: Intrecci e tessuti della preistoria europea. Museo Civico di Riva del Garda-La Roca, 24 maggio-19 ottobre 2003,* Trento, pp. 23-29.

BEAL, J.-CL. (1999): "Remarques sur l'imagerie du pilier funéraire d'Igel", N. Blanc & A. Buisson (eds.): *Imago antiquitatis. Religions et iconographie du monde romain. Mélanges offerts à Robert Turcan,* Paris.

BEBKO, W. (1971): "Les epaves antiques du sud de la Corse", *CC* 1-3.

BECATTI, G. (1961): *Scavi di Ostia IV,* Roma.

BEDINI, A., FERRO, D. & RAPINESSI, I.A. (2004): "Testimonianze di filati e ornamenti in oro nell'abigliamento di età romana", C. Alfaro, J.P. Wild & B. Costa (eds.): *Purpureae Vestes: Textiles y Tintes del Mediterráneo en época romana, Actas del I Symposium Internacional (Ibiza, 8-10 November 2002),* Valencia, pp. 77-88.

BEECH, M.J. (2004): *In the land of the Ichthyophagi: Modelling fish exploitation in the Arabian Gulf and Gulf of Oman from the 5th millennium BC to the Late Islamic period,* (*BAR* (IS) 1217), Oxford.

BEKKER-NIELSEN, T. (2002a): "Nets, boats and fishing in the Roman World", *C&M* 53, pp. 215-233.

BEKKER-NIELSEN, T. (2002b): "Fish in the Ancient Economy", K. Ascani *et alii* (eds.): *Ancient History Matters. Studies presentend to Jens Erik Skydsgaard on His Seventieth Birthday*, (*Analecta Romana Instituti Danici Supplementum* 30), Rome, pp. 29-37.

BEKKER-NIELSEN, T. (2004): "Source materials for fishing in Antiquity and the early middle ages", *Report to first HMAP Mediterranean/Black Sea Workshop, Barcelona 2004,* http://hmap.ruc.dk/projects/m&b/Documents/Bekker-Nielsen.pdf.

BEKKER-NIELSEN, T. (2005): "The Technology and Productivity of Ancient Sea Fishing", T. Bekker-Nielsen (ed.): *Ancient Fishing and Fish Processing in the Black Sea Region,* (*Black Sea Studies* 2), Aarhus, pp. 83-95.

BEKKER-NIELSEN, T. (ed.) (2005): *Ancient Fishing and Fish Processing in the Black Sea Region,* (*Black Sea Studies*) 2, Aarhus.

BEKKER-NIELSEN, T. (2008). "Lo studio della fauna antica del Mar Nero (fino al 1200 d.C.): Aspetti quantitativi e qualitativi", R. Gertwagen *et alii: Il mare, com'era: le interazioni tra uomo ed ambiente nel Mediterraneo dall'Epoca Romana al XIX secolo: una visione storica ed ecologica delle attività di pesca,* Rome, pp. 78-93.

BEKKER-NIELSEN, T. (2009): "La industria pesquera en la región del Mar Negro en la Antigüedad", D. Bernal Casasola (ed.): *Arqueología de la pesca en el Es-*

trecho de Gibraltar: De la Prehistoria al fin del Mundo Antiguo, Cádiz, pp. 286-311.

BELLIN, J.N. (1764): *Carte du Détroit de Gibraltar,* Paris.

BELTRAME, C. (2002): *Vita di bordo in età romana,* Roma.

BELTRAME, C. & GADDI, D. (2005): "The Rigging and the Hydraulic System of the Roman Wreck of Grado (Gorizia, Italy)", *IJNA* 34.1, pp. 79-87.

BELTRAME, C. & GADDI, D. (2008): "Reply to a Comment on an Article Concerning the Hydraulic System of the Roman Wreck ay Grado, Gorizia, Italy", *IJNA* 37.2, pp. 390-392.

BELTRAME, C., GADDI, D. & PARIZZI, S. (in press): "A presumed hydraulic system in the Roman wreck at Grado (Italy): reconstruction and hypothesis", W. Harris (ed.): *Proceedings of Maritime Technology and the Ancient Economy: Ship-design and Navigation Conference,* Rome, 2009.

BELTRÁN DE HEREDIA, J. (2007): "*Cetariae* bajoimperiales en la costa catalana: el caso de *Barcino*", L. Lagóstena, D. Bernal & A. Arévalo (eds.): *Cetariae 2005. Salsas y salazones en Occidente en la Antigüedad,* (*BAR* (IS) 1686), Oxford, pp. 277-284.

BEN LAZREG, N., BONIFAY, M., DRINE, A. & TROUSSET, P. (1995): "Production et commercialisation des *salsamenta* de l'Afrique ancienne", P. Trousset (ed.): *L'Afrique du Nord antique et médiévale: Productions et exportations africaines, actualités archéologiques. VIe Colloque International sur l'Histoire et l'Archéologie de l'Afrique du Nord, Pau, octobre 1993,* Aix-en-Provence, pp. 103-141.

BENDER JØRGENSEN, L. (1990): "Stone-Age Textiles in North Europe", P. Walton & J.P. Wild (eds.): *North European Symposium for Archaeological Textiles,* (*NESAT* 3), York, pp. 1-10.

BENNIS, A. (1996): "Le pêche au Maroc a la première moitie du XVIe siècle", *Revue Maroc Europe* 9, pp. 51-58.

BENOÎT, F. (1958): "Nouvelles èpaves de Provence", *Gallia* 16, pp. 5-39.

BENOÎT, F. (1961): *Fouilles sous-marines. L'épave du Grand Congloué à Marseille, (Gallia* Supplément 14), Paris.

BERGER, L. & M. JOOS (1971): "Das Augster Gladiatorenmosaik", *Römerhaus und Museum Augst: Jahresbericht 1969-70,* August, pp. 25-28, 30-32.

BERNAL CASASOLA, D. (2007): "Algo más que *garum.* Nuevas perspectivas sobre la producción de las *cetariae* hispanas al hilo de las excavaciones en c/ San Nicolás (Algeciras, Cádiz)", L. Lagóstena, D. Bernal & A. Arévalo (eds.): *Cetariae 2005. Salsas y salazones en Occidente en la Antigüedad,* (*BAR* (IS) 1686), Oxford, pp. 93-107.

BERNAL, D., BLANQUEZ, J., ROLDÁN, L., DÍAZ, J.J. & PRADOS, F. (2008): "Un taller de púrpura tardorromano en *Carteia (Betica, Hispania).* Avance de las excavaciones preventivas en el conchero de Villa Victoria –2005–", C. Alfaro & L. Karali (eds.): *Purpureae Vestes. II Symposium Internacional sobre Textiles y Tintes del Mediterráneo en el Mundo Antiguo (Atenas, 2005),* Valencia, pp. 211-228.

BERNAL CASASOLA, D. (2009a): "Arqueología de las redes de pesca. Un tema crucial de la economía marítima hispanorromana", *Mainake* XXX, pp. 181-215.

BERNAL CASASOLA, D. (2009b): "Roma y la pesca de ballenas. Evidencias en el *Fretum Gaditanum*", D. Bernal (ed): *Arqueología de la pesca en el Estrecho de Gibraltar. De la Prehistoria al fin del Mundo Antiguo*, Universidad de Cádiz, Cádiz, pp. 259-285.

BERNAL CASASOLA, D. (ed.) (2009c): *Las factorías de salazón de Traducta. Primeros resultados de las excavaciones arqueológicas en la c/ San Nicolás (Algeciras, Cádiz)*, Universidad de Cádiz & Ayuntamiento de Algeciras, Algeciras, in press.

BERNAL CASASOLA, D. (ed.) (2009d): *Arqueología de la pesca en el Estrecho de Gibraltar: De la Prehistoria al fin del Mundo Antiguo,* Cádiz.

BERNAL, D., ARÉVALO, A., EXPÓSITO, J.A. & DÍAZ, J.J. (2007): "Reocupaciones del espacio y continuidad habitacional en el Bajo Imperio (ss. III y IV d.C.)", A. Arévalo & D. Bernal (eds.): *Las cetariae de Baelo Claudia. Avance de las investigaciones arqueológicas en el barrio industrial (2000-2004)*, Cádiz, pp. 455-486.

BERNAL, D., BUSTAMANTE, M., DÍAZ, J.J., GARCÍA VARGAS, E., HERNANDO, J., LAGÓSTENA, J., RAMOS, J., SÁEZ, A.M., SORIGUER, M. & ZABALA, C. (2009a): "Proyecto SAGENA. Artes de pesca en la *Baetica* en la Antigüedad Clásica", *XVII Internacional Congress of Classical Archaeology. Meeting between Cultures in the Ancient Mediterranean (Rome, 2008),* Rome, in press.

BERNAL, D., COTTICA, D. & ZACCARIA, A. (2009b): "Primera campaña del proyecto *De la pesca al garum Explotación de recursos marinos en Pompeya y Herculano*", (*Noticias de la Escuela Española de Historia y Arqueología en Roma* 3), p. 31.

BERNAL, D., COTTICA, D., ZACCARIA, A., ACQUA, C., ARÉVALO, A., BERMEJO, J., BUSTAMANTE, M., CAPPELLETTO, E., DÍAZ, J.J., JIMÉNEZ-CAMINO, R., LAGÓSTENA, J., LARA, M., LORENZO, L., SÁEZ, A.M., VARGAS, J. M. & VILLADA, F. (2009c): "El *garum* de Pompeya y Herculano (2008-2012). Síntesis de la primera campaña del proyecto hispano-italiano", *Proyectos Arqueológicos en el Exterior 2008*, Ministerio de Cultura, in press.

BERNAL, D. & SÁEZ, A.M. (2006): "*Infundibula gaditana.* Acerca de los vasos troncocónicos perforados para filtrar *garum* y otros usos industriales en la Bahía de Cádiz", *Romula* 5, Sevilla, pp. 167-218.

BERTI, F. (1990): *Fortuna Maris. La nave romana di Comacchio,* Ferrara.

BESNIER, M. (1877): "Salsamentum", *DS* 4.2.2, pp. 1022-1025.

BEUKEMA, J.J. (1970): "Angling experiments with carp (Cyprinus carpio) II. Decreasing catchability through one-trial learning", *NJZ* 20, pp. 81-92.

BIANCHINI CITTON, E., GAMBACURTA, G., RUTA SERAFINI, A. (eds.) (1998): *..."Presso l'Adige ridente"... Recenti rinvenimenti archeologici da Este a Montagnana,* Padova.

BIBBY, G. (1958): *The Testimony of the Spade,* London.

BIETTI SESTIERI, A.M. (ed.) (1992): *La necropoli laziale di Osteria dell'Osa,* Roma.

BINFORD, L. (1968): "Post-Pleistocene adaptations", S.R. Binford & L.R.Binford (eds.): *New perspectives in archaeology,* Chicago, pp. 313-341.

BLANCHARD-LEMÉE, M. (1975): *Maisons à mosaïques du quartier central de Djemila (Cuicul),* Aix-en-Provence.

BLANCO FREIJEIRO, A. (1972): *La Sevilla romana. Colonia Iulia Romula Hispalis,* Sevilla.

BLANCO FREIJEIRO, A. (1978): *Mosaicos romanos de Mérida,* CMRE 1, Madrid.

BLANCO FREIJEIRO, A. (1979): *Historia de Sevilla. I. La ciudad antigua,* Sevilla.

BLANCO FREIJEIRO, A. & LUZÓN, J.M. (1974): *El mosaico de Neptuno de Itálica,* Sevilla.

BLÁZQUEZ, J.M. (1981): *Mosaicos romanos de Córdoba, Jaén y Málaga* (*CMRE* 3), Madrid.

BLÁZQUEZ, J.M. (1982a): *Mosaicos romanos de Sevilla, Granada, Cádiz y Murcia* (*CMRE* 4), Madrid.

BLÁZQUEZ, J.M. (1982b): *Mosaicos romanos de La Real Academia de la Historia, Ciudad Real, Toledo, Madrid y Cuenca* (*CMRE* 5), Madrid.

BLÁZQUEZ, J.M. *et alii* (1989): *Mosaicos romanos de Lérida y Albacete* (*CMRE* 8). Madrid.

BLÁZQUEZ, J.M. *et alii* (1993): *Mosaicos romanos de León y Asturias* (*CMRE* 10), Madrid.

BLUE, L. (2006a): "The sedimentary history of the harbour area", D.P.S. Peacock & L. Blue (eds.): *Myos Hormos – Quseir al-Qadim. Roman and Islamic ports on the Red Sea Volume 1: Survey and Excavations 1999-2003,* Oxford, pp. 43-61.

BLUE, L. (2006b): "Trench 12", D.P.S. Peacock & L. Blue (eds.): *Myos Hormos – Quseir al-Qadim. Roman and Islamic ports on the Red Sea Volume 1: Survey and Excavations 1999-2003,* Oxford, pp. 81-84.

BLUE, L., HOCKER, F. & ENGLERT, A. (eds.) (2006): *Connected by the Sea. Proceedings of the Tenth International Symposium on Boat and Ship Archaeology, Roskilde 2003,* Oxford.

BLUE, L. & PEACOCK, D.P.S. (2006): "Trench 7A", D.P.S. Peacock & L. Blue (eds.): *Myos Hormos – Quseir al-Qadim. Roman and Islamic ports on the Red Sea Volume 1: Survey and Excavations 1999-2003,* Oxford, pp. 68-74

BLÜMNER, H. (1869): *Die Gewerbliche Thätigkeit der Völker des Klassischen Alterthums,* Leipzig.

BOCKIUS, R. (2002): "Die Schiffsfunde von Herculaneum", A. Mees & B. Pferdehirt (eds.): *Römerzeitliche Schiffsfunde in der Datenbank "Navis I" (Kataloge vor- und Frühgeschichtlicher Altertümer 29),* Mainz, pp. 164-166.

BØDKER ENGHOFF, I. (1999): "Fishing in the Baltic region from the 5th century BC to the 16th century AD: Evidence from fish bones", *Archaeofauna* 8, pp. 41-85.

BØDKER ENGHOFF, I. (2000): "Fishing in the southern North Sea region from the 1st to the 16th century AD: Evidence from fish bones", *Archaeofauna* 9, pp. 59-132.

BØDKER ENGHOFF, I. (2004): "Vikinger på smeltfiskeri den 21. april i Viborgsøerne", *Viborg-bogen*, pp. 39-44.

BØDKER ENGHOFF, I. (2005): *Viking Age freshwater fishing at Viborg. Research from an interdisciplinary research project with a special focus on methods of excavation. Paper presented at the 13th ICAZ Fish Remains Working Group Meeting*, Basel.

BOETTO, G. (2001): "Les navires de Fiumicino", J.-P. Descœudres (ed.): *Ostia: port et porte de la Rome antique*, Geneva, pp. 121-130.

BOETTO, G. (2002a): "Fiumicino 4", A. Mees & B. Pferdehirt (eds.): *Römerzeitliche Schiffsfunde in der Datenbank "Navis I" (Kataloge vor- und Frühgeschichtlicher Altertümer 29)*, Mainz, pp. 152-155.

BOETTO, G. (2002b): "Fiumicino 5", A. Mees & B. Pferdehirt (eds.): *Römerzeitliche Schiffsfunde in der Datenbank "Navis I" (Kataloge vor- und Frühgeschichtlicher Altertümer 29)*, Mainz, pp. 156-159.

BOETTO, G. (2005): "Le navi romane di Napoli", Giampaola, D. *et alii: La scoperta del porto di Neapolis: dalla ricostruzione topografica allo scavo e al recupero dei relitti*, (*AMM* 2), pp. 63-76.

BOETTO, G. (2006a): "Roman Techniques for the Transport and Conservation of Fish: the Case of the Fiumicino 5 Wreck", L. Blue, F. Hocker, & A. Englert (eds.): *Connected by the Sea. Proceedings of the Tenth International Symposium on Boat and Ship Archaeology, Roskilde, 2003*, Oxford, pp. 123-129.

BOETTO, G. (2006b): *Les navires de Fiumicino (Italie): architecture, matériaux, types et fonctions. Contribution à l'étude du système portuaire de Rome à l'époque impériale*, PhD Thesis, University of Aix-Marseille 1, Aix-en-Provence.

BOETTO, G. (2008): "Le imbarcazioni vivaio: uno studio etnoarcheologico", F. Lugli & A.A. Stoppiello (eds.): *Atti del III Convegno Italiano di Etnoarcheologia, Mondaino, 2004*, (*BAR* (IS) 1841s), Oxford, pp. 167-171.

BOETTO, G. (2009): "New archaeological evidence of the Horeia-type vessels: the Roman Napoli C shipwreck from Naples (Italy) and the boats of Toulon (France) compared", R. Bockius (ed.): *"Between the seas": Transfer and exchange in Nautical Archaeology. Proceedings of the 11th International Symposium of Boat and Ship Archaeology, Mainz, 2006*, (*RGZM, Tagungen* Bd. 3), Mainz, pp. 289-296.

BOETTO, G. (in press): "L'épave romaine Fiumicino 4 (fin du II-III siècle ap. J.-C.): navire de pêche ou petit caboteur?", H. Tzalas (ed.): *Proceedings of the 8th International Symposium on Ship Construction in Antiquity, Hydra, 2002*, (*Tropis* 8).

BOHLEN, D. (1937): *Die Bedeutung der Fischerei für die antike Wirtschaft. Ein Beitrag zur Geschichte der antiken Fischerei*, Hamburg.

BOHÓRQUEZ JIMÉNEZ, D. (1999): *El ducado de Medina Sidonia en la Edad Moderna: Chiclana de la Frontera (Demografía, Economía, Sociedad e Instituciones)* 1-2, Cádiz.

BONINO, M. (2006): "Il gabbiano. Una barca a remi di età augustea", A. Camilli, A. De Laurenzi & E. Setari (eds.): *Alkedo: Navi e commerci della Pisa romana,* Pisa, pp. 21-24.

BORREANI, M., BRUN, J.-P., LECACHEUR, P. & PASQUALINI, M. (1988): "Découverte de cinq épaves dans le port antique de Toulon (Var)", *Les nouvelles de l'archéologie* 34, pp. 33-34.

BORTOLETTO M., SPAGNOL S. & TONIOLO A., (2000): "Isole di San Francesco del Deserto e di Torcello", Soprintendenza per i Beni Ambientali e Architettonici di Venezia (ed.): *Ritrovare restaurando. Rinvenimenti e scoperte a Venezia e in Laguna,* Cornuda.

BORTOLIN, R. (2005): "Pesi da telaio e fusaiole" A. Zaccaria Ruggiu, M. Tirelli & G. Gambacurta (eds.): *Fragmenta Altino tra Veneti e Romani,* Venezia, pp. 145-147.

BOSCH I LLORET, A., CHINCHILLA SÁNCHEZ, J. & TARRÚS I GALTER, J. (eds.) (2000): *El poblat lacustre neolític de la Draga: excavacions de 1990 a 1998.* Girona.

BOUND, M. (1991): *The Giglio Wreck,* (*Enalia* supplement 1), Athens.

BOURGEOIS, A. & MAYET, F. (1991): *Belo VI: Les Sigillées,* Fouilles de Belo, Madrid.

BRANDT, A. VON (1984): *Fish Catching Methods of the World,* third edition, Farnham.

BRAUND, D. (1995): "Fish from the Black Sea: Classical Byzantium and the Greekness of trade", J. Wilkins, D. Harvey & M. Dobson (eds.): *Food in Antiquity,* Exeter, pp. 162-168.

BRESC, H. (1981): "La pêche et les madragues dans la Sicilie Médiévale", *Actes du 3e Congrès International d'Étude des cultures de la Méditerranée Occidentale,* Jerba, 1981, pp. 13-26.

BREWER, D.J. & R.F. FRIEDMAN (1990): *Fish and Fishing in Ancient Egypt,* Warminster.

BRINKHUIZEN, D.C. (1994): "Some notes on fish remains from the late 16th century merchant vessel Scheurrak SO1", W. van Neer (ed.): *Fish Exploitation in the Past. Proceedings of the 7th meeting of the ICAZ Fish Remains Working Group,* (*Annales du Musée Royal de l'Afrique Centrale, Sciences Zoologiques* 274), Tervuren, pp. 197-205.

BRUCE-MITFORD, R.L.S. (1970): "Ship's Figure-heads in the Migration Period and Early Middle Ages", *Antiquity* 44, pp. 146-148.

BRUN, J.-P. (1999): *Le Var (Carte archéologique de la Gaule),* Paris.

BRUNI, S. (ed.) (2000): *Le navi antiche di Pisa. Ad un anno dall'inizio delle ricerche,* Firenze.

BRUSCHI, T. & WILKENS, B. (1996): "Conserves de poisson á partir de quatre amphores romaines", *Archaeofauna* 5, pp. 165-169.

BUCHI, E. (1987): "Assetto agrario, risorse e attività economiche, E. Buchi(ed.): *Il Veneto nell'età romana. Storiografia, organizzazione del territorio, economia e religione* I, Verona, pp. 103-184.

BULLOCK, A. (2008): "Lo *Scarus* degli antichi. la storia dello *Sparisoma Cretense* nel I secolo d.C.", R. Gertwagen *et alii: Il mare, com'era: le interazioni tra uomo ed ambiente nel Mediterraneo dall'Epoca Romana al XIX secolo: una visione storica ed ecologica delle attività di pesca,* Rome, pp. 94-105.

BÜLOW-JACOBSEN, A. (2003): "The traffic on the road and provisioning of the stations", H. Cuvigny (ed.): *La Route de Myos Hormos. L'armée romaine dans le desert oriental d'Egypte,* Cairo, pp. 399-426.

BÜLOW-JACOBSEN, A., CUVIGNY, H. & FOURNET, J. (1994): "The identification of Myos Hormos: new papyrological evidence", *Bulletin de l'Institut Francais d'Archéologie Orientale* 94, pp. 27-42.

BUNAMANN, L. (1910): *De Piscatorum in Graecorum atque Romanorum Litteris Usu,* Aschendorf.

BURGOS-MADROÑERO, M. (2003): *Hombres de Mar, Pesca y Embarcaciones en Andalucía. La Matrícula de Mar en los Siglos XVIII y XIX (1700-1850),* Sevilla.

BUSTAMANTE ÁLVAREZ, M. (2007): *El Comercio de Terra Sigillata altoimperial en el Círculo del Estrecho. Balance historiográfico y líneas de investigación,* Unpublished dissertation, Universidad de Cádiz, Septiembre de 2007.

BUSTAMANTE ÁLVAREZ, M. (2008): "Cerámica y poder: el papel de la *terra sigillata* en la política romana", *Anales de Arqueología Cordobesa* 19, pp. 185-202.

BUTLER, A.J. (1931): *Sports in Classic Times,* London.

CAMILLI, A. (2002): *Cantiere delle navi di Pisa. La nave C "Giuditta" dallo scavo al laboratorio,* Florence.

CAMILLI, A. (2004): "Il cantiere delle navi antiche di Pisa. Note sull'ambiente e sulla periodizzazione del deposito", *AMM* 1, pp. 53-75.

CAMILLI, A. (2005): "The minor boats", A. Camilli & E. Setari (eds.): *Ancient shipwrecks of Pisa. A guide,* Venice, pp. 64-65.

CAMILLI, A., DE LAURENZI, A. & SETARI, E. (2006): *Catalogo della mostra, Pisa: Un viaggio nel mare dell'Antichità,* Pisa.

CAMPS, G. (ed.) (1998): *L'Homme préhistorique et la mer, 120e Congrès National des Sociétés Historiques et Scientifiques (Aix-en-Provence, 1995),* Paris.

CANAL, E. (1995): "Le Venezie sommerse: quarant'anni di archeologia lagunare", G. Caniato, E. Turri & M. Zanetti (eds.): *La laguna di Venezia,* Verona, pp. 193-225

CANAL, E. (1998): *Testimonianze archeologiche nella Laguna di Venezia - L'età antica,* Cavallino di Venezia (Venezia).

CAO, I. (2003): "*Pectines* altinati e *ostrea* nell'alto adriatico: suggestioni letterarie", G. Cresci Marrone & M. Tirelli (eds.): *Produzioni, merci e commerci in Altino preromana e romana, Roma,* pp. 319-329.

CARANDINI, A. *et alii* (1982): *Filosofiana. La villa de Piazza Armerina,* Roma.

CARLSON, D. (1999): "Roman fishing boats: form and function", R.F. Docter & E.M. Moormann (eds.): *Proceedings of the XVth International Congress of Classical Archaeology, Amsterdam, July 12-17, 1998,* Amsterdam, pp. 107-109.

CARLSON, D. (2002): "Roman fishing boats and the transom prow", H.E. Tzalas (ed.): *7th International Symposium on Ship Construction in Antiquity, Pylos, August 1999,* (*Tropis* 7), Athens, pp. 211-218.

CARMAN, J. (1996): *Valuing ancient things: archaeology and law,* Leicester.

CARRE, B. 1994: "Cap Camarat, Cap Camarat 2", *Bilan Scientifique du DRASSM* 1992, pp. 50-51.

CARRERA RUIZ, J.C., DE MADARIA ESCUDERO, J.L. & VIVES-FERRANDIZ SÁNCHEZ, J. (2000): "La pesca, la sal y el comercio en el Círculo del Estrecho. Estado de la cuestión", *Gerión* 18, pp. 43-76.

CARRERAS, C., AGUILERA, A., BERNI, P., GARROTE, E., PARIMON, P., MORAIS, R., MOROS, J., NIETO, X., PUIG, A., REMESAL, J., ROVIRA, R. & VIVAR, G. (2004): *Culip VIII I les àmfores Haltern 70, (Monografies del CASC* 5), Gerona.

CARRIAZO RUBIO, J.L. (2001): "Pesca, frontera y señorío: Ayamonte, de la Edad Media a la Edad Moderna", *Huelva en su Historia* 8, pp. 41-66.

CASSON, L. (1989): *The Periplus Maris Erythraei,* Princeton.

CASTANYER, P. (2006): "Les arts de pesca a Empuries", X. Aquilué & J. Monturiol (eds.): *Pescadors de l'antiga Empuréis, Catálogo de la Exposición,* Gerona, pp. 20-22.

CASTANYER, P. (2007): "Anzuelos de bronce", *Senatus Populus que Romanus. Catálogo de la Exposición,* Madrid, p. 219.

CAVALLO, D., CIAMPOLTRINI, G. & SHEPHERD, J. (1992): "La pesca nell'agro di Cosa in età romana: prospettive di ricerca e nuove acquisizioni", *V Rassegna di Archeologia Subacquea, Giardini Naxos 1990,* Messina, pp. 103-114.

CAVAZZONI, S. (1995): "La laguna: origine ed evoluzione", G. Caniato, E. Turri & M. Zanetti (eds.): *La laguna di Venezia,* Verona, pp. 41-75.

CECCHINI, F. (ed.) (1990): *Sorella anguilla. Pesca e manifattura nelle valli di Comacchio,* Bologna.

CEDERLUND, C.O. (2006): *Vasa I: The Archaeology of a Swedish Warship of 1628,* Stockholm.

CELUZZA, M.G. (1985): "Terracotta. I pesi da telaio", A. Ricci (ed.): *Settefinestre. Una villa schiavistica nell'Etruria romana. 2. La villa e i suoi reperti,* Modena, p. 69.

CERDÀ JUAN, D. (1980): *La Nave romana-republicana de la Colonia de Sant Jordi,* Palma de Mallorca.

CERRI, L. (2004-2005): *La produzione e il comercio dei salsamenta della Mauretania Tingitana,* Tesi di Dottorato, Università degli Studi di Siena, Siena.

CHABERT, J. (1972): "The Archéonaute", *UNESCO: Underwater Archaeology: A Nascent Discipline,* Paris, pp. 169-174.
CHARLIN, G., GASSEND, J.M. & LEQUÉMENT, R. (1978): "L'èpave antique de la Baie de Cavalière (Le Lavandou, Var)", *Archaeonautica,* 2, pp. 9-93.
CHARPENTIER, V., MERY, S. & PHILLIPS, C.S. (2004): "Des coquillages... outillages des Ichtyophages? Mise en evidence d'industries sur Veneridae, du Neolithique a l'age du Fer (Yemen, Oman, E.A.U.)", *AAE* 15, pp. 1-10.
CHAUCHAT, C., PELEGRIN, J., GÁLVEZ MORA, C., BECERRA URTEAGA, R. & ESQUERRE ALVA, R. (2004): *Projectile point technology and economy, a case study from Paiján, North coastal Peru,* Austin.
CHAVES TRISTÁN, F. (1979): *Las monedas hispano-romanas de Carteia,* Barcelona.
CHEVAL, C. (2008): "Protohistoric weaving, the Minoans' loom-weights: a first approach", C. Alfaro & L. Karali (eds.): *Purpureae vestes II: 2nd Iinternational Symposium on Textiles and Dyes in the Ancient Mediterranean World, Athens, 24-26 November 2005,* Valencia, pp. 19-24.
CHEVALIER, Y. & SANTAMARIA, C. (1971): "L'épave de l'Anse Gerbal à Port Vendres (Pyrénées-Orientales)", *Rivista di Studi Liguri* 37, pp. 7-32.
CHIC GARCÍA, G. (2005): "Marco Aurelio y Cómodo. El hundimiento de un Sistema Económico", *II Congreso Internacional de Historia Antigua. La Hispania de los Antoninos (98-180),* Valladolid, pp. 567-586.
CHRISTENSEN, A.E. (1995): "Boat fragments from Mangersnes", O. Olsen, J.S. Madsen & F. Rieck (eds.): *Shipshape. Essays for Ole Crumlin-Pedersen on the occasion of his 60th anniversary February 24th 1995,* Roskilde, pp. 73-80.
CIAMPOLTRINI, G. & ANDREOTTI, A. (2003): "Pesca e navigazione fluviale lungo l'Auser/Serchio in età romana. I Materiali dalla piana di Lucca", *Atti del II Convegno Nazionale di Archeologia Subacquea (Bari),* pp. 209-224.
CLARK, J.G.D. (1936): *The Mesolithic Settlement of Northern Europe,* Cambridge.
CLARK, J.G.D. (1948): "The Development of Fishing in Prehistoric Europe", *AJ* 28.
CLARK, J.G.D. (1965): *Prehistoric Europe: the economic basis,* London.
CLARKE, J.T. (1888): "On the fish Orphos", *The Classical Review* 2.4, pp. 97-99.
CLEMENTE, I. (1997): *Los instrumentos líticos del Túnel VII: una aproximación etnoarqueológica,* (*Treballs d'Etnoarqueologia* 2), Madrid.
CLEMENTE, I. & GARCÍA, V. (2008): "Yacimientos arqueológicos de la costa Atlántica de la Bahía de Cádiz; aplicación del Análisis Funcional a los instrumentos de trabajo líticos del Embarcadero del Rio Palmones, La Mesa y La Esparragosa", J. Ramos (ed.): M*emoria del Proyecto de Investigación La ocupación prehistórica de la campiña litoral y banda atlántica de Cádiz,* (*Arqueología Monografías*), Sevilla, pp. 185-198.
CLEMENTE, I., GYRIA, E.Y., LOZOVSKA, O.V. & LOZOVSKI, V.M. (2002): "Análisis de instrumentos en costilla de alce, mandíbulas de castor y caparazón de tortuga de Zamostje 2 (Rusia)", I. Clemente, J.F. Gibaja & R. Risch (eds.):

Análisis Funcional: su aplicación al estudio de sociedades prehistóricas, (*BAR* (IS) 1073), Oxford, pp. 187-196.

CLEYET-MERLE, J.J. (1990): *La Prehistoire de la Pêche,* Paris.

COLLS, D. *et alii* (1977): *L'èpave Port-Vendres II et le commerce de la Bétique à l'époque de Claude,* (*Archaeonautica* 1), Paris.

COMPÁN VÁZQUEZ, D. (1988): "La pesca marítima en Andalucía", *Geografía de Andalucía* 5, Sevilla, pp. 201-315.

CORAZZA, C., (1990): "I reperti malacologici", F. Berti (ed.): *Fortuna Maris. La nave romana di Comacchio,* Ferrara, pp. 116-117.

CORCORAN, T.H. (1957): *The Roman Fishing Industry of the Late Roman Republic and Early Empire,* PhD thesis, Northwestern University.

CORCORAN, T.H. (1964): "Fish Treatises in the Early Roman Empire", *CJ* 59, pp. 271-274.

CORREIA, V. *et alii* (1942): *Coimbra,* Coimbra.

COSTANZA, S. (1999): "Gli Uomini, il lavoro, l'economia", N. Ravazza (ed.): *La Terra delle tonnare. Atti del Convegno di San Vito Lo Capo, 3-5 settembre 1999,* Trapani, pp. 32-40.

COSTE, J.-J. (1855): *Voyage d'exploration sur le littoral de la France et de l'Italie,* Paris, (It. ed., 1989: *Industria della Laguna di Comacchio,* Bologna).

COTTE, J. (1944): *Poissons et animaux aquatiques au temps de Pline: Commentaires sur le Livre IX de l'Histoire Naturelle de Pline,* Paris.

COTTICA, D. (2003): "Dalla lana altinata al prodotto finito: filatura e tessitura in Altino romana alla luce dei resti della cultura materiale", G. Cresci & M. Tirelli (eds.): *Produzioni, merci e commerci in Altino preromana e romana,* Roma, pp. 261-283.

COTTICA D., FOZZATI L. & TIRELLI M. (eds.) (2010): "Progetto Costanziaco: un nuovo progetto per il recupero e la valorizzazione di un patrimonio (quasi) scomparso nella Laguna Nord di Venezia", *Quaderni di Archeologia del Veneto* XXV, pp. 66-75.

COTTICA, D., FOZZATI, L., TRAVIGLIA A. & GOTI VOLA, V. (2008): "Nuove ricerche sulla Laguna di Venezia in età romana", S. Gelichi (ed.): *Missioni archeologiche e progetti di ricerca e scavo dell'Università Ca' Foscari –Venezia,* Venezia, pp. 151-158.

COTTICA D., TRAVIGLIA A. & BUSATO D. (2009): "Dalla ricerca d'archivio al remote sensing: metodologie integrate per lo studio del paesaggio antico. Il caso di Costanziaco, Laguna Nord di Venezia", *Agri Centuriati* 5, pp. 33-66.

COUSTEAU, J.-Y. & DUMAS, F. (1953): *The Silent World,* London.

CROWFOOT, G.M. & CROWFOOT, E. (1961): "The Textiles and Basketry", P. Benoit, J.T. Milik & R. de Vaux (eds.): *Les grottes de Murabba Jât, Discoveries in the Judaean Desert II,* Oxford, pp. 51-63 and plates XV-XVIII.

CRUMLIN-PEDERSEN, O. & OLSEN, O. (eds.) (2002): *The Skuldelev Ships I: Topography, History, Conservation and Display,* Roskilde.

CURTIS, R.I. (1984): " *"Negotiatores Allecarii"* and the Herring", *Phoenix* 38, pp. 147-158.

CURTIS, R.I. (1984-86): "Product identification and advertising on Roman commercial amphorae", *Ancient Society* 15-17, pp. 209-228.

CURTIS, R.I. (1991): *Garum and Salsamenta. Production and Commerce in Materia Medica,* Leiden.

CURTIS, R. I. (2001): *Ancient Food Technology,* Brill, Leiden-Boston-Colonia.

CURTIS, R.I. (2005): "Sources for Production and Trade of Greek and Roman Processed Fish", Bekker-Nielsen (ed.): *Ancient Fishing and Fish Processing in the Black Sea Region,* (*Black Sea Studies* 2), Aarhus, pp. 31-46.

CUSCITO, G., (1974): *Immagini di Aquileia,* Trieste.

CUSCITO, G. (1989): *Aquileia,* Aquileia.

CUTTING, C.L. (1955): *Fish Saving. A history of fish processing from ancient to modern times,* London.

CUVIER, G.L. & VALENCIENNES, M. (1832): "Account of the common mackerel *(Scomber scombrus, Lin.)* and the *garum* of the Ancients", *Edinburgh Journal of Science n.s.* 12, pp. 286-294.

CUVIGNY, H. (2003a): "Le fonctionnement du réseau", H. Cuvigny (ed.): *La Route de Myos Hormos. L'armée romaine dans le desert oriental d'Egypte,* Cairo, pp. 295-360.

CUVIGNY, H. (ed.) (2003b): *La Route de Myos Hormos. L'armée romaine dans le desert oriental d'Egypte,* Cairo.

CUVIGNY, H. (2006): "Quelques dipinti amphoriques", V.A. Maxfield & D.P.S. Peacock (eds.): *Survey and excavation: Mons Claudianus, 3: ceramic vessels and related objects,* Cairo, pp. 176-80.

D'AGOSTINO, M. (1995-1996): "Il relitto del vetro", *Bollettino di Archeologia Subacquea* 1-2, pp. 29-89.

D'AGOSTINO, M. & FOZZATI, L. (1997): "Venezia: territorio sommerso e tutela", *Atti del Convegno Nazionale di Archeologia Subacquea,* Bari, pp. 287-298.

D'AGOSTINO, M. & MEDAS, S. (2005): "La navigazione nella laguna di Venezia in epoca romana: nuove evidenze dall'archeologia subacquea", *Rivista di Topografia Antica* XV, pp. 37-54.

D'ONOFRIO, C. (1980): *Il Tevere: l'isola tiberina, le inondazioni, i molini, i porti, le rive, i muraglioni, i ponti di Roma,* Roma.

DAGRON, G. (1995): "Poissons, pêcheurs et poissoniers de Constantinople", C. Mango & G. Dagron (eds.): *Constantinople and its hinterland. Papers from the 27th Spring Symposium of Byzantine Studies, Oxford, April 1993,* Oxford, pp. 57-73.

DAGRON, G. (2002): "The Urban Economy. Seventh-Twelfth Centuries", A.E. Laiou (ed.): *The Economic History of Byzantium: From the Seventh through the Fif-*

teenth Century (*Dumbarton Oaks Research Library and Collection* 39), Washington D.C., pp. 393-461.

DARBY, W.J., GHALIOUNGUI, P. & GRIVETTI, L. (1977): *Food – the gift of Osiris* 1-2, London.

DAVIDSON, A. (1997): *Poissons de la Méditerranée. Comment les reconnaître et les cuisiner,* Aix-en-Provence.

DAVIDSON, G.R. (1952): *Corinth. Volume XII The Minor Objects,* Princeton.

DE GROSSI MAZZORIN, J. (2000): "État de nos connaissances concernant le traitement et la consommation du poisson dans l'Antiquité à la lumière de l'archéologie; l'exemple de Rome", *MEFRA* 112, pp. 155-167.

DE HOMAN, H. (1756): *Carte Topographique des pays et côtes maritimes qui forment le Détroit de Gibraltar,* Nuremberg.

DE ROMANIS, F. (2003): "Between the Nile and the Red Sea. Imperial trade and barbarians", M. Liverani (ed.): *Arid Lands in Roman Times: Papers from the International Conference in Rome, July 9th-10th, 2001,* (*Arid Zone Archaeology* 4), Florence, pp. 117-22.

DE SAINT-DENIS, E. (1947): *Le vocabulaire des animaux marins en Latin classique,* Paris.

DE VINGO, P. & FOSSATI, A. (2001): "Gli utensili da pesca", Mannoni, T. & Murialdo, G. (eds.): *S. Antonino: Un insediamento fortificato nella Ligura bizantina,* Bordiguera, pp. 657-660.

DEAN, M. *et alii* (1992): *Archaeology Underwater, The NAS guide to Principles and Practice,* London.

DÉCHELETTE, J. (1910): *Manuel d'archéologie Prehistorique, Celtique et Gallo-Romaine,* Paris.

DEEBEN, J., HALLEWAS, D.P. & MAARLEVELD, TH.J. (2002): "Predictive modelling in Archaeological Heritage Management of the Netherlands: the Indicative Map of Archaeological Values (2nd Generation)", *Berichten van de Rijksdienst voor het Oudheidkundig Bodemonderzoek,* (*Proceedings of the National Service for Archaeological Heritage in the Netherlands* 45), pp. 9-56.

DEGANI, E. (1982): "Appunti di poesia gastronomica greca", *Prosimetrum e Spoudogeloion,* Genova, pp. 29-54.

DEISS, J.J. (1995): *The Town of Hercules: a buried treasure trove,* Malibu.

DELUSSU, F. & WILKENS, B. (2000): "Le conserve di pesce; alcuni dati da contesti italiani", *MEFRA* 112, pp. 53-65.

DESSE, J. & AUDOUIN-ROUZEAU, F. (eds.) (1993): *Exploitation des animaux sauvages à travers le temps: XIIIe Rencontres Internationales d'Archeologie et d'Histoire d'Antibes: Actes des rencontres 15-16-17 octobre 1992,* Juan-les-Pins.

DESSE, J. & DESSE-BERSET, N. (1993): "Pêche et surpêche en Méditerranée: le témoignage des os", J. Desse & F. Audouin-Rouzeau (eds.): *Exploitation des animaux sauvages à travers le temps: XIIIe Rencontres Internationales d'Archeologie et d'Histoire d'Antibes: Actes des rencontres 15-16-17 octobre 1992,* Juan-les-Pins, pp. 327-340.

DESSE-BERSET, N. (1993): "Contenus d'amphores et surpêche: l'exemple Sud-Perduto (Bouches de Bonifacio)", J. Desse & F. Audouin-Rouzeau (eds.): *Exploitation des animaux sauvages à travers le temps: XIIIe Rencontres Internationales d'Archeologie et d'Histoire d'Antibes: Actes des rencontres 15-16-17 octobre 1992,* Juan-les-Pins, pp. 341-346.

DESSE-BERSET, N. & DESSE, J. (2000): "*Salsamenta, garum,* et autres préparations de poissons; ce qu'en disent les os", *MEFRA* 112, pp. 73-97.

DI STEFANO, G. (2002): "Marmi africani e *garum* spagnolo nel Mediterraneo centrale: tracce di alcune rotte commerciali di età romana", *AR* 14, pp. 627-642.

DÖLGER, F.J. (1928): *Das Fisch-Symbol in frühchristlicher Zeit. Ichthys als Kürzung des Namen Jesu,* Münster.

DONATI, A. & PASINI, P. (ed.) (1997): *Pesca e pescatori nell'antichità,* Milan.

DORIGO, W. (1994): *Venezie sepolte nella terra del Piave. Duemila anni fra il dolce e il salso,* Roma.

DORIGO, W. (1995): "Fra il dolce e il salso: origini e sviluppi della civiltà lagunare", G. Caniato, E. Turri & M. Zanetti (eds.): *La laguna di Venezia,* Verona, pp. 137-191.

DOS SANTOS DURAN KREMER, M.J. (1999): "Contribuçâo para o estudo de alguns mosaicos romanos da *Gallaecia* e da *Lusitania*", *Actas do V Congreso Internacional de Estudios Galegos (Tréveris 1997),* Trier, pp. 509-519.

DRAGENDORFF, H. (1895-96) "*Terra Sigillata,* ein Beitrag zur Geschichte der griechischen und römischen Keramik", *Bonner Jahrbucher* 96, pp. 18-155.

DRAGENDORFF, H. & C. WATZINGER (1948): *Arretinische Reliefkeramik, mit Beschreibung der Sammlung in Tübingen,* Reutlingen.

DUHAMEL DE MONCEAU, H.-L. (1771): *Traité general des pêches et histoire des poisons qu'elles fournissent, tant pour la subsistence des homes que pour plusieurs autres usages qui ont rapport aux arts et au commerce 1,* Paris (reprinted 1998).

DUMONT, J. (1976-77): "La peche du thon a Byzance a l'époque hellenistique", *REA* 78-79, pp. 96-119.

DUMONT, J. (1977): "La pêche dans le Fayoum hellénistique: traditions et nouveautés d'après le Papyrus Tebtynis 701", *Chronique d'Égypte* 52, pp. 125-142.

DUMONT, J. (1981): *Halieutika. Recherches sur la pêche dans l'antiquité grecque,* PhD thesis, University of Limoges.

DUNBABIN, K.M.D. (1978): *The Mosaics of Roman North Africa,* Oxford.

DUNBABIN, T.J. (1948): *The Western Greeks,* Oxford.

EBERL, G. (1892): *Die Fischkonserven der Alten,* Stadtamhof.

ECKMAN, S. (1953): *Marine Zoogeography,* London.

EDMONDSON, J.C. (1987): *Two Industries in Roman Lusitania: Mining and Garum Production,* (*BAR* (IS) 362), Oxford.

EERDEN, R. VAN (2004): "Uitgeest, tunnel onder spoorbaan", *Holland* 36, pp. 48-49.

EISEMAN, C.J. & RIDGWAY, B.S. (1987): *The Porticello Shipwreck. A Mediterranean Merchant Vessel of 415-385 B.C.*, College Station, Texas.

EKROLL, Ø. (1989): "Farvel til båten?", *Arkeo* 1, pp. 19-22.

EL SHAHAWY, A. (2005): *El Museo Egipcio de El Cairo. Un paseo por los rincones del antiguo Egipto*, Farid Atiya Press, El Cairo.

ELLMERS, D. (1972): *Frühmittelalterliche Handelsschiffart in Mittel- und Nordeuropa*, Neumünster.

ELVIRA BARBA, M.A. (1981): "Los dioses romanos en la *terra sigillata* hispánica", *La Religión romana en Hispania, Symposio organizado por el Instituto de Arqueología 'Rodrigo Caro' del CSIC del 17 al 19 de diciembre de 1979*, Madrid, pp. 59-68

ERBATI, E. & TRAKADAS, A. (2008): *The Morocco Maritime Survey: a contribution to the history of Mauretania Tingitana*, (*BAR* (IS) 1890) Oxford.

ERLANDSON, J.M. & MOSS, M.L. (2001): "Shellfish feeders, Carrion eaters and the Archaeology of Aquatic Adaptations", *American Antiquity* 66 (3), pp. 413-432.

ÉTIENNE, R. (1970): "À propos de *"garum sociorum"*", *Latomus* 29, pp. 297-313.

ÉTIENNE, R., MAKAROUN, Y. & MAYET, F. (1994): *Un grand complexe industriel à Tróia (Portugal)*, Paris.

ÉTIENNE, R. & MAYET, F. (1998a): "Cartographie critique des établissements de salaisons de poisson dans la Péninsule Ibérique", É. Rieth (ed.): *Méditerranée antique. Pêche, navigation, commerce*, Paris, pp. 33-60.

ÉTIENNE, R. & MAYET, F. (1998b): "Les *mercatores* de saumure hispanique", *MEFRA* 110, pp. 147-165.

ÉTIENNE, R. & MAYET, F. (2002): *Salaisons et sauces de poisson hispaniques*, Paris.

FELDTKELLER, A. & SCHLICHTHERLE, H. (1998): "Flechten, Knüpfen und Weben in Pfahlbausiedlungen der Jungsteinzeit", *Archäologie in Deutschland* 14, pp. 22-27.

FERDI, S. (1998): *Mosaïques des eaux en Algérie*, Alger.

FERNÁNDEZ NIETO, F.J. (2002): "*Hemeroskopeion = Thynnoskopeion.* El final de un problema histórico mal enfocado", *Colonizadores e indígenas en la Península Ibérica*, (*Mainake* 14), Málaga, pp. 231-255.

FERNÁNDEZ NIETO, F.J. (2006): "Titularidad y cesión de los derechos de la pesca marítima en la antigua Grecia", H.A. Rupprecht (ed.): *Vorträge zur griechischen und hellenistischen Rechtsgeschichte. Symposion 2003*, Wien, pp. 207-232.

FERNÁNDEZ PÉREZ, J. (2002): "Consideraciones sobre la pesca romana en *Hispania*", *Artifex. Ingeniería romana en España*, Madrid, pp. 331-352.

FERNÁNDEZ URIEL, P. (1995): "La púrpura en el Mediterráneo Occidental", E. Ripoll Perelló & M.F. Ladero Quesada (eds.): *Actas del II Congreso Internacional "El Estrecho de Gibraltar" Ceuta 1990*, vol. 2., Madrid, pp. 309-327.

FERNÖ, A. & HUSE, I. (1983): "The effect of experience on the behaviour of cod *(Gadus morhua L.)* towards a baited hook", *Fisheries Research* 2 , pp.19-28.

FERREIRA, E. (1988): *Galicia en el comercio marítimo medieval*, Santiago de Compostela.

FERRER ALBELDA, E. & GARCÍA VARGAS, E. (2001): "Producción y comercio de salazones y salsas de pescado de la costa malagueña en épocas púnica y romano-republicana", F. Wulff Alonso, G. Cruz Andreotti & C. Martínez Maza (eds.): *II Congreso de Historia Antigua de Málaga. Comercio y comerciantes en la Historia Antigua de Málaga (siglo VIII a.C.-711),* Málaga, pp. 547-572.

FERRER ANDRADE, J. (1995): "La motorización de la flota pesquera portuense", *Revista de Historia de El Puerto* 11, pp. 53-78.

FERRO, R. (2007): *Ferro e acqua. Oggetti del lavoro in laguna,* Venice.

FIORENTINI, M. (2003): *Fiumi e mari nell'esperienza giuridica romana: profili di tutela processuale e di inquadramento sistematico,* (*Pubblicazioni della Facoltà di Giurisprudenza della Università di Trieste* 53), Milano.

FISCHER, A. (2004): "Submerged Stone Age – Danish examples and North Sea potential", N.C. Flemming (ed.): *Submarine prehistoric archaeology of the North Sea,* (*CBA Research Report* 141), York, pp. 23-36.

FLANNERY, K. (1969) "Origins and ecological effects of early domestication in Iran and the Near East", P.J. Ucko & G.W. Dimbleby (eds.): *The domestication and exploitation of plants and animals,* London, pp.72-100.

FLORIDO DEL CORRAL, D. & MENANTEAU, L. (2006): "Geohistoria de las almadrabas del golfo ibero-marroquí (siglos XVI-XXI)", *Historia de la pesca en el ámbito del estrecho, I conferencia internacional (1-5 de Junio de 2004, Puerto de Santa Maria Cádiz),* Sevilla, pp. 859-925.

FOERSTER LAURES, F. (1989): "The Line or Net 'Free-er' Ring", *IJNA* 14.1, pp. 80-82.

FOKKENS, H., COLES, B.J., VAN GIJN, A.L., KLEIJNE, J.P., PONJEE, H.H. & SLAPPENDEL, C.G. (eds.) (2008): *Between Foraging and Farming, an extended broad spectrum of papers presented to Leendert Louwe Kooijmans,* (*Analecta Praehistorica Leidensia* 40), Leiden.

FORSTER, G.R. (1973): "Line fishing on the continental slope. The selective effects of different hook patterns", *Journal of the Marine Biological Association of the United Kingdom* 53, pp. 749-751.

FOUCHER, L. (1960): *Inventaire des mosaïques de Sousse,* Tunis.

FOUCHER, L. (1961): "Un mosaïque de *triclinium* trouvée à *Thysdrus*", *Latomus* 20, pp. 291-297.

FOUCHER, L. (1965): "Les mosaiques nilotiques africaines", M.G. Picard & H. Stern (eds.): *Colloque international sur la mosaïque gréco-romaine. Paris, 29 août-3 septembre 1963* (= *CMGR* 1), Paris, pp. 137-145.

FOZZATI, L. & GOBBO, V. (2007): "Le indagini archeologiche nell'area urbana e nel territorio di Caorle", L. Fozzati (ed.): *Caorle archeologica tra mare, fiume e terra,* Venezia, pp. 65-121.

FOZZATI, L. & TONIOLO, A. (1998): "Argini-strade nella laguna di Venezia", S. Mattioli Pesavento (ed.): *Bonifiche e drenaggi con anfore in epoca romana: aspetti tecnici e topografici,* Modena, pp. 197-208.

FRADIER, G. (1982): *Mosaïques romaines de Tunisie,* Tunis.

FRANCIOSI, G. (2002): "Il divieto della *piscatio tinnaria:* un'altra servitù prediale?" *Revue Internationale des droits de l'Antiquité* 49, pp. 101-107.

FRANCO MORENO A. & MORENO OLLERO, A. (1981): "Datos sobre el comercio del puerto de Sanlúcar de Barrameda en el primer tercio del siglo XVI", *Actas del II Coloquio de Historia Medieval de Andalucía: Hacienda y Comercio,* Sevilla, pp. 283-296.

FROST, H. (1962): "Submarine Archaeology and Mediterranean Wreck Formations", *Mariner's Mirror* 48, pp. 82-89.

FROST, H. (1976): *Lilybaeum (Marsala) –The Punic ship: Final Excavation Report,* (*Notizie degli Scavi suppl.* 30), Roma.

FUSCO, N. (1941): *Il fondo del mare da Fiumicino a Capo Circeo con annessa carta di pesca n. 2.* Rome.

GALASSO, M. (1997): "Rinvenimenti archeologici subacquei in Sardegna sud-occidentale e nord-occidentale", AIAsub (ed.): *Atti del Convegno Nazionale di Archeologia Subacquea, Anzio 1996,* Bari, pp. 121-133.

GALILI, E. & ROSEN, B. (2008): "Fishing Gear from a 7th-Century Shipwreck off Dor, Israel", *IJNA* 37.1, pp. 67-76.

GALILI, E., ROSEN, B. & SHARVIT, J. (2002): "Fishing-gear sinkers recovered from an underwater wreckage site, off the Carmel coast Israel", *IJNA* 31.2, pp. 182-201.

GALLANT, T. (1985): *A Fisherman's Tale,* (*Miscellanea Graeca* 7), Ghent.

GALLART, L., ESCARICHE, T. & FITO, P. (2004): *La Salazón de pescado, una tradición en la dieta mediterránea,* Valencia.

GALLIAZO, V. (1979): *Bronzi romani del Museo Civico di Treviso,* Roma.

GÁLVEZ MORA, C. & QUEIROZ MORENO, C.E. (2008): "En torno a la hipótesis del uso de puntas de proyectil para cazar peces en el paijiense (*ca.* 11.000 A.P.)", *Archaeobios* 1, pp. 63-74.

GARCÍA ENTERO, V. (2004): "Nueva propuesta interpretativa de la llamada Casa de *Hippolytus* de *Complutum* (Alcalá de Henares. Madrid). Un complejo termal suburbano", *Archivo Español de Arqueología* 77, pp. 143-158.

GARCÍA VARGAS, E. (2001): "Pesca, sal y salazones en las ciudades fenicio-púnicas del sur de Iberia", J. Fernandez & B. Costa (eds.): *De la mar y de la tierra. Producciones y productos fenicio-púnicos. XV Jornadas de Arqueología Fenicio-púnica, Ibiza, 2000,* (*Trabajos del Museo arqueológico de Ibiza y Formentera* 47), Ibiza, pp. 9-66.

GARCÍA VARGAS, E. (2003): "La pesca de especies pelágicas en la antigua Bética", *Actas del III Congreso de Historia de Andalucía, Córdoba 2001,* Cordoba, pp. 473-489.

GARCÍA VARGAS, E. (2004): "Las pesquerías de la Bética durante el Imperio romano y la producción de púrpura", C. Alfaro, J.P. Wild & B. Costa (eds.): *Purpureae Vestes. I Symposium Internacional sobre Textiles y Tintes del Mediterráneo en el Mundo Antiguo (Ibiza, 2002),* Valencia, pp. 219-235.

GARCÍA VARGAS, E. (2006a): "*Garum sociorum*: pesca, salazones y comercio en los litorales gaditano y malacitano (época altoimperial romana)", *Simpósio Internacional Produçao e Comércio de Preparados Piscícolas durante a proto-história e a Época Romana no Occidente da Península Ibérica. Homenagem a Françoise Mayet, Setúbal 2004,* (*Setúbal Arqueológica* 13), Setubal, pp. 39-56.

GARCÍA VARGAS, E. (2006b): "Pesca y salazones en la Bética Altoimperial", *Historia de la pesca en el ámbito del estrecho, I conferencia internacional (1-5 de Junio de 2004, Puerto de Santa Maria Cádiz),* Sevilla, pp. 531-576.

GARCÍA VARGAS, E. (2007): "La distribución interior de las salazones de pescado (época romana)", D. Bernal Casasola & L. Lagóstena Barrios (eds.): *Cetariae 2005. Salsas y salazones de pescado en Occidente durante la Antigüedad, Actas del Congreso Internacional, Cádiz 7-9 de noviembre de 2005,* (*BAR* (IS) 1686), Oxford. pp. 527-532.

GARCÍA VARGAS, E. (2008): "Entre el consumo de lujo y el gusto popular: Las salazones de la Iberia púnica y su romanización (siglos V-I a.C.), Una perspectiva histórica y cultural", J. Napoli (ed.): *Colloque International Ressources et activites maritimes des peuples de l'Antiquite,* Université du Littoral Côte d'Opale. Boulogne, pp. 87-108.

GARCÍA VARGAS, E. & FERRER ALBELDA, E. (2001): "*Salsamenta* y *liquamina* malacitanos en época imperial romana. Notas para un estudio histórico y arqueológico", F. Wulff Alonso, G. Cruz Andreotti & C. Martínez Maza (eds.): *II Congreso de Historia Antigua de Málaga. Comercio y comerciantes en la Historia Antigua de Málaga (siglo VIII a.C.-711),* Málaga, pp. 573-594.

GARCÍA VARGAS, E. & FERRER ALBELDA, E. (2006): "Producción y comercio de salazones y salsas saladas de pescado del litoral andaluz en época fenicio-púnica. Temas y problemas", *Simpósio Internacional Produçao e Comércio de Preparados Piscícolas durante a proto-história e a Época Romana no Occidente da Península Ibérica. Homenagem a Françoise Mayet. Setúbal 2004,* (*Setúbal Arqueológica* 13), Setúbal, pp. 19-38.

GARCÍA VARGAS, E. & MUÑOZ VICENTE, A. (2003): "Reconocer la cultura pesquera de la Antigüedad en Andalucía", *Boletín del Instituto Andaluz del Patrimonio Histórico* 44, pp. 43-53.

GARCÍA VARGAS, E., ROSELLÓ, E., BERNAL CASASOLA, D. & MORALES, A. (in press): "Salazones y salsas de pescado en la Antigüedad. Un primer acercamiento a las evidencias de paleocontenidos y depósitos primarios en el ámbito europeo-mediterráneo", D. Bernal (ed.): *Las factorias de salazón de la calle San Nicolás de Algeciras (Cádiz),* Cádiz.

GARCÍA Y BELLIDO, A. (1954): "El vaso puteolano de Ampurias", *AEspA* 27, pp. 212-226.

GARCÍA Y BELLIDO, A. (1979): *Colonia Aelia Augusta Italica,* Madrid.

GAUCKLER, P. (1897): "Le Mosaique de l'Arsenal à Sousse", *Revue d'Archeologie* 2, pp. 8-22.

GAUCKLER, P. (1910): *Inventaire des mosaïques de la Gaule et de l'Afrique, 2: Afrique proconsulaire (Tunisie),* Paris.

GAUER, W. (1975): *Olympische Forschungen, 8: Die Tongefässe aus den Brunnen unter Stadion Nordwall und im Südöst-Gebiet,* Berlin.

GAZDA, E.K. & MCCANN, A.M. (1987): "Reconstruction and function: port, fishery, and *Villa*", A.M. McCann, J. Bourgeois, E.K. Gazda, J.P. Oleson & E. Lyding Will: *The Roman Port and Fishery of Cosa,* Princeton, pp. 141-155.

GIACOPINI, L., BELELLI MARCHESINI, B. & RUSTICO, L. (1994): *L'itticoltura nell'antichità,* Rome.

GIANFROTTA, P.A. (1999): "Archeologia subacquea e testimonianze di pesca", *MEFRA,* 111.1, pp. 9-36.

GIANFROTTA, P.A. & POMEY, P. (1981): *Archeologia subacquea: storia, tecniche, scoperte e relitti,* Milano.

GIL DE SOLA, L. (1999): *Ictiofauna demersal del Mar de Alborán: distribución, abundancia y espectro de tamaño,* Unpublished Ph.D. dissertation, Universidad Autónoma de Madrid.

GILMAN, A. (1975): *The Later Prehistory of Tangier, Morocco,* Cambridge, MA.

GINELLA, F. & KOCH, P. (2006): "Archäologie der römischen Binnenfischerei", H. Hüster Plogmann (eds.): *Fisch und Fischer aus zwei Jahrtausenden. Eine fischereiwirtschaftliche Zeitreise durch die Nordwestschweiz (Forschungen in Augst 39),* Basel, pp. 109-122.

GIOVANNINI, A. (1985): "Le sel et la fortune de Rome", *Athenaeum* 73, pp. 373-387.

GIOVANNINI, A. (2001): "Les salines d'Ostie", J.-P. Descœudres (ed.): *Ostia: port et porte de la Rome antique,* Geneva, pp. 36-38.

GONZÁLEZ ANTÓN, R. (2004): "Anzuelos", *Fortunatae Insulae. Canarias y el Mediterráneo. Catálogo de la Exposición,* Tenerife, p. 301.

GONZÁLEZ ANTÓN, R. & CHAVES TRISTÁN, F. (2004): *Fortunatae Insulae. Canarias y el Mediterráneo. Catálogo de la Exposición, Museo Arqueológico de Tenerife,* Tenerife.

GOULD, S.J. (1987): *Time's Arrow, Time's Cycle: Myth and Metaphor in the discovery of Geological Time,* Cambridge.

GOW, A.S.F. (1968): "On the *Halieutica* of Oppian", *The Classical Quarterly* 18.1, pp. 60-68.

GRACIA, F. (1981-1982): "Ordenación tipológica del instrumental de pesca en bronce ibero-romano", *Pyrenae* 17-18, pp. 315-328.

GRAMSCH, B. (1987): "Ausgrabungen auf dem mesolithischen Moorfundplatz bei Friesack, Bezirk Potsdam", *Veröffentlichungen des Museum für Ur- und Frühgeschichte Potsdam* 21, p. 89, pl. 25.

GRANDINETTI, G. (2000): "Manufatti in fibre vegetali", S. Bruni (ed.): *Le navi antiche di Pisa. Ad un anno dall'inizio delle ricerche*, Firenze, pp. 109-117.

GRAU ALMERO, E., PÉREZ JORDA, G., IBORRA ERES, P., RODRIGO GARCÍA, J., RODRÍGUEZ SANTANA, C.G. & CARRASCO PORRAS, S. (2001): "Gestión de Recursos y Economía", C. Aranegui Gascó (ed.): *Lixus: Colonia fenicia y ciudad púnico-mauritana anotaciones sobre su occupación medieval*, Valencia, pp. 191-230.

GRAZIANI, S. (2002): "La mattanza de San Pietro", *Chasse-marée* 149, pp. 42-51.

GREENSPAN, R.L. (1998): "Gear Selectivity Models, Mortality Profiles, and the Interpretation of Archaeological Fish Remains: a Case Study from the Harney Basin, Oregon", *Journal of Archaeological Science* 25, pp. 973-984.

GUEST-PAPAMANOLI, A. (1996): "Hunting and Trapping in Prehistory Crete: a Proposal for Ethnoarchaeological Research", D. Reese (ed.): *The Pleistocene and Holocene Fauna of Crete and its First Settlers, (Monographs in World Archaeology* 28), Madison, WI, pp. 337-349.

GUILLAUME-ALONSO, A. (2006): "Conil, año 1563, una almadraba espectacular", *Boletín La Laja* 7, http://www.lalaja.org/articulos/705.html

HAGEN, E. (2001): "Northwest African upwelling scenario", *Oceanologica Acta* 24.1, pp. 113-128.

HALD, M. (1950): *Olddanske Tekstiler, Nordiske Fortidsminder 5,* Copenhagen.

HALD, M. (1980): *Ancient Danish textiles from bogs and burials: a comparative study of costume and Iron Age textiles,* Copenhagen.

HALDANE, D. (1985): "Recent Discoveries about the Dating and Construction of Wooden Anchors", *Thracia Pontica* 3, pp. 416-427, 555-557.

HALDANE, D. (1990): "Anchors of antiquity", *Biblical Archaeologist* 53.1, pp. 19-24.

HAMILTON-DYER, S. (2001): 'The faunal remains", V.A. Maxfield & D.P.S. Peacock (eds.): *Survey and excavation Mons Claudianus, vol. 2: Excavations part 1*, Cairo, pp. 251-312.

HAMILTON-DYER, S. (2003): "Faunal remains", D.P.S. Peacock, L. Blue & S. Moser (eds.): *Myos Hormos – Quseir al-Qadim: A Roman and Islamic port on the Red Sea coast of Egypt,* Southampton, pp. 51-55.

HAMILTON-DYER, S. (2007): "Faunal remains", D.P.S. Peacock & V.A. Maxfield (eds.): *The Roman Imperial quarries: Survey and excavation at Mons Porphyrites 1994-1998, vol. 2: The excavations,* London, pp. 143-75.

HAUSCHILD, T. (1994): "Die Mosaiken am *Podium* des Wasserheiligtums com Milreu, Estoi (Portugal)", *CMGR* 4, pp. 285-291.

HAYWARD, P.J. & RAYLAND, J.S. (1995): *Handbook of the marine fauna of northwest Europe*, Oxford.

HELFMAN, G.S., COLLETTE, B.B. & FACEY, D.E. (1997): *The Diversity of Fishes,* Malden, MA.

HERMET, F. (1934): *La Graufesenque, 1: Vases sigillés,* Paris.

HESNARD, A. (1998): "Le sel des plages (*Cotta* et Tahadart, Maroc)", *MEFRA* 110, pp. 167-192.

HESNARD, A. *et alii* (1999): *Parcours de villes. Marseille: 10 ans d'archéologie, 2600 ans d'histoire,* Aix-en-Provence.

HIEBERT, F.T. (1991): "Commercial organisation of the Egyptian port of Quseir al-Qadim", *Archeologie Islamique* 2, pp 127-59.

HIGGINBOTHAM, J. (1997): *Piscinae. Artificial fishponds in Roman Italy,* Chapel Hill, NC.

HINKS, R.P. (1933): *Catalogue of the Greek, Etruscan and Roman Paintings and Mosaics in the British Museum,* London.

HINZ, K., DOSTMANN, H. & FRITSCH, J. (1982): "The continental margin of Morocco: seismic sequences, structural elements and geological development", U. von Rad, K. Hinz, M. Sarnthein & E. Seibold (eds.): *Geology of the Northwest African Continental Margin,* Berlin, pp. 34-60

HOCKER, F. (2005): "Sampling a Byzantine Vintage: Bozburun, Turkey", G.F. Bass (ed.): *Beneath the Seven Seas,* London, pp. 100-105.

HØJTE, J.M. (2005): "The Archaeological Evidence for Fish Processing in the Black Sea Region", Bekker-Nielsen, T. (ed.): *Ancient Fishing and Fish Processing in the Black Sea Region,* (*Black Sea Studies* 2), Aarhus, pp. 133-160.

HOLTORF, C. (2007): *Archaeology is a brand! The meaning of archaeology in contemporary popular culture,* Oxford.

HONORÉ, T. (1982): *Ulpian,* Oxford.

HÖPPENER, H. (1931): *Halieutica: Bijdrage tot de Kennis der oud-griekshe Visscherij,* Amsterdam.

HORDEN, P. & PURCELL, N. (2000): *The Corrupting Sea. A Study of Mediterranean History,* Oxford.

HORN, R.C. (1929): "P.S.I. 798 Fragments of documents regarding fishing", *CPh* 24, pp. 164-168.

HURST, H.R. (1994): *Excavations at Carthage: The British Mission, vol. 2.1, The Circular Harbour, North Site,* Oxford.

INRH (2002): *Ressources halieutiques: situation et niveau d'exploitation,* Casablanca.

IVERSEN, TH. (1937): "Utviklingen av fiske og fiskemetoder i Norge" *Årsberetning vedkommende Noregs Fiskerier* 1937, 4.

JACKSON, J.B.C., KIRBY, M.X., BERGER, W.H., BJORNDAL, K.A., BOTSFORD, L.W., BOURQUE, B.J., BRADBURY, R.H., COOKE, R.G., ERLANDSON, J., ESTES, J.A., HUGHES, T.P., KIDWELL, S., LANGE, C.B., LENIHAN, H.S., PANDOLFI, J.M., PETERSON, C.H., STENECK, R.S.,

TEGNER, M.J. & WARNER, R.R. (2001): "Historical Overfishing and the Recent Collapse of Coastal Ecosystems", *Science* 293, pp. 629-638.

JARDIN, C. (1961): "*Garum* et sauces de poisson de l'antiquité", *Rivista di Studi Liguri* 27, pp. 70-96.

JENKINS, I. & DIFERÍ WILLIAMS, D. (1985): "Sprang Hair Nets: Their manufacture and use in Ancient Greece", *AJA* 89, pp. 411-418.

JEREZ LINDE, J.M. (2000): "Eros y Psiqué en un fragmento de *TS* Aretina del Museo Nacional de Arte Romano de Mérida", *Anas* 13, pp. 89-100.

JONCHERAY, J.P. (1975): "Un épave du Bas Empire, Dramont F", *CAS* 4, pp. 91-140.

JONCHERAY, J.P. (1975): *L'épave C de La Chrétienne, Supplément au Cahiers d'archéologie Subaquatique.*

JONCHERAY, J.P. (1987): "L'épave G du Dramont. Notes sur six épaves de tuiles", *CAS* 6, pp. 51-84.

JONCHERAY, J.P. (2002): "Chrétienne M. Trois épaves distinctes, entre le cinquième siècle avant et le premier siècle après Jésus-Christ", *Cahiers d'archéologie subaquatique* XIV, pp. 57-130.

JONES, A.H.M. (1964): *The Later Roman Empire (284-602) 1-3,* Oxford.

JURLARO, R. (1972): "Di alcuni bicchieri romani in ceramica *sigillata* dell'officina di Norbano trovati presso il porto di Brindisi", *Faenza* 58, pp. 51-57.

KAGAN, R.L. (ed.) (1989): *Spanish Cities of the Golden Age. The Views of Anton van den Wyngaerde,* Los Angeles and Berkeley.

KANKELEIT, A. (1999): "Représentations de pêcheurs sur des mosaïques en Grece", *CMGR* 7, pp. 69-79.

KAPITÄN, G. (1984): "Ancient anchors – technology and classification", *IJNA* 13.1, pp. 33-44.

ΚΑΘΗΜΕΡΙΝΗ ΖΩΝ ΕΤΟ ΒΥΖΑΝΤΙΟ (2002): *Catalogue of the Exhibition,* Athens.

KATZEV, S.W. (1982a): "Iron Objects", G.F. Bass, & F.H. van Doorninck Jr. (eds.): *Yassi Ada: A Seventh-Century Byzantine Shipwreck,* College Station, Texas, pp. 231-265.

KATZEV, S.W. (1982b): "Miscellaneous Finds", G.F. Bass, & F.H. Van Doorninck Jr. (eds.): *Yassi Ada: A Seventh-Century Byzantine Shipwreck,* College Station, Texas, pp. 265-295.

KEAY, S., MILLET, M., PAROLI, L. & STRUTT, K. (eds.) (2005): *Portus: An Archaeological Survey of the Port of Imperial Rome,* (*Archaeological Monographs of the British School at Rome* 15), Rome.

KELLER, O. (1902-1912): *Die Antike Tierwelt 1-2,* Leipzig.

KNORR, R. (1912): *Südgallische Terra Sigillata von Rottweil,* Stuttgart.

KÖHLER, H.K.E. (1832): *Τάριχος, ou recherches sur l'histoire et les antiquités des pêcheries de la Russie méridionale,* St. Petersburg.

KOLFLÅTH, BÅRD (1986): "Rester frå bronse/jernalderkultur funne på Radøy", *Frå Fjon til Fusa* 1986, pp. 72-81.

KOLFLÅTH, BÅRD (1987): "Mangerkulturen", *Frå Fjon til Fusa* 1987, pp. 68-85

KOLFLÅTH, BÅRD (1988): "Frå Mangerkulturen. Båtfunna ved Nesvågen", *Frå Fjon til Fusa* 1988, pp. 102-105

KÖRBER-GROHNE, U. (1990): "Textiles, fishing nets, wickerwork and rope from the Neolithic sites of Hornstaad and Wangen on Lake Constance (Bodensee): botanical investigations", *NESAT* 3, pp. 11-20.

KRISTIANSEN, K. & LARSSON, T.B. (2006): *La emergercia de la sociedad del Bronce. Viajes, transmisiones y transformaciones,* Barcelona.

KUNIHOLM, P. (1982): "The fishing gear", G.F. Bass, & F.H. Van Doorninck Jr. (eds.): *Yassi Ada: A Seventh-Century Byzantine Shipwreck,* College Station, Texas, pp. 296-310.

KURLANSKY, M. (1998): *Cod: A Biography of the Fish that changed the World,* New York.

LADERO QUESADA, M.A. (1993): "Las almadrabas de Andalucía (siglos XIII-XV)", *Boletín de la Real Academia de la Historia* 190, pp. 345-354.

LAFAYE, G. (1910): *DS,* 4.1, pp. 489-494, *s.v. Piscatio et Piscatus.*

LAGÓSTENA BARRIOS, L. (2005): "Pesquerías en la *Hispania* Altoimperial. Reflexiones y perspectivas para su estudio", J. Molina & M.J. Sánchez (eds.): *III Congreso Internacional de Estudios Históricos. El Mediterráneo: la cultura del mar y de la sal,* Santa Pola, pp. 77-82.

LAGÓSTENA BARRIOS, L., BERNAL CASASOLA, D. & ARÉVALO GONZÁLEZ, A. (eds.) (2007): *Cetariae 2005. Salsas y salazones de pescado en Occidente durante la Antigüedad, Actas del Congreso Internacional, Cádiz 7-9 de noviembre de 2005,* (*BAR* (IS) 1686), Oxford.

LAMOTTA, V.M. & SCHIFFER, M.B. (2001): "Behavioral Archaeology: Toward a New Synthesis", I. Hodder (ed.): *Archaeological Theory Today,* Cambridge, pp. 14-64.

LANFRANCHI STRINA B. (ed.) (2006): *Codex Publicorum (Codice del Piovego),* vol. II, Venezia.

LARJE, R. (1995): "Favourite fish dish of the Romans in Carthage", *Archaeofauna* 4, pp. 7-26.

LASSUS, J. (1965): "Vénus marine", *CMGR* 1, pp. 175-192.

LAWRENCE, M. (1962): "Ships, Monsters and Jonah", *AJA* 66, pp. 289-296.

LE GALL, J. (1953): *Le Tibre. Fleuve de Rome dans l'Antiquité,* Paris.

LE GURUN, G. (2006): "France", S. Dromgoole (ed.): *The Protection of the Underwater Cultural Heritage. National Perspectives in Light of the UNESCO Convention 2001,* Leiden, pp. 59-95.

LEACH, F. (2006): "Fishing in Pre-European New Zealand", *Archaeofauna* 15, pp. 1-267.

LEGUILLOUX, M. (2003): *"Les animaux et l'alimentation d'apres la faune: Les restes de l'alimentation carnee des fortins de Krokodilo et Maximianon"*, H. Cuvigny (ed.): *La Route de Myos Hormos. L'armée romaine dans le desert oriental d'Egypte,* Cairo, pp. 549-588.

LEROI-GOURHAN, A. (1973): *Milieu et techniques,* second edition, Paris.

LIMANE, H. *et alii* (1998): *Volubilis. De mosaïque à mosaïque,* Casablanca.

LOCKER, A. (2007): "*In piscibus diversis,* the Bone Evidence for Fish Consumption in Roman Britain", *Britannia* 38, pp. 141-180.

LOESCHKE, S. (1909): "Keramische Funde in Haltern", *Mitteilungen der Altertumskommission für Westfalen* 5, pp. 101-190.

LØKKEBORG , S. & BJORDAL, Å. (1989): "Responses of cod *(Gadus morhua)* and haddock *(Melanogrammus aeglefinus)* to baited hooks in their natural environments", *CJFAR* 46(9), pp. 1478-1483.

LØKKEBORG, S., BJORDAL, Å. & FERNÖ, A. (1993): "The reliability of studies of fish behaviour in long-line gear research", *ICES Marine Science Symposium* 196, pp. 41-46.

LONG, L. (1987): "L'Épave Antique Bénat 4, Expertise Archéologique d'un Talus d'Amphores à Grande Profondeur", *CAS* 6, pp. 99-108.

LÓPEZ, I. (2003): *Enciclopedia de la pesca,* Tikal ediciones, Madrid.

LÓPEZ MONTEAGUDO, G. (1993): "Representaciones de ciudades en mosaicos del Norte de Africa y de *Hispania*", *AR* 10, pp. 1251-1257.

LÓPEZ MONTEAGUDO, G. (2006): "La pesca en el arte clásico", *Historia de la pesca en el ámbito del estrecho, I conferencia internacional (1-5 de Junio de 2004, Puerto de Santa Maria Cádiz),* Sevilla, pp. 219-268.

LÓPEZ MONTEAGUDO, G. (2008): "Las riquezas de las aguas en los mosaicos. Aspectos de la economía hispano-romana", *AR* 17, pp. 2547-2568.

LÓPEZ PALOMO, L.A. (2007): "El complejo arqueológico de Fuente Álamo (Puente Genil). Excavaciones actualmente en marcha", *Arte, Arqueología e Historia* 14, pp. 145-156.

LOUWE KOOIJMANS, L.P. (1987): "Neolithic Settlement and Subsistence in the Wetlands of the Rhine/Meuse Delta of the Netherlands", J.M. Coles & A.J. Lawson: *European Wetlands in Prehistory,* Oxford, pp. 227-251.

LOUWE KOOIJMANS, L.P., VAN DEN BROEKE, P.W., FOKKENS, H. & VAN GIJN, A. (2005): *The Prehistory of the Netherlands,* Amsterdam.

LOWE, B.J. (1997): *Trade and Production of Garum and its Role in the Provincial Economy of Hispania Tarraconensis,* PhD thesis, University of Edinburgh.

LUZÓN NOGUÉ, J.M. (1988): "Los *hippoi* gaditanos", *I Congreso Internacional sobre el Estrecho de Gibraltar, Ceuta 1987,* vol. 1, Madrid, pp. 445-458.

LYNSLAGER, H. (1726): *Nieuwe Paskaart ant Naauw van de Straat, opgedragen aan den Hoog Edel,* Amsterdam.

MAARLEVELD, T.J. (1998): *Archaeological heritage management in Dutch waters: exploratory studies,* Leiden.

MAARLEVELD, T.J. (2004): "Finding "new" boats: enhancing our chances in heritage management, a predictive approach", P. Clark (ed.): *The Dover Bronze Age Boat in Context. Society and water transport in prehistoric Europe,* Oxford, pp. 138-147.

MAARLEVELD, T.J. (2008): "Maritime Heritage, Mutual Heritage: Research beats Collecting", A. Tripathi (ed.): *A Marine Archaeological Perspective of the Indian Ocean: A Gateway to the Continents,* New Delhi, pp. 305-329.

MACINTOSH TURFA, J. & STEINMAYER Jr., A.G. (1999): "The Syracusia as a giant cargo vessel", *IJNA* 28.2, pp. 105-125.

MAGGIO, T. (2000): *Mattanza: love and death in the sea of Sicily,* London.

MAIA, M. GARCIA PEREIRA (2006): "La pesca, la actividade conserveira e as ânforas de Tavira", *Historia de la pesca en el ámbito del Estrecho. I Conferencia Internacional (El Puerto de Santa María, Cádiz)*, Sevilla, pp. 455-487.

MAIURI, A. (1933): *La Casa del Menandro e il suo tesoro di argenteria,* Roma.

MANNONI, T. & GIANNICCHEDA, E. (1997): *Archeologia della Produzione,* Genova.

MARCUZZI, L. (1993): *Aquileia,* Aquileia.

MAREC, E. (1958): "Trois mosaïques d'Hippone à subjets marines", *Libyca* 6, pp. 109-112.

MARGALEF, R. (ed.) (1989): *El Mediterráneo Occidental,* Barcelona.

MARTÍN DE LA CRUZ, J.C. (1994): *El tránsito del Neolítico al Calcolítico en el litoral del sur-oeste peninsular,* Madrid.

MARTIN, T. (2005) "Présigillées languedociennes de Narbonne et de Bram à Bordeaux : l'apport des fouilles récents", L. Rivet (ed.): *Actes du congrès de Blois de la SFECAG, 5-8 mai 2005,* Marseille, pp.427-447

MARTÍNEZ MAGANTO, J. (1992): "Las técnicas de pesca en la Antigüedad y su implicación en el abastecimiento de las factorías de salazón", *Cuadernos de Prehistoria y Arqueología de la Universidad Autónoma de Madrid* 19, pp. 219-244.

MARTÍNEZ MAGANTO, J. (2000): "Inscripciones sobre ánforas de salazón: interpretación sobre la estructura y significación commercial de los *tituli picti", Congreso internacional Ex Baetica Amphorae. Conservas, aceite y vino de la Bética en el Imperio Romano. Sevilla-Écija, 17 al 20 de diciembre de 1998,* vol. 4, Écija, pp. 1207-1219.

MARTIN-KILCHER, S. (1990): "Fischsaucen und Fischkonserven aus dem römischen Gallien", *Archäologie der Schweiz* 13, pp. 37-44.

MARTINUSSEN, A.O. (1993): "Teknologi i arbeidsliv – innovasjonsprosesser og deres betydning. Hva gjør museene", *Museumsnettverk 3 – Teknologi, arbeidsliv og industrihistorie,* Oslo, pp. 27-36.

MARTINUSSEN, A.O. (2004): "Bruk av open båt. Tradisjonell bruk av Oselvaren" *Havstrilen volume,* Årbok, pp. 34-45.

MARTINUSSEN, A.O. (2005): "Vidareføring av handlingsboren kunnskap" *Maihaugen 100 år – evig ung: Årbok* 2004, pp. 191-200.

MARTINUSSEN, A.O. (2006a): "Introduction", *MAST* 5.1, pp. 7-11

MARTINUSSEN, A.O. (2006b): "Nylon fever. Technological Innovation, Diffusion and Control in Norwegian Fisheries during the 1950s", *MAST* 5.1, pp. 29-44.

MARTINUSSEN, A.O. (2007): "Norsk handverksutvikling si stipendiatordning i tradisjonelle handverk. Prøveordning for høgare utdanning av handverkarar 1995-2006", *Maihaugens Årbok* 2007, pp. 53-65.

MASTON, R.B. (1901): *Supplement to the Bibliotheca Piscatoria,* London.

MASTROMARCO, G. (1998): "La pesca del tonno nella Grecia antica: dalla realtà quotidiana alla metafora poetica", *Rivista di Cultura Classica e Medioevale* 1-2, pp. 229-236.

MAXFIELD, V.A. & PEACOCK, D. P. S. (1996): "The Archaeology of an Industrial Landscape: An Interim report on the work of the Imperial Quarries *(Mons Porphyrites)* project", O.E. Kaper (ed.): *Life on the Fringe: Living in the Southern Egyptian Deserts During the Roman and Early Byzantine Periods,* Leiden, pp. 181-196.

MAYET, F. & TAVARES DA SILVA, C. (1998): *L'atelier d'amphores de Pinheiro (Portugal),* Paris.

MAYET, F. & TAVARES DA SILVA, C. (2002): *L'atelier d'amphores d'Abul,* Paris.

MAZARAKIS, A. (2002): "Recent excavations at *Oropos* (northern Attica)", M. Stamatopoulou & M. Yeroulanou (eds.): *Excavating Classical Culture: Recent Archaeological Discoveries in Greece,* Oxford, pp. 149-178.

MÉDARD, F. (2003): "La produzione di filo nei siti lacustri del Neolitico", M. Bazzanella, A. Mayor, L. Moser & A. Rast-Eicher (eds.): *Textiles: Intrecci e tessuti della preistoria europea. Museo Civico di Riva del Garda-La Roca, 24 maggio-19 ottobre 2003,* Trento pp. 79-86.

MEDEROS, A. & ESCRIBANO, G. (1999): "Pesquerías gaditanas en el litoral atlántico norteafricano", *Rivista di Studi Fenici* 27, pp. 93-113.

MEDRI, M. (1993): *Terra Sigillata Tardoitálica decorata,* Roma.

MEES, A. & PFERDEHIRT, B. (eds.) (2002): *Römerzeitliche Schiffsfunde in der Datenbank "Navis I" (Kataloge vor- und Frühgeschichtlicher Altertümer 29),* Mainz.

MEIGGS, R. (1973): *Roman Ostia,* Oxford.

MENASANCH DE TOBARUELA, M. (2007): "*Baria* tardoantigua. Cambios sociales y económicos del siglo V al siglo X", *Actas de las Jornadas sobre la Zona Arqueológica de Villaricos (Almería, 2005),* Sevilla, pp. 131-167.

MERINO, J.M. (1997): *La Pesca,* Vitoria.

MERLIN, A. (1930): "Submarine Discoveries in the Mediterranean", *Antiquity* 4, pp. 405-414.

MEZQUÍRIZ DE CATALÁN, M.A. (1961): *Terra sigillata hispánica,* Valencia.

MOLINA FAJARDO, F. (2000): "Las factorías de salazón de pescado", *Almuñécar romana,* Granada, pp. 129-185.

MOMBER, G. (2004): "The inundated landscapes of the western Solent", N.C. Flemming (ed.): *Submarine prehistoric archaeology of the North Sea, (CBA Research Report* 141), York, pp. 37-42.

MONTAÑÉS, M., PÉREZ, M., GARCÍA, M.E. & RAMOS, J. (1999): "Las primeras sociedades campesinas. Las sociedades comunitarias y los comienzos de la jerarquización social", J. Ramos, M. Montañés, M. Pérez, V. Castañeda, N. Herrero, M.E. García & I. Cáceres (eds.): *Excavaciones arqueológicas en La Mesa (Chiclana de la Frontera, Cádiz). Campaña de 1998. Aproximación al estudio del proceso histórico de su ocupación, (Arqueología en Chiclana de la Frontera* 1*)*, Chiclana de la Frontera, pp. 111-134.

MORALES, A. (1996): "Algunas consideraciones teóricas en torno a la fauna como indicadora de espacios agrarios en la Prehistoria", *Trabajos de Prehistoria* 53.2, pp. 5-17.

MORALES, A., ANTIPINA, Y., ANTIPINA, A. & ROSELLÓ, E. (2007): "An Ichthyoarchaeological Survey of the Ancient Fisheries from the Northern Black Sea Coast", *Archaeofauna* 16, pp. 117-172.

MORALES, A. & ROSELLÓ, E. (1988): "La riqueza del Estrecho de Gibraltar como inductor potencial del proceso colonizador en la Península Ibérica", *I Congreso Internacional El Estrecho de Gibraltar, Tomo IV*, UNED, Madrid, pp. 447-457.

MORALES, A. & ROSELLÓ, E. (2005/2006): "Ictiofaunas Musterienses de la Península Ibérica: ¿Evidencias de pesca Neanderthal?", *Munibe (Antropología-Arkeologia)* 57.1, pp. 183-195.

MORALES, A. & ROSELLÓ, E. (2006): "La pesca en la antigüedad andaluza: una perspectiva biológica", *Historia de la pesca en el ámbito del estrecho, I conferencia internacional (1-5 de Junio de 2004, Puerto de Santa Maria Cádiz)*, vol. 1, Sevilla, pp. 41-79.

MORALES, A. & ROSELLÓ, E. (2007): "Los atunes de *Baelo Claudia* y Punta Camarinal (s. II a.C.). Apuntes Preliminares", A. Arévalo & D. Bernal (eds.): *Las Cetariae de Baelo Claudia,* Sevilla, pp.491-500.

MORALES, A. & ROSELLÓ, E. (2008): "20,000 years of fishing in the Strait", T.C. Rick & J.M. Erlandson (eds.): *Archaeology, Historical Ecology, and Human Impacts on Marine Environments,* Berkeley, pp. 243-277.

MORALES, A., ROSELLÓ, E. & HERNÁNDEZ, F. (1998): "Late Upper Paleolithic Subsistence Strategies in Southern Iberia: Tardiglacial Faunas from Cueva de Nerja (Málaga, Spain)", *European Journal of Archaeology* 1.1, pp. 9-50.

MORALES, D.C., ROSELLÓ, E. & MORALES, A. (2009): "Pesquerías medievales hispanas: las evidencias arqueofaunísticas", *La Pesca en la Edad Media,* Madrid, pp. 145-156.

MORAN, E. (1996): "An agenda for anthropology", E. Moran (ed): *Transforming Societies, Transforming Anthropology,* Ann Arbor, pp. 1- 24.

MOREL, J-P., RONDI-COSTANZO, C. & UGOLINI, D. (2000): *Corallo di ieri, corallo d'oggi, Atti del Convegno (Ravello, 1996), (Scienze e Materiali del Patrimonio Culturale* 5), Bari.

MORELLI, C., CARBONARA, A., FORTE, V., GROSSI, C. & ARNOLDUS-HUYZENDVELD, A. (in press): "La topografia romana dell'agro portuense alla luce delle nuove indagini", S. Keay & L. Paroli (eds.): *Recent research at Portus and in its hinterland: Workshop organised at the BSR, Rome, 2008,* in press.

MORELLI, C., OLCESE, O. & ZEVI, F. (2004): "Scoperte recenti nelle saline portuensi *(Campus salinarum romanarum)* e un progetto di ricerca sulla ceramica di area ostiense in età repubblicana", A. Gallina Zevi & R. Turchetti (eds.): *Méditerranée occidentale antique: les échanges, III Seminario ANSER, Marseille, 2004,* Soveria Mannelli, pp. 43-55.

MORENO, R. (1994): "Los moluscos", Roselló, E. & A. Morales (eds.): *Castillo de Doña Blanca: Archaeo-environmental investigations in the Bay of Cádiz. Spain (750-500 B.C.),* (*BAR* (IS) 593), Oxford, pp. 143-182.

MORENO PÁRAMO, A. & ABAD CASAL, L. (1971): "Aportaciones al estudio de la pesca en la antigüedad", *Habis* 2, pp. 209-222.

MOTTA, R. (1986): "Il fiume e l'assetto topografico urbano e territoriale nel Medio Evo", *Tevere un'antica via per il Mediterraneo, catalogo mostra,* Rome, pp. 120-126.

MÜLLER, S. (1897): *Nordische Altertumskunde,* Strasbourg.

NADEL, D. *et alii* (1994): "19,000-Year-Old Twisted Fibres from Ohalo II", *Current Anthropology* 35, pp. 451-458.

NIETO, X., JOVER, A., IZQUIERDO, P., PUIG, A., ALAMINOS, A., MARTIN, A., PUJOL, M., PALOU, H. & COLOMER, S. (1989): *Excavacions Arqueològiques subaquàtiques a Cala Culip I,* Gerona.

NIETO, X. & PUIG, A.M. (2001): *Excavacions arqueològiques subaquàtiques a Cala Culip, 3. Culip IV: la Terra Sigil·lata decorada de La Graufesenque* (*Monografies del CASC* 3), Gerona.

NIETO, X. & PUJOL, M. (1989): "Objectes del vaixell i de la tripulació", AA.VV.: *Excavacions Arqueològiques subaquàtiques a Cala Culip I,* Gerona, pp. 209-219.

NIMA (1994): *Sailing directions (en route), West Coast of Europe and Northwest Africa,* Bethesda, MD.

NOCETE, F. (ed.) (2004): *Odiel. Proyecto de investigación arqueológica para el análisis del origen de la desigualdad social en el Suroeste de la Península Ibérica,* Sevilla.

NORDGAARD, O. (1928): "Utviklingsfaser i norsk fiskeri", special edition of Nidaros 1928, pp. 6-7

NORDSTRAND, L. (1980): *Fiskeridirektoratet og den tekniske utvikling i norsk fiske på 1900-talet,* unpublished paper.

OLESON, J.P. & STEIN, R. (2007): "Comment on an Article Concerning the Hydraulic System of the Roman Wreck at *Grado,* Gorizia, Italy", *IJNA* 36.2, pp. 415-417.

OLIVER, A. (1994): *El poblado ibérico del Puig de la Misericòrdia de Vinaròs,* Vinaròs (Alicante).

OLIVER NARBONA, M. (1982): *Las almadrabas de la costa alicantina,* Alicante.

OLSEN, O.M. (1998): *Bymann og fiskar? Ei analyse av fiskereiskap frå mellomalderen funne i Bergen,* Master's dissertation, University of Bergen.

ORDÓÑEZ AGULLÁ, S. (2003): "El puerto romano de *Hispalis.* Puertos fluviales antiguos: ciudad, desarrollo e infraestructuras", *IV Jornadas de Arqueología Subacuática,* Valencia, pp. 59-79.

ORSI, J.A., WERTHEIMER, A.C. & JAENICKE, W. (1993): "Influence of selected hook and lure types on catch, size and mortality of commercially troll-caught Chinook salmon", *North American Journal of Fisheries Management* 13, pp. 709-722.

ØRSTED, P. (1998): "Salt, fish and the sea in the Roman Empire", I. Nielsen & H. Sigismund Nielsen (eds.): *Meals in a social context: Aspects of the Communal Meal in the Hellenistic and Roman World,* Aarhus, pp. 13-35.

OSSOWSKI, W. (1999a): *Studia nad Lodziami jednopiennymi z obszaru polski,* Gdansk.

OSSOWSKI, W. (1999b): "Some results of the study of logboats in Poland", J. Litwin (ed.): *Down the river to the sea. Proceedings of the 8th International Symposium of Boat and Ship Archaeology, Gdansk, 1997,* Gdansk, pp. 59-66.

OSWALD, F. (1931): *Index of Potter's Stamps on Terra Sigillata (Samian Ware),* London.

OSWALD, F. (1936-37): *Index of figure-types on Terra Sigillata "Samian Ware",* Liverpool.

OSWALD, F. & PRYCE, D. (1920): *An introduction to the Study of Terra Sigillata,* London.

OUT, W.A. (2009): *Sowing the seed?: human impact and plant subsistence in Dutch Wetlands during the Late Mesolithic and Early and Middle Neolithic (5500-3400 cal BC),* Dissertation, Leiden University Press.

OWEN, J.F. & MERRICK, J.R. (1994): "Analysis of coastal middens in South eastern Australia: selectivity of angling and other fishing techniques related to Holocene deposits", *JAS* 21, pp. 11-16.

PÄFFGEN, B. & ZANIER, W. (1994): "Kleinfunde aus Metall", G. Hellenkemper Salies, H.-H. von Prittwitz und Gaffron & G. Bauchhenß: *Das Wrack: Der antike Schiffsfund von Mahdia,* Köln, pp. 111-130.

PÄLSI, S. (1920): "Ein steinzeitlicher Moorfund", *Finska Fornminneföreningens Tidsskrift* 28.2, 1-19.

PANVINI, R. (2001): *The Archaic Greek Ship at Gela,* Caltanissetta.

PAPPALARDO, U. (1990): "L'eruzione pliniana del Vesubio nel 79 d.C.: ercolano", Albore Libadie, C. & Widemann, F. (eds.): *Volcanology & Archeology, PACT* 25, pp. 197-215.

PARKER, A.J. (1992): *Ancient Shipwrecks of the Mediterranean and the Roman Provinces,* (*BAR* (IS) 580), Oxford.

PASSELAC, M. (1993) : "Céramique présigillée sud-gauloise", *Lattara* 6, pp. 532-535.

PEACOCK, D.P.S. (1993): "The Site of Myos Hormos: A View from Space", *JRA* 6, pp. 226-32.

PEACOCK, D.P.S. & BLUE, L. (eds.) (2006): *Myos Hormos – Quseir al-Qadim. Roman and Islamic ports on the Red Sea Volume 1: Survey and Excavations 1999-2003,* Oxford.

PEACOCK, D.P.S. & BLUE, L. (eds.) (forthcoming): *Myos Hormos – Quseir al-Qadim. Roman and Islamic ports on the Red Sea Volume 2: Maritime aspects,* Oxford.

PEACOCK, D.P.S. & MAXFIELD, V.A. (eds.) (1997): *Survey and Excavations. Mons Claudianus 1987-93, vol. 1: Topography and Quarries,* Cairo.

PEDERSEN, R.K. (1996): *Waterschip ZN 42': a clenched-lap fishing vessel from Flevoland, the Netherlands,* Lelystad.

PELLANDRA, I. (1997): "Due poco note peschiere romane a Santa Severa e a Santa Marinella", *ASubacq* 2, pp. 21-33.

PELLIZZATO, M. & GIORGIUTTI, E. (1997): *Attrezzi e sistemi di pesca nella provincia di Venezia,* Venezia.

PÉREZ, M. (2004): *Primitivas comunidades aldeanas en Andalucía,* Electronic publication, ProQuest Information and Learning, Spain.

PÉREZ, M. (2005): "Sociedades cazadoras-recolectoras-pescadoras y agricultoras en el Suroeste: una propuesta para un cambio social", *Arqueología y Territorio* 2, pp. 153-168.

PÉREZ, M., RAMOS, J., VIJANDE, E. & CASTAÑEDA, V. (2005): "Informe preliminar de la excavación arqueológica de urgencia en el asentamiento prehistórico de La Esparragosa (Chiclana de la Frontera, Cádiz)", *Anuario Arqueológico de Andalucía 2002,* vol. 3, pp. 93-103.

PÉREZ DE MESSA, D. (1595): *Primera y segunda parte de las cosas memorables de España,* Alcalá de Henares.

PIERCY, G.V & BASS, G.F. (2004): "Fishing Gear", G.F. Bass, S.D. Matthews, J.R. Steffy & F.H. van Doorninck Jr. (eds.): *Serçe Limani: An Eleventh-Century Shipwreck, 1: The Ship and its Anchorage, Crew, and Passengers,* College Station, Texas, pp. 399-435.

PINEDO REYES, J. (2004): "Plomos de pesca", *Scombraria. La Historia oculta bajo el mar. Arqueología submarina en Escombreras, Cartagena, Catálogo de la Exposición,* Murcia, p. 172.

PITCHER, T.J. (1993): *Behaviour of Teleost fishes* (second edition), London.

PLAYÀ GUIRADO, R.M. (1998): "Exploitation des domaines marins par les Ibères de la côte catalane (VI-II s. av. J.-C.)", *L'Homme préhistorique et la mer, 120 Congrès National des Sociétés Historiques et Scientifiques (Aix-en-Provence, 1995),* Paris, pp. 395-406.

POINSSOT, C. (1965): "Quelques remarques sur les mosaïques de la Maison de Dionysos et d'Ulysse à Thugga (Tunisie)", *CMGR* 1, Paris, pp. 219-230.

POMATA, S. & SANNA, A. (1998): "La tradizione cantieristica di Carloforte", M. Marzari (ed.): *Navi di legno: evoluzione tecnica e sviluppo della cantieristica nel Mediterraneo dal XVI secolo ad oggi, Atti del Convegno internazionale, Grado 1997,* Trieste, pp. 273-280.

POMEY, P. (1995): "Les épaves grecques et romaines de la place Jules-Verne à Marseille", *CRAI,* pp. 459-484.

POMEY, P. (1998): "Les épaves grecques du VIe siècle av. J.-C. de la Place Jules-Verne à Marseille", P. Pomey & E. Rieth (eds.): *Construction navale maritime et fluviale. Approches archéologique, historique et ethnologique, Proceedings of the 7th International Symposium of Boat and Ship Archaeology, Île Tatihou, 1994,* (*Archaeonautica* 14), Paris, pp. 147-154.

POMEY, P. (2000): "Un témoignage recent sur la pêche au corail à Marseille à l'époque archaïque", J.-P. Morel, C. Rondi-Costanzo & D. Ugolini (eds): *Corallo di ieri. Corallo di oggi, Atti del Convegno, Ravello, 1996,* Bari, pp. 37-39.

POMEY, P. (2003): "Reconstruction of Marseilles sixth century BC Greek ships", C. Beltrame (ed.): *Boats, Ships and Shipyards, Proceedings of the nineth International Symposium of Boat and Ship Archaeology, Venice, 2000,* Oxford, pp. 57-65.

POMEY, P. (2009): "A new approach of Mediterranean nautical archaeology: harbour, river and river-sea boats", R. Bockius (ed.): *Between the seas: Transfer and exchange in Nautical Technology, Proceedings of the 11th International Symposium of Boat and Ship Archaeology, Mainz 2006,* (*RGZM Tagungen,* Bd. 3), Mainz, pp. 267-280.

POMEY, P. (in press): "L'épave Jules-Verne 8 et autres données en archéologie navale", H. Tzalas (ed.): *Proceedings of the 10th International Symposium on Ship Construction in Antiquity, Hydra,* (*Tropis* 10), Athens.

POMEY, P. & TCHERNIA, A. (2006): "Les inventions entre l'anonimat et l'exploit: le pressoir à vis et la Syracusia", E. Lo Cascio (ed.): *Innovazione tecnica e progresso economico nel mondo romano, Atti degli Incontri capresi di storia dell'economia antica, Capri, 2003,* Bari, pp. 81-99.

PONSICH, M. (1964a): "Exploitations agricoles romaines de la région de Tanger", *BAM* 5, pp. 235-252.

PONSICH, M. (1964b): "Contribution à l'Atlas archéologique du Maroc: Région de Tanger", *BAM* 5, pp. 253-290.

PONSICH, M. (1966): "Contribution à l'Atlas archéologique du Maroc: Région de *Lixus,*" *BAM* 6, pp. 377-423.

PONSICH, M. (1970): *Recherches archéologiques à Tanger et dans sa region,* Paris.

PONSICH, M. (1981): *Lixus: le quartier des temples,* Rabat.

PONSICH, M. (1988): *Aceite de oliva y salazones de pescado. Factores geo-económicos de Betica y Tingitania,* Madrid.

PONSICH, M. & TARRADELL, M. (1965): *Garum et industries antiques de salaison dans la Méditerranée occidentale,* Paris.

POSAC MON, C. (1998): "*Sigillata* itálica hallada en Ceuta", *AA* 34, pp. 45-56.

POTIER, E. (1918): *DS,* 4.2, pp. 850-853, *s.v. Rete.*

POWELL, J. (1996): *Fishing in the Prehistoric Aegean,* Jonsered.

PROWSE, T., SCHWARCZ, H.P., SAUNDERS, S., MACCHIARELLI, R. & BONDIOLI, L. (2004): "Isotopic paleodiet studies of skeletons from the Imperial Roman-age cemetery of *Isola Sacra.* Rome, Italy", *JAS* 31, pp. 259-272.

PULAK, C. (1998): "The Uluburun shipwreck: an overview", *The International Journal of Nautical Archaeology* 27.3, pp. 188-224.

PURCELL, N. (1995): "Eating fish. The paradoxes of sea-food", J. Wilkins, D. Harvey & M. Dobson (eds.): *Food in Antiquity,* Exeter, pp. 132-149.

PURPURA, G., (2007): " *"Servitus thynnos non piscandi"* (D. 8.4.13 pr.)", F. D'Ippolito (ed.): *Filia scritti per Gennaro Franciosi,* Napoli, pp. 2163-2174.

PURPURA, G. (2008): "*Liberum mare,* acque territoriali e riserve di pesca nel mondo antico", J. Napoli (ed.): *Ressources et activités maritimes des peuples de l'Antiquité. Actes du Colloque International de Boulogne-sur-Mer, 12, 13 et 14 Mai 2005,* Boulogne, pp. 533-554.

RADCLIFFE, W. (1921): *Fishing from the earliest times,* Chicago.

RADIC ROSSI, I. (2006): "Due testimonianze particolari sull'economia marinara nella Dalmazia romana", I. Radic Rossi (ed.): *Archeologia subacquea in Croazia: Studi e ricerche,* Venezia, pp. 46-57.

RAMON, J. (1995): *Las ánforas fenicias y púnicas del Mediterráneo central y occidental,* Barcelona.

RAMON, J., SÁEZ, A., SÁEZ, A.M. & MUÑOZ, A. (2007): *El taller alfarero tardoarcaico de Camposoto (San Fernando, Cádiz), Arqueología Monografías,* Sevilla.

RAMOS, J. (2004): "El poblamiento calcolítico en la Banda atlántica de Cádiz. Aproximación a la sociedad clasista inicial del III[er.] milenio a.n.e.", *Las primeras sociedades metalúrgicas en Andalucía: III Simposio de Prehistoria, Cueva de Nerja,* Málaga, pp. 352-360.

RAMOS, J. (ed.) (2008): *Memoria del Proyecto de Investigación La ocupación prehistórica de la campiña litoral y banda atlántica de Cádiz,* Sevilla.

RANKE, L. VON (1885): *Geschichte der romanischen und germanischen Völker von 1494 bis 1514,* third edition, Leipzig.

RASCÓN, S. (2007): "La así llamada Casa de *Hippolytus:* la fundación de los Anios y la *schola* de una agrupación colegial de la ciudad romana de *Complutum*", *AEspA* 80, pp. 119-152.

RASCÓN, S. *et alii* (1995-1997): "*Hippolytus:* estudio de un nuevo mosaico del género de pesca y con inscripción procedente de *Complutum,* Alcalá de Henares. Madrid", *Lucentum* 14-16, pp. 39-62.

RAST-EICHER, A. (1992): "Neolitische Textilien im Raum Zürich", L. Bender Jørgensen, & E. Munksgaard (eds.): *Archaeological Textiles in Northern Europe: Report from the 4th NESAT Symposium 1-5 May 1990 in Copenhagen,* (*Tidens Tand* 5), Copenhagen, pp. 9-19.

RAST-EICHER, A. (2003): "Scheda n. 36, 37, 44", M. Bazzanella, A. Mayor, L. Moser & A. Rast-Eicher (eds.): *Textiles. Intrecci e tessuti della preistoria europea. Museo Civico di Riva del Garda-La Roca, 24 maggio-19 ottobre 2003,* pp. 221 and 237.

RAVAZZA, N. (ed.) (1999): "La terra delle Tonnare", *Atti del Convengo di San Vito Lo Capo, 3-5 settembre 1999,* Trapani, pp. 32-40.

RAVAZZA, N. (2000): *L'ultima muciara,* Trapani.

REGUEIRA, J. & REGUEIRA, E. (1993): *Túnidos y tunantes en las almadrabas de las costas gaditanas,* Algeciras.

REINDERS, H.R., VAN VEEN, H., VLIERMAN, K. & ZWIERS, P.B. (1986): *Flevobericht nr. 140. Het wrak van een 16e eeuws vissersschip in flevoland, het onderzoek van een visserschip gevonden op kavel W 10 in Flevoland,* Lelystad.

RENDINI, P. (1997): "Vasi per la pesca del polpo?", *Atti del Convegno Nazionale di Archeologia Subacquea,* Bari, pp. 75-78.

RICARD, R. (1927): "La côte atlantique du Maroc au début du XVIe siècle d'après des instructions nautiques portugaises", *Hespéris* 7, pp. 229-258.

RICARD, R. (1955): *Études sur l'histoire des Portugais au Maroc,* Coimbra.

RIMANTIENË, R. (1995): *Lietuva iki Kristaus,* Vilnius.

ROBERT, L. (1950): "Pécheurs de Parion", *Hellenica* 9, pp. 81-97.

RODRIGO, M.J. (1994): "Remains of *Melanogrammus aeglefinus* (Linnaeus, 1758) in the Pleistocene-Holocene passage in the cave of Nerja, Málaga/Spain", *Offa* 51, pp. 348-351.

RODRÍGUEZ RODA, J. (1964): "Biología del atún (*Thunnus thynnus L.*) de la costa sudatlántica española", *Investigaciones Pesqueras* 25, pp. 33-146.

RODRÍGUEZ RODA, J. (1973): "Descripción de la pesquería de atún rojo *Thunnus thynnus (L.)* de almadraba", *ICCAT, Colección de documentos científicos* 11, Madrid, pp. 401-404.

RODRÍGUEZ SANTANA, C.G. & RODRIGO GARCÍA, M.J. (2005): "Las ictiofaunas arqueológicas", C. Aranegui & M. Habibi (eds.): *Lixus-2. Ladera Sur,* Valencia, pp. 241-252

ROLAND, H. (1958): "Information archéologique. Aix-en-Provence (Partie Nord)", *Gallia* 16, pp. 406-412.

ROSADA, G. (1992): "Aggregazioni insediative e strutture urbane" L. Cracco Ruggini *et alii* (eds.): *Storia di Venezia, I. Origini, età ducale,* Roma, pp. 209-268.

ROSELLÓ, E. (1989): *Arqueoictiofaunas ibéricas. Aproximación metodológica y biocultural.* Madrid.

ROSELLÓ, E. & BRINKHUIZEN, D. (1994): "Laminak II/Spain: Alternative taxonomies as approaches to the interpretation of a fish fauna", *Offa* 51, pp. 401-409.

ROSELLÓ, E. & MORALES, A. (1988): "Ictiofaunas de yacimientos costeros ibéricos: patrones de agrupamiento con ayuda de técnicas multivariantes e implicaciones paleoculturales", *I Congreso Internacional Internacional sobre el Estrecho de Gibraltar, Ceuta 1987*, vol. 4, Madrid, pp. 459-472.

ROSELLÓ, E. & MORALES, A. (1992): "Grouping patterns in Iberian ichthyoarchaeological assemblages from coastal sites", *Archaeofauna* 1, pp. 11-22.

ROSELLÓ, E. & MORALES, A. (1994): "The fishes", E. Rosello & A. Morales (eds.): *Castillo de Doña Blanca: Archaeo-environmental investigations in the Bay of Cádiz. Spain (750-500 B.C.)*, (*BAR* (IS) 593), Oxford, pp. 143-182.

ROSELLÓ, E. & MORALES, A. (eds.) (1994): *Castillo de Doña Blanca: Archaeo-environmental investigations in the Bay of Cádiz. Spain (750-500 B.C.)*, (*BAR* (IS) 593), Oxford.

ROSELLÓ, E. & MORALES, A. (2008): *Evidencias de Pesca en Santimamiñe*, unpublished interim report, Laboratorio de Arqueozoologia de la UAM, Madrid.

ROSELLÓ, E., MORALES, A. & POPOV, S.V. (2005): "Gihayu: A Late Stone Age Fishing Station in the coast of Yemen", *Paléorient* 31.1, pp. 116-125.

ROSSI, R. (1990): "Gli attrezzi da pesca", Berti, F. (ed.): *Fortuna Maris. La nave romana di Comacchio*, Bolonia, pp. 114-115.

ROUILLARD, P. (1992): "Le commerce grec du Ve et du IVe siècle av. J.-C. dans les régions de *Lixus* et *Gadés*", *Lixus. Actes du colloque organize par l'Institut des Sciences de l'Archéologie et du Patrimoine de Rabat avec le concours de l'École française de Rome, Larache, 8-11 novembre 1989*, Rome, pp. 207-215.

ROUX, J. (1764): *Carte de la mer Mediterranée en douze feuilles*, Marseille.

ROYO, J.I. & ACÍN, J.L. (eds.) (1991): *Arqueología Aragonesa 1988-1989*, Zaragoza.

RUIZ MATA, D., RUIZ GIL, J.A. & LÓPEZ AMADOR, J.J. (2006): "La pesca en época prerromana en la Bahía de Cádiz. Apéndice sobre las factorías de salazones en el Puerto de Santa María", *Historia de la pesca en el ámbito del Estrecho. I Conferencia Internacional (El Puerto de Santa María, Cádiz)*, Sevilla, pp. 273-337.

RUSSELL, M. (ed.) (2002): *Digging Holes in Popular Culture*, Oxford.

RUSSO, F. (2005): "Di sangue e d'oro. Alternanze storico-militari nella pesca del corallo sulle coste nordafricane", C. Del Mare & F. Russo (eds.): *Il corallo nel gioiello étnico di Marocco e Algeria*, Torre del Greco, pp. 11-21.

RUSTICO, L. (1999): "Peschiere romane", *MEFRA* 111.1, pp. 51-66.

SÁEZ, A.M. (2008): *La producción cerámica en Gadir en época tardopúnica (ss. –III/-II)*, (*BAR* (IS) 1812), Oxford.

SAGLIO, M.E. (1918): *DS* 4.2, p. 855, *s.v. Reticulum, Retiolum.*

SAHRHAGE, D. (1998): *Fischfang und Fischkult im alten Ägypten*, Mainz.

SAHRHAGE, D. & LUNDBECK, J. (1992): *A History of Fishing*, Berlín.

SALADINO, L. (2001): "La pesca nel lago Fucino", A. Campanelli: *Il tesoro del lago. L'archeologia del Fucino e la collezione torlonia*, Avezzano, pp. 72 and 292-293.

SALAS ALMELA, L. (2006): "La agencia en Madrid del VIII duque de Medina Sidonia, 1615-1636", *Hispania: Revista española de Historia* 66 (224), pp. 909-958.

SAN NICOLÁS PEDRAZ, M.P. (2004-2005): "Seres mitológicos y figuras alegóricas en los mosaicos romanos de *Hispania* en relación con el agua", *Espacio, Tiempo y Forma, Serie II, Historia Antigua* 17-18, pp. 301-333.

SÁÑEZ REGUART, A. (1791-1795): *Diccionario histórico de las artes de la pesca nacional,* Madrid (reprinted Madrid 1988).

SÁÑEZ-REGUART, A. (1796): *Colección de Producciones de los Mares de España,* Tomo I, Madrid (reprinted Madrid 1993).

SANGRISO, P. (1998): "*Terra Sigillata* e politica augustea: alcune note su Cn. Ateivs", *Studi Classici e Orientali* 46.3, pp. 919-932.

SANTINI, P. (1775): *Carte de la Barbarie contenant les Royaumes de Maroc, de Fez, d'Alger, de Tunis et de Tripoli, avec les deserts limitrophes de l'intérieur de l'Afrique,* Venice.

SANTOS, A. (2003): "La Torre de Guzmán en 1411. El origen del pueblo de Conil en un documento del Archivo Parroquial de Santa Catalina", *Boletín La Laja* 3, http://www.lalaja.org/articulos/305.html

SANTOS, A. (2007): "El Siglo de Oro de la pesca del atún", *Boletín La Laja* 8, 28-34, http://www.lalaja.org/articulos/806.html

SARMIENTO, M. (2007): "Mosaico imperial. Sensacional hallazgo en Noheda Cuenca", *La Aventura de la Historia* 104, pp. 56-62.

SCAGLIARINI CORLAITA, D. (1992): *Villa romana Desenzano,* Roma.

SCHIFFER, M.B. (1987): *Formation Processes of the Archaeological Record*, Albuquerque.

SCHLABOW, K. (1950): *Trachten der Eisenzeit aus Moorfunden in Schleswig-Holstein,* Neumünster.

SCHUBART, H. (2006): *Morro de Mezquitilla. El asentamiento fenicio-púnico en la desembocadura del río Algarrobo, (Mainake*, Anejos 1), Málaga.

SCIALLANO, M. (1997): *Poissons de L'Antiquité*, Catalogue, Musée d'Istres, Istres.

SÉGUR, J./ZED (UNESCO 2007): *Scope and content*, http://www.unesco.org/culture/ich/index.php?pg=00057

SEILER-BALDINGER, A. (1973): *Systematik der Textilen Techniken,* (*Basler Beiträge zur Ethnologie* 14), Basel.

SEILER-BALDINGER, A. (2003): "Reperti tessili ingannevoli" *Textiles: Intrecci e tessuti della preistoria europea. Museo Civico di Riva del Garda-La Roca, 24 maggio-19 ottobre 2003,* Trento, pp. 55-63.

SELLA, M. (1928): "Biologia e pesca del tonno (*Thunnus thynnus L*)", *Atti del Convegno di Biologia Marina Applicata alla Pesca*, Messina, pp. 1-32.

SELLA, M. (1929): *Migrazioni e habitat del tonno studiati col metodo degli ami, con osservazioni su l'accrescimento, sul regime delle tonnare, ecc.*, (*Reale Comitato Talassografico Italiano, Memorie* 156), pp. 1-24.

SEMENOV, S.A. (1964): *Prehistoric Technology*, London.

SERNA, J.M. DE LA, ALOT, E., MAJUELOS, E. & RIOJA, P. (2004): "La migración trófica post-reproductiva del atún rojo", *ICCAT Collective Volume of Scientific Papers* 56.3, Madrid, pp. 1196-1209.

SIDEBOTHAM, S.E. (1991): "University of Delaware archaeological project at 'Abu Sha'ar: the 1990 season", *American Research Center in Egypt Newsletter* 153, pp. 1-6.

SIDEBOTHAM, S.E. (2002): "From Berenike to Coptos: Recent results of the desert route survey", *TopOO Supplement* 3, pp. 415-438.

SIDEBOTHAM, S.E. & WENDRICH, W.Z. (2007): *Berenike 1999/2000, Report on the Excavations at Berenike, including Excavations in Wadi Kalalat and Siket and the Survey of the Mons Smaragdus Region,* Los Angeles.

SIRET, L. (1906): *Villaricos y Herrerías. Antigüedades púnicas, romanas, visigóticas y árabes,* Madrid.

SMIDTH, J.K. (1876): *Historical observations on the condition of the fisheries among the ancient Greeks and Romans, and on their mode of salting and pickling fish. Report of the U.S. Commission of Fish and Fisheries for 1873-74 and 1874-75, III,* Washington D.C.

SNOUSSI, M. (2000): "Implications of Accelerated Sea-Level Rise (ASLR) for Morocco", A.C. de la Vega-Leinert, R.J. Nicholls, A. Nasser Hassan & M. El-Raey (eds.): *Proceedings of SURVAS Expert Workshop on 'African Vulnerability and Adaptation to Impacts of Accelerated Sea-Level Rise (ASLR)' Cairo, Egypt, 5th-8th November 2000,* Enfield, pp. 20-22.

SOFFER, O. (1998): "Science-week", August 21, University of Illinois, http://www.science-week.com/1998/SW980821.htm

SOTOMAYOR, M., ROCA, M. & FERNÁNDEZ, M.I. (1999): "Centro de producción de los Villares, Andújar (Jaén)" M. Roca & M.I. Fernández (eds.): *Terra Sigillata Hispánica. Centros de fabricación y producciones alto-imperiales,* Málaga.

STATENS FISKEREDSKAPSIMPORT (1972): *Syntetfibre i fiskeredskap*, 7, Bergen.

STEFANI, G. (1990): *Guida all'Antiquarium di Boscoreale: Uomo e ambiente nel territorio Vesuviano,* Boscoreale.

STEFANI, G. (2001): *Uomo e ambiente nel territorio vesubiano. Guida all'Antiquarium di Boscoreale,* Herculano.

STEFFY, J.R. (1985): "The *Herculaneum* Boat: Preliminary Notes on Hull Details", *AJA* 89.3, pp. 519-521.

STEFFY, J.R. (1994): *Wooden Shipbuilding and the Interpretation of Shipwrecks,* College Station, Texas.

STEINGRÄBER, S. (ed.) (1984): *Catalogo ragionato della pittura etrusca,* Milano.

STERNBERG, M. (1995): *La pêche à Lattes dans l'Antiquité à travers l'analyse de l'ichtyofaune,* (*Lattara* 8), Lattes.

STERNBERG, M. (1998): "Les produits de la pêche et la modification des structures halieutiques en Gaule Narbonnaise du IIIe siècle av. J.-C. au Ier siècle ap. J.-C. Les données de Lattes (Hérault), Marseille (Bouches-du-Rhône) et Olbia-de-Provence (Var)", *MEFRA* 110.1, pp. 81-109.

STERNBERG, M. (2000a): "Données sur les produits fabriqués dans une officine de *Neapolis* (Nabeul, Tunisie)", *MEFRA* 112, pp. 135-153.

STERNBERG, M. (2000b): "État des connaissances sur la pêche dans le monde ibérique (VIème s.-IIIème s. av. J.-C.)", C. Mata Parreño & G. Pérez Jordà (eds.): *Ibers, agricultors, artesans i comerciants, III Reunió sobre Economia en el Món Ibèric,* (*Saguntum* suppl. 3), València, pp. 93-97.

STERNBERG, M. (2002): "La mer nourricière. Pêche et infrastructures portuaires du IIe s. av. J.-C. au IIe s. après J.-C. Le cas de Lattes", *Lattara* 15, pp. 189-202.

STERNBERG, M. (2005a): "Les zones de pêche exploitées en Gaule méditerranéenne pré-romaine et romaine", E. Barré, E. Ridel & A. Zysberg (eds.): *Ils vivent avec le rivage: pêche côtière et exploitation du littoral, Colloque du Musée maritime de l'île Tatihou. Tatihou 2000,* Caen, pp. 7-14.

STERNBERG, M. (2005b): "Les restes de poisson des puits de Lattes : témoignages d'activité halieutique, d'exploitation, de production et de consommation alimentaire", *Lattara* 18, pp. 277-291.

STEWART, H. (1977): *Indian Fishing,* Seattle.

STONE, M. & STONE, J.C. (eds.) (1968): *Africa on maps dating from the twelfth to the eighteenth century,* Lepizig.

STRASBURGER, E. *et alii* (1994): *Tratado de botánica,* sixth Spanish edition, Barcelona.

STRÖMBERG, R. (1943): *Studien zur Etymologie und Bildung der griechischen Fischnamen,* Göteborg.

STRØMSHEIM, M. (1970): *Norsk fiskeredskapsindustri,* unpublished paper, Archive: Museum Vest, Bergen.

SWINY, H. & KATZEV, M.L. (1973): "The Kyrenia shipwreck: a fourth-century BC Greek merchant ship", D. Blackman (ed.): *Marine Archaeology,* (*Colston Papers* 23), London, pp. 339-359.

SYMINGTON, D. & COLLON, D. (2007): "Spindle Whorls", N. Postgate & D. Thomas (eds.): *Excavations at Kilise Tepe 1994-98. Volume 1: Text,* Cambridge/London, pp. 481-497.

TAILLIEZ, P. (1961): "Travaux de l'été 1958 sur l'épave du "Titan" à l'île du Lévant (Toulon)", *Atti del II Congresso Internazionale di Archeologia Sottomarina, Albenga 1958,* Bordighera, pp. 175-198.

TEICHNER, F. (1997): "Die römischen Villen von Milreu (Algarve/Portugal). Ein Beitrag zur Romanisierung der südlichen Provinz *Lusitania*", *MM* 38, 1997, pp. 106-162.

TESTAGUZZA, O. (1970): *Portus,* Rome.

THOMAS, R.I. (2006): "Trench 15", D.P.S. Peacock & L. Blue (eds.): *Myos Hormos – Quseir al-Qadim. Roman and Islamic ports on the Red Sea Volume 1: Survey and Excavations 1999-2003,* Oxford, pp. 87-94.

THOMAS, R.I. (2007): "The *Arabaegypti Ichthyophagi:* Cultural connections with Egypt and the maintenance of identity", J.C.M. Starkey, P. Starkey & T. Wilkinson (eds): *Natural resources and cultural connections of the Red Sea,* Oxford, pp. 149-60.

THOMAS, R.I. (forthcoming a): "Fishing equipment", D.P.S. Peacock & L. Blue (eds.): *Myos Hormos – Quseir al-Qadim Roman and Islamic ports on the Red Sea Volume 2: Maritime aspects,* Oxford.

THOMAS, R.I. (forthcoming b): "Hull maintenance", D.P.S. Peacock & L. Blue (eds.): *Myos Hormos – Quseir al-Qadim Roman and Islamic ports on the Red Sea Volume 2: Maritime aspects,* Oxford.

THOMAS, R.I. (forthcoming c): "The vessel stoppers", D.P.S. Peacock & L. Blue (eds.): *Myos Hormos – Quseir al-Qadim Roman and Islamic ports on the Red Sea Volume 2: Maritime aspects,* Oxford.

THOMAS, R. I. & MASSER, P. (2006): "Trench 8", D.P.S. Peacock & L. Blue (eds.): *Myos Hormos – Quseir al-Qadim. Roman and Islamic ports on the Red Sea Volume 1: Survey and Excavations 1999-2003,* Oxford, pp. 127-140.

THOMPSON, D.W. (1947): *A Glossary of Greek Fishes,* London.

TIMOULE, A. (1985): *Evolution et potentialitiés de la pêche au Maroc, 1914-1985,* Casablanca.

TIRELLI, M., BALISTA, C., GAMBACURTA, G. & RAVAGNAN G.L. (1988): "Altino (Venezia): proposta di articolazione in fase della necropoli "Le Brustolade" attraverso l'analisi di un settore (trincea I 1985-1987)", *Quaderni di Archeologia del Veneto* 4, pp. 348-394.

TOMBER, R. (2001): "The pottery", V. Maxfield & D.P.S. Peacock (eds.): *The Roman Imperial quarries: survey and excavation at Mons Porphyrites 1994-1998, vol. 1: topography and quarries,* London, pp. 241-303.

TOMBER, R. (2005) "Trogodytes and Troglodytes: Exploring interaction on the Red Sea during the Roman period", J.C.M. Starkey (ed.): *People of the Red Sea: Proceedings of Red Sea Project II held in the British Museum October 2004,* (*BAR* (IS) 1395), Oxford, pp. 41-50.

TOMBER, R. (2006): "The pottery", V.A. Maxfield & D.P.S. Peacock (eds.): *Survey and excavation Mons Claudianus. Volume 3: ceramic vessels and related objects,* Cairo, pp. 3-238.

TONIOLO, A. (2007): "Anfore dall'area lagunare", S. Gelichi & C. Negrelli (eds.): *La circolazione delle ceramiche nell'Adriatico tra tarda antichità e altomedioevo,* Mantova, pp. 91-106.

TONIOLO, A. (2008): *Anfore. Vino, olio, pesce lavorato, spezie, profumi nella Laguna di epoca antica,* Venezia.

TRAKADAS, A. (2003): "The Morocco maritime survey: the 2002 season", *INA Quarterly* 30.1, pp. 12-21.

TRAKADAS, A. (2004): "Morocco maritime survey: 2003 season", *INA Quarterly* 31.4, pp. 3-9.

TRAKADAS, A. (2005): "The archaeological evidence for fish processing in the Western Mediterranean", T. Bekker-Nielsen (ed.): *Ancient Fishing and Fish Processing in the Black Sea Region,* (*Black Sea Studies* 2), Aarhus, pp. 47-82.

TRAKADAS, A. (2006): ""Exhausted by fishermen's nets": Roman sea fisheries and their management", *Journal of Mediterranean Studies* 16.1/2, pp. 259-272.

TRAKADAS, A. (2009): *Piscationes in Mauritania Tingitana: marine resources exploitation in a Roman North African Province,* PhD Thesis, unpublished, University of Southampton.

TROTTA, F. (1996): "La pesca nel mare di Magna Grecia e Sicilia", F. Prontera (ed.): *La Magna Grecia e il Mare. Studi di Storia Marittima,*Taranto, pp. 227-250.

TROUSSET, P. (1992): "La vie littorale et les ports dans la Petite Syrte à l'époque romaine", *Afrique du Nord antique et médiévale. Spectacles, vie portuaire, religions. Actes du Ve colloque international sur l'histoire et l'archéologie de l'Afrique du Nord (Avignon, 9-13 avril 1990),* Paris, pp. 317-332.

TROUSSET, P. (1998): "La pêche et ses techniques sur les côtes de l'Africa", É. Rieth (ed.): *Méditerranée antique. Pêche, navigation, commerce,* Paris, pp. 13-32.

TUCCINARDI, M. (1998): "I nuovi scavi. Il nucleo edilizio all'angolo nord-occidentale della città", A. De Simone, F. Ruffo, M. Tuccinardi & U. Cioffi*: Ercolano 1992-1997. La villa dei Papiri e lo scavo della cittá,* (*Cronache ercolanesi* 28), pp. 47-59.

TURRI, E. (1995): "La valva di Venezia", G. Caniato, E. Turri & M. Zanetti (eds.): *La laguna di Venezia,* Verona, pp. 3-39.

UGGERI, G. (1992): "La laguna e il mare", L. Cracco Ruggini *et alii* (eds.): *Storia di Venezia, I. Origini, età ducale,* Roma, pp. 149-173.

UNESCO (2007): "The Intangible Heritage" *Courier,* February 2007, p. 5.

VAN KEULEN, J. (1694): *Nieuwe Paskaart van de Kust van Hispania. De Groote Nieuwe Vermeerderde Zee-Atlas ofte Water-Werelt,* Amsterdam.

VAN NEER, W. & ERVYNCK, A.M.H. (1998): "The faunal remains", S.E. Sidebotham & W.Z. Wendrich (eds.): *Berenike 1996: Report of the 1996 excavations at Berenike (Egyptian Red Sea coast) and the survey of the Eastern Desert,* Leiden, pp. 349-388.

VAN NEER, W. & ERVYNCK, A.M.H. (1999): "The faunal remains", S.E. Sidebotham & W.Z. Wendrich (eds.): *Berenike 1997,* Leiden, pp. 325-348.

VAN NEER, W., HAMILTON-DYER, S., CAPPERS, R., DESENDER, K. & ERVYNCK, A.M.H. (2007): "The Roman trade in salted Nilotic fish products: some examples from Egypt", *Documenta Archaeobiologiae* 4, pp. 173-188.

VAN NEER, W. & LENTACKER, A. (1996): "The faunal remains", S.E. Sidebotham & W.Z. Wendrich (eds.): *Berenike 1995,* Leiden, pp. 337-356.

VAN NEER, W., LERNAU, O., FRIEDMAN, R., MUMFORD, G., POBLOME, J. & WAELKENS, M. (2004): "Fish remains from archaeological sites as indicators of former trade connections in the Eastern Mediterranean", *Paléorient* 30, pp. 101-148.

VAN NEER, W. & MORALES, A. (1992): "Fish Middens: Anthropogenic Accumulations of Fish Remains and their Bearing on Archaeoichthyological Assemblages", *JAS* 19, pp. 683-695.

VAN NEER, W. & PARKER, S.T. (2008): "First Archaeozoological Evidence for *Haimation,* the "invisible" *garum*", *JAS* 35, pp. 1821-1827.

VAN RENGEN, W. (2002): "Sebakh excavations and the written material", D.P.S. Peacock, L. Blue & S. Moser (eds.): *Myos Hormos – Quseir al-Qadim: A roman and Islamic port on the Red Sea coast of Egypt,* Southampton, pp. 53-54.

VAN RENGEN, W. (forthcoming): "The Roman written evidence", D.P.S. Peacock & L. Blue (eds.): *Myos Hormos – Quseir al-Qadim: Roman and Islamic ports on the Red Sea, vol. 2: The finds from the 1999-2003 excavations,* Oxford.

VAN RENGEN, W. & THOMAS, R.I. (2006): "The sebakh excavations", D.P.S. Peacock & L. Blue (eds.): *Myos Hormos – Quseir al-Qadim. Roman and Islamic ports on the Red Sea Volume 1: Survey and Excavations 1999-2003,* Oxford, pp. 146-54.

VARGAS COSTA, M.L. (1985): "Contribuiçâo para o estudo de alguns dos mosaicos da villa romana de Pisôes", *ABeja, 2nd series* 2, pp. 95-135.

VÁZQUEZ DE LA PARGA, L. (1934): "De *terra sigillata.* Un vaso inédito de Germanus", *Anuario Cuerpo Facultativo de Archiveros, Bibliotecarios y Arqueólogos* 1, pp. 105-113.

VELDMEIJER, A. (2004): "Fishing nets from Berenike (Egyptian Red Sea coast)", *Papers on Ancient Egypt* 3, pp. 99-110.

VELDMEIJER, A. (2005A): "Archaeologically attested cordage. Terminology on the basis of the material from Ptolemaic and Roman Berenike (Egyptian Red Sea coast)", *Eras Journal* 7, pp. 1-32.

VELDMEIJER, A. (2005b): "Indentifiable and associated cordage. Examples from Berenike (Egyptian Red Sea coast)", *AntO* 3, pp. 65-87.

VELDMEIJER, A. (2006): "The cordage from the 2001 season of the excavations at Berenike (Egyptian Red Sea coast): Preliminary results", *AntO* 4, pp. 119-35.

VELDMEIJER, A. & VAN RODE, S.M. (2004): "Carrier netting from the Ptolemaic Roman harbour town of Berenike (Egyptian Red sea coast)", *AntO* 2, pp. 9-25.

VENDITTELLI, M. (1992): "Diritti e impianti di pesca degli enti ecclesiastici romani tra X e XIII secolo", *MEFRM* 104. 2, pp. 387-430.

VERNHET, A. (1991): *La Graufesenque, céramiques gallo-romaines,* Millau.

VIANELLO, R. (2004): *Pescatori di Pellestrina. La cultura della pesca nell'isola veneziana,* Treviso/Verona.

VIVAR, G. (2003): "Els objectes de la tripulació", *Culip VIII u les àmfores Haltern 70, (Monografies del CASC 5)* Gerona, pp. 147-154.

VIZCAINO, J. (2005): "Utensilios de pesca", *Bizancio en Cartago Spartaria. Aspectos de la vida cotidiana, Catálogo de la Exposición*, Cartagena, p. 63.

VOGT, E. (1937): *Geflechte und Gewebe der Steinzeit,* Basel.

WACHSMANN S. (1990a): *The excavation of an Ancient Boat in the Sea of Galilee (Lake Kinneret), (Atiqot, English series* 19), Jerusalem.

WACHSMANN, S. (1990b): "The Kinneret boat: The discovery and excavation", H. Tzalas (ed.): *Proceedings of the 2nd International Symposium on Ship Construction in Antiquity, Delphi 1987,* (*Tropis* 2), Athens, pp. 371-384.

WACHSMANN, S. (1995): *The Sea of Galilee Boat: a 2000 Year Old Discovery from the Sea of Legends*, Cambridge.

WACKE, A. (1993). "Freedom of Contract and Restraint of Trade Clauses in Roman and Modern Law", *Law and History Review* 11-1, 1-19.

WEERD DE, M.D. (1988): *Schepen voor Zwammerdam,* Phd Thesis, University of Amsterdam.

WENDRICH, W.Z. & W. VAN NEER (1994): "Preliminary notes on fishing gear and fish at the late Roman fort at 'Abu Sha'ar (Egyptian Red Sea coast)", W. Van Neer (ed.): *Fish exploitation in the past: Proceedings of the 7th meeting of the ICAZ fish remains working group, Tervuren, (Annales du Musee Royal de l'Afrique Centrale)*, pp. 183-189.

WESTWOOD, T. & SATCHELL, T. (1883): *Bibliotheca Piscatoria,* London.

WHEELER, A. & A.J.K. JONES (1989): *Fishes* (*Cambridge Manuals in Archaeology), *Cambridge.

WHITCOMB, D.S. & JOHNSON, A.C. (1982*):* "1982 season of excavation at Queseir al-Qadim", *American Research Center in Egypt Newsletter* 120, pp. 24-30.

WHITEHEAD, P.J.P., BAUCHOT, M.L., HUREAU, J.C., NIELSEN, J. & TORTONESE, E. (1984-1989): *Fishes of the North-Eastern Atlantic and the Mediterranean 1-3,* Paris.

WHITEWRIGHT, J. (2007): "Roman Rigging Material from the Red Sea Port of Myos Hormos", *IJNA* 36.2, pp. 282-292.

WHITTAKER, C.R. (1983): "Late Roman Trade and Traders", P. Garnsey, K. Hopkins & C.R. Whittaker (eds.): *Trade in the Ancient Economy,* London, pp. 163-180.

WILD, J.P. (2001): "Textiles et activités relatives au textil sur le monument d'Igel", *Annales de l'Est* 2, pp. 83-92.

WILKINS, J. (2005): "Fish as a Source of Food in Antiquity", T. Bekker-Nielsen (ed.): *Ancient Fishing and Fish Processing in the Black Sea Region,* (*Black Sea Studies* 2), Aarhus, pp. 21-30.

WILKINS, J. & HILL, S. (1994): *Archestratus: The Life of Luxury,* Totnes.

WILLIAMS, C.K. (1979): "Corinth, 1978. Forum Southwest", *Hesperia* 4, pp. 105-144.

WILSON, A. (1999): "Commerce and industry in Roman Sabratha", *Libyan Studies* 30, pp. 29-52.

WILSON, A. (2002): "Marine resource exploitation in the cities of coastal Tripolitania", *AR* 14, pp. 429-436.

WILSON, A. (2006): "Fishy business: Roman exploitation of marine resources", *JRA* 19, pp. 525-537.

WITTEYER, M. (1982): "Ausgewählte Kleinfunde", G. Rupprecht (ed.): *Die Mainzer Römerschiffe. Berichte über Entdeckung, Ausgrabung und Bergung, (Archälogische Berichte aus Rheinhessen und dem Kreis Bad Kreuznach* 1). Mainz, pp. 134-156, figure 6-11.

WOOD, F.A. (1927): "Greek fish names: part I", *AJPh* 48.4, pp. 297-325.

WOOD, F.A. (1928a): "Greek fish names: part II", *AJPh* 49.1, pp. 36-56.

WOOD, F.A. (1928b): "Greek fish names: part III", *AJPh* 49.2, pp. 167-187.

WOOTTON, R.J. (1990): *Ecology of Teleost fishes,* London.

YACOUB, M. (1993): *Le Musée du Bardo,* Tunis.

YACOUB, M. (1995): *Splendeurs des mosaïques de Tunisie,* Tunis.

YADIN, Y. (1962): "The Cave of the Letters", *IEJ* 12, pp. 227-257.

ZAHN, R. (1910): "*Garum*", *RE* 7.1, pp. 841-849.

ZAPHIROPOULOU, A. (2005): *Rhodes from the 4th c. AD to its capture by the Ottoman Turks (1522), Palace of the Grand Master,* Athens.

ZELENIN D.K. (1989): "Tabù linguistici nelle popolazioni dell'Asia settentrionale (III)", *Quaderni di Semantica* 10, pp. 183-276.

ZEVI, F. (2001): "Le invenzioni di Archimede e le grandi navi", M. Giacobelli (ed.): *Lezioni Fabio Faccenna,* Bari, pp. 95-114.

ZITTERKOPF, R.E. & S.E. SIDEBOTHAM (1989): "Stations and Towers on the Quseir-Nile Road", *JEA* 75, pp. 155-189.

Geographical Index

Index of Species

Index of Fishing Gear

Index of Ancient Sources